Water Engineering in Ancient Civilizations

5,000 Years of History

Water Engineering in Ancient Civilizations

5,000 Years of History

Pierre-Louis Viollet
2005

translated into English by
Forrest M. Holly Jr.
2007

IAHR International Association of Hydraulic Engineering and Research, 2007
www.iahr.org

Water Engineering in Ancient Civilizations
5,000 Years of History

Published by:
IAHR International Association of Hydraulic Engineering and Research
Pº Bajo Virgen del Puerto, 3
28005 Madrid, Spain
www.iahr.org

ISBN 978-90-78046-05-9

Original edition in French, *L'Hydraulique dans les Civilisations Anciennes,*
Presses des Ponts et Chaussées, 2000, 2005

Printed by Intigraf in Spain

Other IAHR Monographs in the series:

(published by IAHR- for more information and to order visit www.iahr.org or contact the Secretariat at iahr@iahr.org)

Fluvial Processes monograph	M. Selim Yalin A.M. Ferreira da Silva, 2001, 197 pages.
Fluvial Processes Solutions Manual	M. Selim Yalin A.M. Ferreira da Silva, 2001, 83 pages.
Hydraulicians in Europe 1800-2000	W. H. Hager, 2003, 774 pages
Flow and Sediment Transport in Compound Channels: the experiences of Japanese and UK research	I. K. McEwan and S. Ikeda, 2007

(published by Taylor and Francis - Balkema, www.taylorandfrancis.com/)
IAHR Members receive a 10% discount on all IAHR publications

Physical Processes and Chemical Reactions in Liquid Flows	F.S Rhys/A. Gyr, 1998, 200 pages.
Mudflow Rheology and Dynamics	Ph. Coussot, 1997, 270 pages.
Hyperconcentrated Flow	Zhaohui Wan/Zhaoyin Wang, 1994, 320 pages,.
Turbulence in open-channel Flows	I. Nezu/H.Nakagawa, 1993, 286 pages.
Turbulence Models and Their Application in Hydraulics	W. Rodi, 1993 3rd edition, 116 pages.
Mobile Barrages and Intakes on Sediment Transporting Rivers	M. Bouvard, 1992, 322 pages.
Debris Flow	T. Takahashi, 1991, 156 pages.

The International Association of Hydraulic Engineering and Research (IAHR), founded in 1935, is a worldwide independent organisation of engineers and water specialists working in fields related to hydraulics and its practical application. Activities range from river and maritime hydraulics to water resources development and eco-hydraulics, through to ice engineering, hydroinformatics and continuing education and training. IAHR stimulates and promotes both research and it's application, and by so doing strives to contribute to sustainable development, the optimisation of world water resources management and industrial flow processes. IAHR accomplishes its goals by a wide variety of member activities including: working groups, research agenda, congresses, specialty conferences, workshops and short courses; Journals, Monographs and Proceedings; by involvement in international programmes such as UNESCO, WMO, IDNDR, GWP, ICSU, The World Water Forum; and by co-operation with other water-related (inter)national organisations.

For Dominique-Marie, who joined me in this voyage.

Translator's Note

Pierre-Louis Viollet published the first edition of his book as I was finishing my academic career at the University of Iowa. I decided then that a challenging first project in my new life would be to spend a few months translating his work into English. A few months curiously inflated to nearly three years as other projects caught my attention, and meanwhile Pierre-Louis published the second edition of his book. I am pleased to have been able to complete this most enjoyable and educational task, and thank Pierre-Louis for having given me the opportunity to make his wonderful work available to a wider audience.

To the extent possible, I have attempted to preserve Pierre-Louis' organization of material at the paragraph level. Although the names of ancient people and places do not always have unique representations in French or English, I have tried to adopt consistent English names from among those that appear in the literature. In translating several ancient texts for which it was not possible to cite published English translations, I preserved the ancient modes of expression captured in Pierre-Louis' translations in adapting them to the English edition. Responsibility for any inconsistencies or mistranslations is mine and mine alone.

I was fortunate to be able to do much of this work in the Iowa City Public Library, and also in the University of Iowa's main library – both are marvelous facilities of which the City and University can be extremely proud. I completed the draft translation while in residence at the University of Nice Sofia-Antipolis, France, as a guest instructor in the Erasmus Mundus EuroAquae Hydroinformatics Masters program. I would especially like to thank Philippe Gourbesville and his colleagues and staff, especially Annie Vahramain, for their support and collegiality during this period. During this time Pierre-Louis and Dominique Viollet graciously hosted me for a final working session in Paris.

Finally, this project would have been impossible without the encouragement and moral and editorial support of my wife Joyce, who suffered through several proofreadings, was never shy about pointing out the need for rewriting, and patiently tolerated my endless hours at the keyboard.

Forrest M. Holly Jr.

Acknowledgements

I would like to thank the archaeologists, specialists in the ancient civilizations, and engineers interested in the history of hydraulics who so generously shared their unpublished work with me. In particular, I would like to thank Gilbert Argoud, Frank Braemer, Corinne Debaine-Francfort, Bernard Geyer, and Philippe Leveau. I would especially like to thank Günther Garbrecht, with whom I have had a continuing correspondence, and who provided me with abundant documentation of his own work in Egypt, Palestine and Anadolu. I would also like especially to thank Jean-Claude Margueron who opened his personal library on Mesopotamia to me, and with whom I had lively discussions. Felipe Martinez and Cristobal Mateos graciously shared several works and articles on ancient hydraulic works in Spain, for which I thank them. I would also like to thank the staff of the Center of Contemporary and Historic Documentation of l'École Nationale des Ponts et Chaussées, who helped me in my searches of the ancient archives. This work is a synthesis, and therefore I must recognize all those who, by their field work and study of ancient texts, have put together the body of knowledge without which this book would not have been possible. Finally, I would like to thank Forrest and Joyce Holly for their teamwork in bringing this English translation to fruition.

Pierre-Louis Viollet

Preface

This work of Pierre-Louis Viollet on the history of hydraulics in the ancient civilizations, more generally in the civilizations of the classical era and the Middle Ages, is important for several reasons.

First of all, the author is the first scholar who has attempted, with success, a complete synthesis of techniques in hydraulics, from the birth of agriculture in Syria-Palestine up to the beginning of the modern period. He gives due consideration to the role of the Mediterranean world and the Near and Far East, as well as the Indian and Chinese worlds, as precursors to this development of techniques. There has been no comparable effort of this scale to present, and to explain in a concrete manner, the diversity and evolution of hydraulic knowledge and techniques over such a vast geographical space and over the long expanse of several millennia, taking into account the historical context.

The second reason, one that is fundamental to the importance of this marvelous book, is that the author is neither a historian nor an archaeologist. He is, rather, an engineer whose background gives him a unique ability to understand the operation of and interest in hydraulic works, installations that had been previously known only through vague descriptions, imprecise representations, or physical remains in an extreme state of deterioration. The author's contribution would be fundamental for this reason alone. Despite the originality and quality of previous works on archaeological hydraulics, their impact has often been compromised by a weak knowledge of the physical principles that are indispensable to an understanding of the workings, importance, and innovation of hydraulic projects. The present work is rich in such technical analyses of ancient innovations, providing thoughtful explanations and commentary on both the nature of these discoveries and their technical pertinence. Therefore this is much more than an ordered compilation of facts – which would be of great interest in its own right – it is a true synthesis that is focused on the importance of ancient discoveries, giving them texture and richness through the author's scientific and technical perspective.

This book is important for yet another reason. It always presents hydraulic developments and innovation in their historical and intellectual context. Even though his primary objective is the historical development of hydraulics, Pierre-Louis Viollet has endowed his book with an overview of world history in general. His work is aligned with traditional notions of historical periods, but he gives these periods fresh significance in highlighting the number and importance of technical innovations associated with them. In this respect, the decisive changes are those that occur after the conquests of Alexander, in particular at Alexandria but also in the rest of the Hellenistic world. The originality of Chapter 5 is not only in its presentation of a broad panorama of these innovations and inventions; but it is also in its demonstration that these developments not only represent a natural continuation of the classical age's tradition of technical thought, but also reflect the application of analytical methods elaborated by philosophers from the empirical developments of Oriental civilizations. This demonstration deflates the importance of works that see these inventions as simple inspirations of thought, unconnected to any context of reality, and that expect to find "cultural obstacles" to the exploitation of these inventions. The lack of any practical application or development of "Heron's steam ball" invention, which demonstrates the principle of a steam engine, has undoubtedly led to erroneous conclusions in this regard. The consequences of techniques invented by the Alexandrians resulted in numerous and very important applications during the Roman period. From all perspectives imperial Rome is a civilization of water, as seen in its technology for the transport and distribution of water, as well as in its thermal installations.

This example is but one of many showing the richness of a book that succeeds in not only presenting an inventory of the state of knowledge in hydraulic techniques, but also in enrichening this knowledge in many respects, some of which are of considerable significance. Study of this book is indispensable for specialists in the history of technology, economy, and thought.

Georges Tate
Professor of Ancient History at the
University of Versailles – Saint-Quentin
Former Deputy Scientific Director of the
Department of Humanities and Social Sciences of CNRS
Former Director of IFAPO at Damascus

Table of Contents

Introduction

Water is the key to all civilization. The diverse and often competing uses of water inevitably lead to tensions and conflicts in its management and allocation. Water technology has progressed from the primitive to the advanced, but this progress has not changed man's continuous responsibility for careful and fair management of this precious resource.

Initially, my objective was to give my students at l'École Nationale des Ponts et Chaussées a historical perspective before getting into the craft of the engineer and modern techniques for flow modeling. But as I tried to travel back in time to reconstruct the historical relations between mankind and water, I quickly realized what a distant horizon this quest represented. The wealth of knowledge to be mined from the past quickly became apparent - not only in the descriptions of hydraulic works and analysis techniques developed by our distant ancestors, but also in the relation between the development of hydraulics and of civilization itself. It seems to me that it is as important to understand the context and circumstances of innovations, and their entry into the knowledge base of civilization, as it is to describe the innovations themselves.

This book does not pretend to be a comprehensive catalog of hydraulic works. I have tried to be reasonably complete, while limiting the scope of my studies to the vast and continuous landmass extending from the Atlantic Ocean to the China Sea. My historical perspective extends in time from the ancient Near East, to Antiquity (the historic period preceding the Middle Ages in Europe), and then to the medieval world; from the known origins of Neolithic water management, up until the Renaissance and the advent of modern fluid mechanics. One could legitimately criticize this work for having ignored certain civilizations, for example the pre-Columbian world in the Americas. But the objective of the book – to describe ancient works and processes and situate them in the melding of the East and the West in a unified manner – led me to limit my attention to the Eastern hemisphere.

To give the reader a feel – a bit of taste and smell - of the ancient civilizations, I liberally include citations from ancient authors themselves - scribes and chroniclers, travelers such as Xenophon and Ibn Battûta, historians such as Herodotus and Sima Qian, geographers or architects like Strabo and Vitruvius. I make an effort to complement the text with numerous maps, plans, and sketches, for nothing is more annoying than to read the description of a site without being able to see where it is.

This work is presented in two parts. The first covers the period prior to the 3rd century BC. It deals with the land bounded by Mesopotamia, Egypt and the Aegean Sea, where the stage was set for the meeting of the geometry of the Greeks and the hydraulic know-how of the East at Alexandria and elsewhere. The second part begins with the hydrostatics of Archimedes and the earliest devices based on the use of water pressure. This part of the book broadens the perspective to include the main developments of the Roman, Chinese, and Arab empires, and finally, those of the medieval world. This perspective does not necessarily lead to a strictly chronological presentation of the material; the chronological table at the end of the book serves this purpose for the interested reader.

Part I

4,000 YEARS OF HYDRAULIC DEVELOPMENT IN THE EAST

From the era of the early cultivators to the conquests of Alexander the Great

From the beginning of history up to the conquests of Alexander the Great, continuous and rapid development of civilization occurred in the valleys of the Tigris, the Euphrates, and the Nile Rivers, as well as on the shores of the Aegean Sea. Each of these regions has its own particular historical context, and each would be worthy of its own detailed description. But the regions were also closely linked together and unified by extensive trade and military ventures, and by the transfer of technology that came with them.

The historical period of Part I of this book ends in the era of Alexander the Great, whose conquests marked the end of a civilization in the Orient.

To understand this period, one must keep in mind that it includes the Bronze Age, from the IVth millennium BC to the end of the IInd millennium BC, and the transition to the Iron Age. In some areas, such as Egypt, this transition occurred smoothly and continuously. In other areas, such as Greece, the transition represented profound ruptures with the past.

1. Hydraulics and The Birth of Civilization

Water and the infrastructure for its conveyance are ever-present needs of civilization, whether for irrigation or flood protection, for water supply or for wastewater drainage from the earliest cities. Added to these needs are those of waterborne commerce, canals, and ports.

This story begins in the East with the great Neolithic revolution, humanity's fundamental stride into an economic system of production, of agriculture and its accompanying development of the first cities.

From the birth of agriculture to the development of irrigation: the origins of the great Mesopotamian civilizations

In the near-East there is a zone of hills called the "fertile crescent" extending from Syria-Palestine to the foot of Mounts Taurus and Zagros. On these hills, blessed by ample rainfall, naturally grow wild grains such as barley and wheat. The natural fertility of this zone began to develop around 16000 BC, when the climate began to become warmer and moister. This occurred first in the western portions of the area where, after the interruption of a dry and cool period from 10500 to 9000 BC, the climate stabilized to become more or less as it is today, albeit somewhat more humid.[1] The change continued into the eastern portions of the zone, in the foothills of Mount Zagros in present-day Iraq, and finally ended about 7000 BC with the permanent inundation of the Persian Gulf.

In about 12500 BC, in this fertile land in the middle and to the west of the present fertile crescent, the harvesting of wild grains led the hunter-gatherers to begin to settle. Toward 9500 BC they began to take charge of their means of subsistence. They came down from the hills to begin early cultivation of domestic grains and cereals in the sedimentary corridor from the Jordan Valley to the upper valley of the Euphrates, and in the oasis of Damas. For example, in the IXth millennium BC, Jericho is a village located near a spring; a rather large settlement of two hectares and probably having several hundred inhabitants, surrounded by a thick wall possibly designed to protect the inhabitants from floods. Prehistorians call this corridor the Levantine core (Figure 1.1).[2]

Subsequently, about 8000 BC, this seat of early agriculture moved north to take root in the Syrian interior and in the south of Anadolu. This is when animal husbandry first appeared, as well as the rectangular dwelling, a major architectural innovation

1 Sanlaville (1996).

2 For a recent synthesis of the birth of agriculture and the Neolithic revolution, see the book of the prehistorian and archaeologist Jacques Cauvin, Naissance des divinités, naissance de l'agriculture, la révolution des symbols au Néolithique, revised in 1996.

(prior to this, the semi-underground huts were round). The population began to increase, and this in effect marked the beginning of the *Neolithic revolution* which occurred from about 7500 BC to 4500 BC. [3] Toward the west, this movement gradually reached the Syrian coast, then the West through two parallel paths: the Mediterranean and the Danube. The population spread also extended toward the east, and it is this eastern spread that interests us here. The Neolithic human tide reaches the Zagros mountains, where it has now become possible for men to live. Indeed, since 7000 BC the Persian Gulf had been flooded, and was even deeper than it is today. The human tide spreads into some hospitable niches of the arid Syrian desert: the oases of the regions of Palmyra and El-Kown, and the site of Bouqras, at the confluence of the Euphrates and the Khabur.

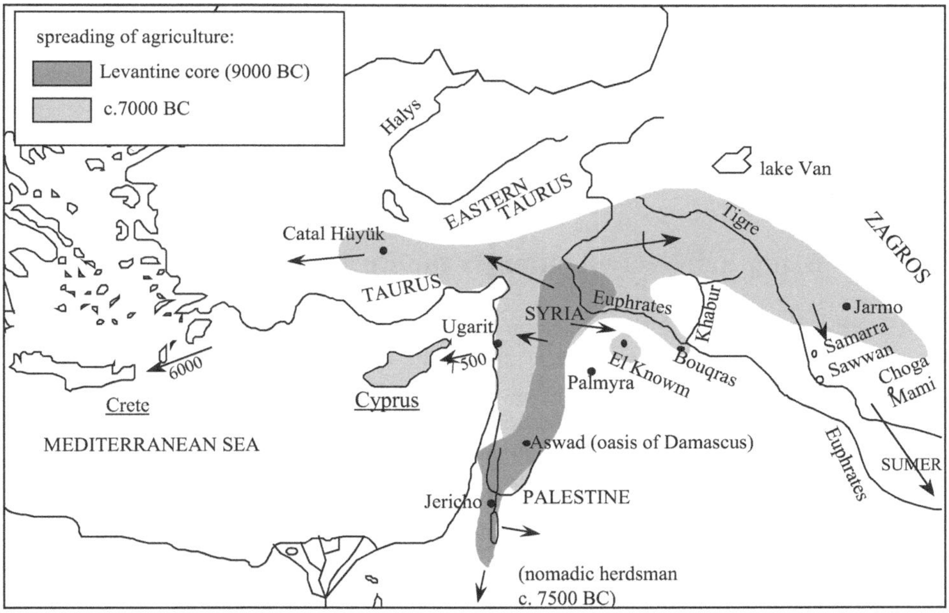

Figure 1.1 From the Neolithic revolution in the East to the first irrigation canals: birth and spreading of agriculture from the Levantine core toward the culture of Samarra and the land of Sumer, from 9000 to 6000 BC.

The rain and natural runoff are at first sufficient to provide enough water for grains and vegetables. Then, the needs of individual, isolated farmers or perhaps small groups of them, led to the advent of irrigation through small ditches. This made it possible to improve yields and cultivate new land that lacked sufficient rainfall. It is very

3 Some have proposed that population pressure explains the exodus which clearly accompanied the Neolithic spread (each generation seems to have migrated 20 kilometers or so). But according to Jacques Cauvin, this explanation is inadequate; he sees in addition a profound change of mentality, responding to the call of the "new frontier".

difficult to date these very first and rudimentary "hydraulic works". It is possible that when settlers began to occupy the oases mentioned above, as at Bouqras where, between 7400 BC and 6800 BC, the flood plain of the Euphrates valley was cultivated,[4] or at El-Kowm where, in the first half of the Vth millennium BC, artesian springs were available, some management of water was already occurring. Moreover, the earliest evidence of drainage of water from dwellings was found at El Kowm (Figure 1.4).[5]

The first definite evidence of irrigation dates from the VIIth millennium BC – in the middle Tigris valley, at the foot of the Zagros mountains, and at the sites known as the Samarran civilization (Samarra, Sawwan, Choga Mami). At Choga Mami, remains of what are thought to be two-meter wide canals have been found, canals that connect to the rivers and follow elevation contours for hundreds of meters before distributing the spring flood waters into the fields.[6] The first wells appeared in the VIth millennium BC.

Also in the VIth millennium BC, the irrigated cereal-growing know-how began to spread to the Euphrates delta, a potentially fertile area thanks to its silt deposits, but an arid one. This is the Ubaid culture that may have been the inheritor of the Samarran culture. The multiple channels of the delta surely facilitated irrigated farming in this semi-marshy alluvial region. The first great urban civilizations, as we recognize them today, took root here.

Irrigation and urban civilization in the Euphrates delta

The driving force of this urban development is likely the significant population growth, as seen in a proliferation of villages and small towns in the IVth millennium BC. Subsequently, it may be that as the climate became drier, some of the villages were abandoned causing market towns to grow and evolve into cities. As some branches of the river became dry, it was necessary to dig canals and establish a complex system of water distribution, and also to bring more land under cultivation by draining swampy areas and irrigating dry land. The accompanying need to organize a work force and coordinate the construction gave birth to the Sumerian civilization, the first to have a hierarchical organization.[7] Studies of human settlements in certain regions of lower Mesopotamia, performed by the American archaeologist Robert Adams, show a decrease in the number of villages, and a concomitant increase in the number of cities and population increases in existing cities, during the period between 3000 and 2500 BC. These really were cities in the true sense of the word: Uruk, one of the largest and oldest, occupies 550 hectares

---

4 According to Geyer and Besançon (1997), the Euphrates was in a sedimentation phase until the VIIth millennium BC, having a braided morphology which favored early seasonal, non-irrigated agriculture. From the VIth millennium BC, the river entered a new phase of erosion of its bed into its own alluvia. The terraces that were formed as a consequence, protected from flooding a dozen or so meters above the riverbed, became favorable to permanent settlements. Irrigation then became necessary.

5 Cauvin (1969), p. 239. Evidence of such drainage may also have been observed at Bouqras.

6 See for example Huot (1994), describing the explorations performed by the American archaeologist John Oates in 1967-68.

7 See reference books like Roux (Ancient Iraq, Allen & Unwin, 1964); Oates (Babylon, Thames & Hudson, 1986).

with a wall of circumference 9.5 km. The reconstitution of Uruk's urbanization in 2500 BC is shown in Figure 2.3 of the following chapter.

The notion of *writing* first appeared in this urban civilization, in particular in Uruk about 3300 BC (and perhaps also in Suse to the east). One of the oldest texts describes the creation of man, vegetation and animals, and the first five cities (Eridu, Bad Tibira, Larak, Sippar, and Shuruppak). It goes on to argue for the vital need to maintain the hydraulic system, mentioning the necessity of "the cleaning of the small ditches."[8]

Another account contains the following:

> "At this time, water was short in Lagaš, there was famine in Girsu. Canals were not dug, vast lands were not irrigated by a shadoof *(shaduf)*,[9] abundant water was not used to dampen meadows and fields, because humanity counted on rainwater. Ašnan did not bring forth dappled barley, no furrow was plowed nor bore fruit! No land was worked nor bore fruit! No country or people made libations of beer or wine, [...] sweet wine [...], to the gods. No one used the plow to work the vast lands. (...) In order to dig the canals, in order to dredge the irrigation ditches, in order to irrigate the vast lands by a shadoof, in order to utilize abundant water so that the meadows and fields were moistened, An and Enlil put a spade, a hoe, a basket, a plow, the life of the land, at the disposal of the people. After this time, human beings gave all their attention to making the barley grow." *(there follows a list of numerous canals dug by the leaders of Lagash).*[10]

A third text describes periods of famine, caused by a conflict between the "waters of the primordial sea" having invaded the earth and the beneficial water of the Tigris.[11] Does this perhaps refer to the conflict, common to all deltaic and estuarine zones of large rivers, between fresh and salt water, the latter useless for both cultivation and human consumption?

> "Famine was severe, nothing was produced, The small rivers *were not cleaned, the dirt was not carried off*, On the steadfast fields no water was sprinkled, there was no digging of ditches, In all the lands there were no *crops*, only weeds grew."

The River: menace or blessing?

All of the great early civilizations were born in alluvial valleys – notably the Tigris-Euphrates delta, then the valleys of the Nile, the Indus, and the Yellow River (Figure 1.2). The soil in those valleys is fertile, but must be irrigated and drained, and this requires the establishment of coordinated water management. But there is another con-

---

8 This is from a tablet that also carries the very first account of the Flood. The tablet is from the Nipur collection at the University Museum of Philadelphia; the translation was published in 1914 by A. Poebel (Kramer, History Begins at Sumer, Chapter 23).

9 Balancing device making it possible to use manpower to lift water for irrigation of fields; see Figure 2.4 in the following chapter.

10 This account, called Royal Chronicle of Lagash, is thought to be somewhat satirical. It probably dates from the middle of the 2nd millennium BC. Livius, Articles on Ancient History, http://www.livius.org/cg-cm/chronicles/cm/lagash.html.

11 Extract of a poem entitled "The Feats and Exploits of Ninurta" (Kramer, 1961, Chapt. 3).

sideration: populations near the river are exposed to the deadly threat of flooding. Flood protection is therefore another fundamental need, demanding societal organization of significant manpower, and attracting population agglomeration to sites that can readily be protected. All of this implies a certain collective and hierarchical organization.

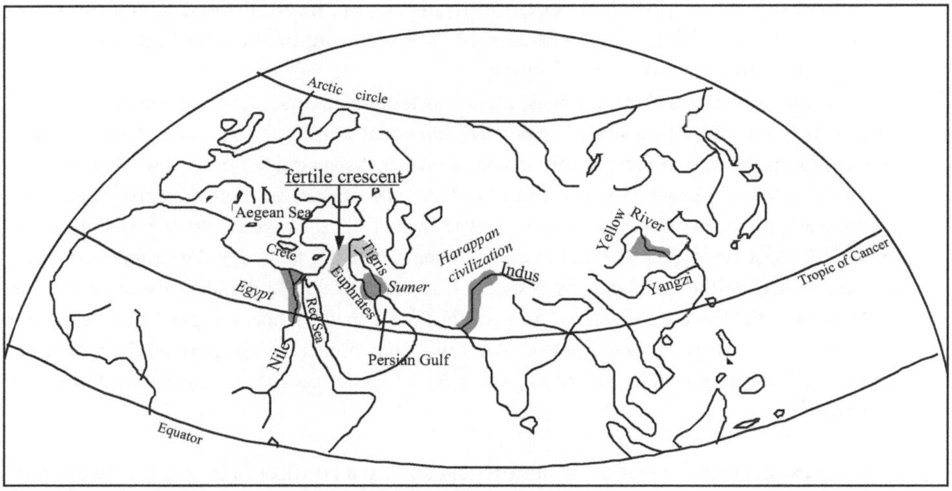

Figure 1.2. Alluvial valleys, nurseries of the first great civilizations.

The myth of the Flood, a reflection of the ancestral threat of floods and inundations?

The Euphrates River, although less capricious than the great rivers of China, is subject to major floods and changes in course. The *Flood* is a common myth in all of Mesopotamia, existing in several versions. The oldest written version, written in Sumerian and discovered at Nippur, is unfortunately in very poor condition.[12] The most well-known version, the one that very likely inspired the biblical account in Genesis, is part of *The Epic of Gilgamesh*, a Babylonian account written in the first half of the IInd millennium BC in which the description of the event is situated in the Sumerian city of Shuruppak. Here is an extract:

> "That stated time had arrived. In the morning he let loaves of bread shower down, and in the evening a rain of wheat. I watched the appearance of the weather—the weather was frightful to behold! I went into the boat and sealed the entry….
> Just as dawn began to glow there arose from the horizon a black cloud. Adad rumbled inside of it, before him went Shullat and Hanish, heralds going over mountain and land. Erragal pulled out the mooring poles, forth went Ninurta and made the dikes overflow. The Anunnaki lifted up the torches, setting the land ablaze with their flare. Stunned shock over Adad's deeds overtook the heavens, and turned to blackness all that had been light. The... land shattered like

12 Kramer, History Begins at Sumer, Chapt. 23.

a... pot. All day long the South Wind blew ..., blowing fast, submerging the mountain in water, overwhelming the people like an attack....
When the seventh day arrived, the storm was pounding, the flood was a war—struggling with itself like a woman writhing (in labor). The sea calmed, fell still, the whirlwind (and) flood stopped up.
I looked around all day long—quiet had set in and all the human beings had turned to clay! The terrain was as flat as a roof."[13]

Initially attracted by the desire to find traces of the great events of the Bible in the soil of Mesopotamia, archaeologists have searched for the ruins of the Sumerian villages in the signatures of the geological strata. Their studies have revealed significant and long-duration flood deposits, but only in some cities, and from different eras: At Ur from between 4500 and 4000 BC and then another episode dating from 2800 to 2600 BC; at Kish from three periods between 2800 and 2600 BC; and at Shuruppak, actually about 2900 BC.[14] This city, which had been quite populous before, never again attained such status. No trace was found in the very ancient city of Eridu, despite its proximity to Ur (see Figure 1.3 or Figure 2.1 of Chapter 2 for the geography of these cities.)

The Indus and the Harappa civilization
Bactria and the Margiana, on the banks of the Oxus

It is somewhat frustrating to write of the great civilization of the Indus Valley. Its origins, near the beginning of the IIIrd millennium BC, are unknown; and the reasons for its demise, a thousand years later, hardly less so. What is known results from archaeological digs at the two large sites of the twin cities Harappa and Mohenjo-Daro. The civilization is thought to be the country called *Meluha.* The towns of the Indus civilization are built on terraces raised above the flood level, their perimeters protected from erosion by brick structures. One of the most notable aspects of these towns is, as we will see later, the large number of wells and the integrated water use in the housing, including wastewater drainage. The writing of the Indus civilization has not yet been deciphered. The civilization had maritime trade with Sumer, the two cultures having some elements in common.

It is known that there were also land links in this period between the Indus civilization and Mesopotamia, passing to the north by Bactria and the Caspian Sea, supporting the trade in lapis lazuli (a semi-precious stone).[15] Indeed, at Shortughai, in the eastern portions of the Oxus basin (present-day Amou Daria) the remains of a substantial establishment of Harappean culture have been found. Here, in the middle of the IIIrd millennium BC, extensive irrigation was practiced, as evidenced by 25-km long canals (see

---

13 The Epic of Gilgamesh, Translation of Timothy R.(Wolf) Carnahan http://www.ancienttexts.org/index.html.
14 Contenau (1927), Volume 3, p. 1507; Roux (1964), Chapter 7.
15 A 1977 investigation in the northeast of Afghanistan uncovered ceramic remnants of the Indus civilization: see Lyonnet (1981).

Figures 2.20 and 7.2). Moreover, Harappean relics have been found in towns that flourished to the west in the same period, such as Altyn, proof of long-distance commerce. Should one then consider the Indus as the origin of irrigated agriculture, key to the subsequent prosperity of central Asia, whereas conventional wisdom considers Mesopotamia to be the origin? The answer is not simple.

In the VIIth or VIth millennium BC the Neolithic revolution reached the regions to the southeast of the Caspian Sea, to the plains where the Gorgan and Atrek rivers flow, and to the foothills of the Kopet Dag mountains (Figure 1.3). The Jeitun culture that developed at the foot of these mountains had already likely developed modest irrigation, even if there is no formal proof of this. Somewhat later, population increase led to new

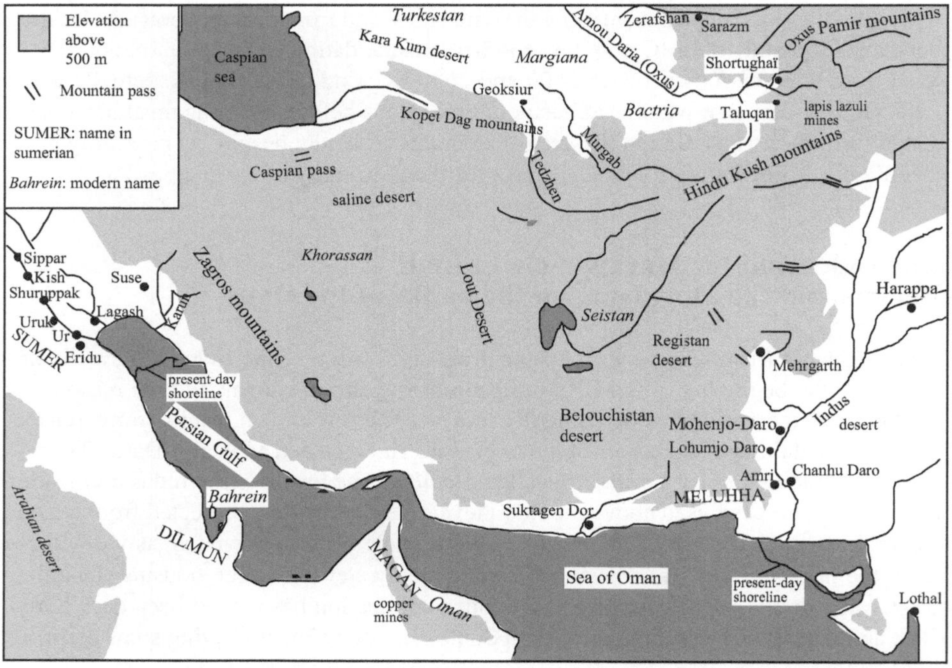

Figure 1.3 From Sumer to the sites of the Indus Valley (Harappa civilization), and the Oxus Civilization, c. 2500 – 2200 BC. The oasis of Geoksiur is shown because of its importance to the history of irrigation, but its settlement dates from much earlier (5th and 4th millennia BC)

settlements to the east, first where the Tedzhen River forms a branched delta that vanishes into the desert of Kara Kum, in the Geoksiur oasis. Artificial irrigation appears in this oasis during the IVth millennium BC. Russian archaeologists have studied derivation canals extending perpendicular to the Tedzhen. These canals are 3 – 5 m wide, 0.8 – 1.2 m deep, and can be identified along a distance of 3 km. It is interesting to note, moreover, that the most active of this irrigation activity, supposedly the first to occur outside of Mesopotamia, apparently coincides with the concentration of population into a single site of the oasis. The settlements had previously been dispersed around the oasis, suggesting that the irrigation efforts were a response to the gradual weakening of the water supply, a prelude to complete abandonment of the oasis. The neighboring cities of Namagza and Altyn, whose existence would be inconceivable without intensive irrigation, surely were the destinations of the dislocated peoples. This culture and its techniques, conveniently called the Oxus civilization, come together later in the IIIrd and IInd millennia BC in all of the Bactria-Margiana region. Traces of this civilization have even been found to the north at Sarazm, on the Zeravchan River (later to become the sites of Boukhara and Samarcand).[16]

Despite the decline of urbanization in central Asia, including the disappearance of the cities of Altyn and Namagza in about 2200 BC, the destiny of Bactria and Margiana is to remain civilized, an obligatory passageway between Mesopotamia and the Far East. The destiny of the large cities of the Indus, on the other hand, is to vanish forever. After several centuries of existence, they become uninhabited, for reasons unknown. Around 1750 BC, Harappa is destroyed, and Mohenjo Daro is abandoned.

The Yellow River valley and its catastrophic floods

In China, agriculture first appears about 6000 BC along the Yellow River. Omnipresent in the beginnings of the Chinese civilization is the legend of its founding hero, Yu the Great. In about 2000 BC Yu was apparently "master of River Control."

> "In ancient times Emperor Yü deepened the rivers and saved the empire from flood, bringing relief and security to the nine provinces."[17]

During this time, according to legend, "the flood waters rose as high as the sky".[18]

Yu's father, Kouen, had earlier been given the responsibility to curb the floodwaters, and constructed dikes over a nine-year period. But he failed in his assignment and thus was banished. Yu then decided to sacrifice himself to the river, taking the sins of all upon himself. The river, it was said, took half of Yu (tradition has it that half his face was shriveled and that he dragged one leg). He took a different approach from his dike-building father. During a period of 13 years, during which he never once returned to his home (according to legend), Yu dug canals and channels and dredged the rapids.

> "He led the Rivers to the sea, as lords file to the court".

Tradition credits Yu with a superhuman exploit: digging the channel of the Door of the Dragon (*Longmen* pass, see Figure 8.2) to provide an outlet for the waters of the Yellow River. Yu succeeded where Kouen had failed, and became the founder of the legendary first dynasty of Xia.

---

16 This view reflects discoveries by Russian archaeologists in recent decades. See Kohl (1984), Masson (1992), Sarianidi (1992).

17 Quote from the historian Sima Qian, who lived about 100 BC, Sima Qian, Shi Ji 29, transl Burton Watson.

18 The China scholar Marcel Granet has collected the legends and traditions of the ancient Chinese writings. For the legend of Yu the Great, see "Danses et legends de la Chine anciènne", first published in 1926 (pages 244 and 468 in the re-publication of 1994).

In ancient China, animals and even humans were sacrificed to the Yellow River before the followers of Confucius put an end to this practice. The battle to control the rivers, and to provide protection from their floods, is a constant theme in the hydraulic history of China. We return to this theme in Chapter 8.

In Egypt, the Nile flood is a blessing

Agriculture developed in Egypt about 5000 BC, perhaps under the influence of Mesopotamia and Syria. Subsequently, the need to take maximum advantage of the flood for land fertilization and irrigation led naturally to the organization of human resources for this purpose. Flood risk on the Nile is less than on the Euphrates and Yellow Rivers. The Nile has a relatively regular annual flood cycle, but still has sufficient variability, from one year to the next, to cause plenty or famine. The importance of the use of the flood in Egypt appears in numerous texts. In the *Book of the Dead*, in the heart of a long litany in which a person, embarking on a final voyage, proclaims his (or her) purity, are the following verses:

> "I have not stopped water when it should flow. I have not made a cutting in a canal of running water."[19]

or again, from the *Book of the Dead:*

> "O Osiris I am your son Horus and I come to work your fields for you.
> "O Osiris I am your son Horus and I come to irrigate your land"
> "O Osiris I am your son Horus and I come to work the land according to your intention"
> "O Osiris I am your son Horus and I come to dig canals for you"[20]

One can see in this hymn to the Nile, dating from about 1365 BC, a rather moving parallel between the blessings of the flood in Egypt and the benefits of the rain in other lands:

> "Thou createst the Nile in the nether world below, and thou bringest it at thy will to provide life to the men of Egypt, the men thou created for thyself. [...] Thou also giveth life to the most distant foreign lands, for thou givest them the Nile descending from the heavens *(i.e. rain)*. [...] The Nile in the heavens is for the foreigners, and for all animals of foreign lands who walk on their feet. The Nile that comes from the world below belongs to the beloved Land."[21]

Blessings and calamities

A flood in an alluvial valley is two-faced. It is always a threat when it is more severe than usual. But when well synchronized with the cycles of agriculture, the flood can be used to fertilize and water the soil before plowing and planting. This is the case with the

---

19 Papyrus of Ani; Egyptian Book of the Dead Translated by E.A. Wallis Budge http://www.sas.upenn.edu/African_Studies/Books/Papyrus_Ani.html.
20 Lalouette (1984), II, Chapt. 3, adapted.
21 Text engraved in a tomb at Amarna, after Lalouette (1984), II, Chapt. 4, adapted.

Figure 1.4 The deity "Nile" (Hapy), God of nourishment worshipped during the flood (note its full breasts), temple of Philae at Aswan (photo by the author).

summer flood of the Nile, which extends from June through August. The Tigris and Euphrates Rivers, on the other hand, flood in the spring (March through June), which coincides with the ripening of grains. Therefore in these regions, it is essential to complete the harvest before the flood - whose early arrival is always possible - can wash out the crops, especially from unprotected lands. In stark contrast to the hymns that laud the floods of the Nile, one can see a sense of urgency in a letter from a senior officer of the kingdom of Mari, located on the middle course of the Euphrates just downstream of the Neolithic site of Bouqras (see the map of Figure 2.1), written in about 1800 BC. The letter pleads to the king for help in completing the harvest before the early flood arrives:

> "My Lord: this is Kibri-Dagan, your servant. I realized that the river was in flood. It is worsening for three days I have undertaken to harvest the palace grain. But, the river is in flood. [....]. The remaining grain in my district exceeds my strength. If my Lord agrees, it is necessary to obtain help from Dumtan, Zurubban, and Hishamta *(three villages in the region).* These workers must set out now for the water is already into the fields of Zurubban."[22]

The following letter from the same official shows the need to mobilize significant manpower to prevent flood damage:

> "The Khabur *(a tributary of the Euphrates)* has flooded and Yaqqim-Addu sent me a message asking for help. I have called upon the people of Terqa *(the city of which the author is the governor)* and the people of my own district and I hastened to the aid of the flood gates of Khabur. My Lord need not worry."[23]

22 After Jean-Marie Durand (1998), Documents épistolaires du palais de Mari, II, 806.
23 Durand (1998), II, 81.

Water in the early cities

Another dimension of hydraulics appears in the early cities: that of wastewater drainage. Many medieval and modern civilizations will come to treat this problem casually, and as a result endow their cities with an atmosphere of filth. Yet the early civilizations of the East were precocious in their concern for urban drainage. Evidence of the oldest known systems for draining water from houses can be found as early as the end of the Neolithic period, around 6500 BC, at El-Kowm.[24] These comprise plaster-lined gutters dug into the ground and crossing the doorsills, as well as holes pierced through walls, and even passages below the hard surface of the ground (Figure 1.4).

Mohenjo-Daro, in the Indus valley, gives us a particularly striking example of such hydraulic works at the scale of an entire city. Here there are about 700 cylindrical wells more than 15 m deep, often located within the houses themselves. Such houses are provided with bathing rooms and often with latrines. The wastewater (including from the rooms above the ground floor) is drained through clay pipes which obliquely pass through the massive walls to connect to gutters covered with slabs, water being then conveyed into brick-covered passages dug underneath the walks between houses, and finally into larger collectors. In these drainage systems, settling basins prevent blockage by debris. In alleys and passageways that are not on the drainage system, large bottomless urns serve as cesspools.[25]

A number of the early cities of Mesopotamia, at the end of the IVth or beginning of the IIIrd millennium BC, are similarly endowed with networks of wastewater and stormwater drainage.[26] These include Habuba Kebira, Mari, Eshnunna, then Ugarit in the IInd millennium BC (see the maps of Figures 2.1 and 4.1 for the geography of these cities). At Habuba Kebira, which is a Sumerian establishment of a thousand inhabitants founded about 3500 BC and occupied for only a century and a half, various systems are used to drain wastewater. The streets, well maintained and paved with an aggregate of gravel, are equipped with U-shaped gutters made from 64-cm long sections of clay, or sometimes from conduits covered with slabs of stone, draining wastewater and stormwater outside the city walls. Even true conduits made from interlocking sections of clay pipe have been found.[27] At Mari, pipe networks whose total length can be greater than a hundred meters service multiple sanitary installations in the same house. But in Mesopotamia, this preoccupation with drainage seems to have faded away in time.

It is in Crete, in the IInd millennium BC, that we rediscover elaborate systems of water drainage – and for the first time also systems of water supply. We return to these in Chapter 4.

---

24 Stordeur (1989); Cauvin (1997), Chapt. 16, p. 239; we have adopted the dating indicated by Cauvin, recently recalibrated.
25 After Michaael Janssen (1988).
26 Margueron (1991), Volume II, p. 29.
27 Vallet (1997).

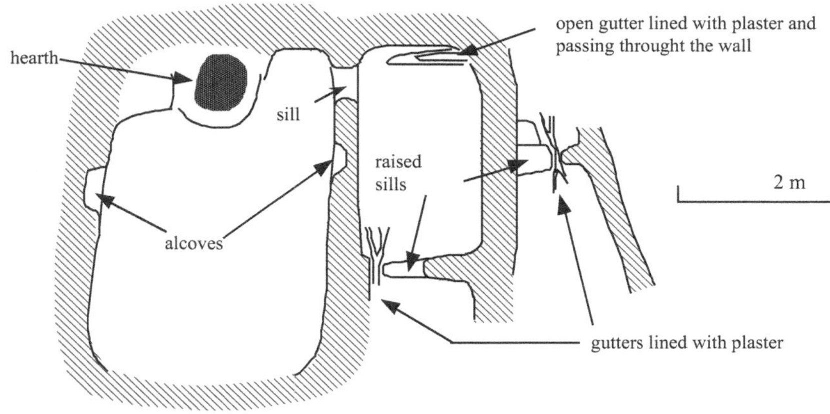

Figure 1.5 House with adjoining rooms at El-Kowm (about 6500 BC), with the remnants of drainage systems: the oldest known traces of water conveyance? (adapted from Stordeur, 1989).

First of the maritime civilizations

We complete this overview of the very first hydraulic works with a brief look at early navigation, which has prehistorical origins. The migrations that accompanied the Neolithic spread toward the western Mediterranean are thought to have been by sea. The island of Cyprus, already populated in the IXth millennium BC from Palestine or from southeast Europe, experienced the neolithic migration about 7500 BC, including the arrival of cattle from the continent. Crete was populated about 6000 BC from Anatolia. The Sicilian Neolithic was populated through seaborne migration from the Near East, thought to have passed through Greece on the way.[28]

A fundamental invention for the development of long-distance maritime commerce appeared in the IVth millennium BC: the sailboat. The oldest evidence is a boat model found in a tomb at Eridu, one of the oldest Sumerian cities, located to the south of Ur near the Persian Gulf (Figure 1.6). This model, dating from the first half of the IVth millennium BC,[29] includes what is in effect a mast socket and attach holes for the stays. In addition, two other sailboat models that date from the IIIrd millennium and are similar to the one found at Eridu, have been recovered at the mouth of the Indus.

28 Rachet (1993).
29 De Graeve (1981), Chapt. 4, F; Roaf (1990).

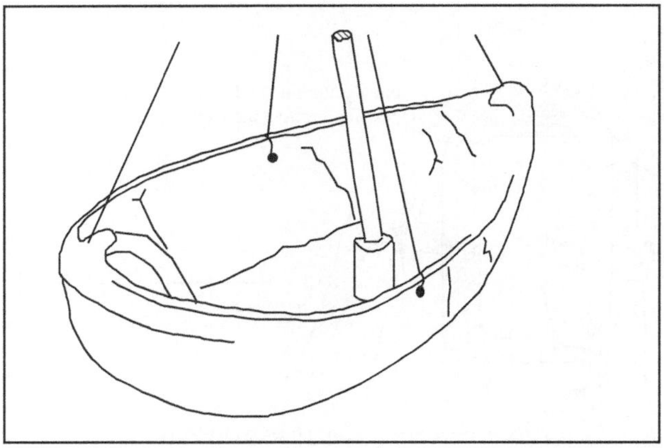

Figure 1.6 Clay model, found at Eridu, one of the first Sumerian cities, located on an arm of the Euphrates, near the Persian Gulf. This is the oldest known evidence of use of the sail (note the mounting of the mast and the holes to which the stays are attached), first half of the IVth millennium BC. (Sketch from a photograph)[30]

Important maritime routes (Figure 1.3) provided very early links among the Euphrates, the region of Bahrain (*Dilmun* of the Sumerians), the Oman peninsula[31] (*Magan*) and the Indus (likely the region called *Meluhha*). The Indus civilization founded a kind of trading base facing the Persian Gulf, at Suktagen Dor, as well as a port to the east, at Lothal. On this site there is a large rectangular basin made of clay bricks, and whose purpose is still subject to debate: freshwater reservoir, or basin of a port?

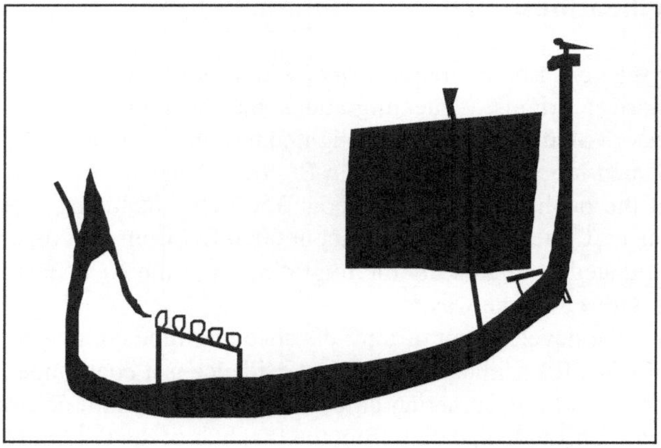

Figure 1.7 The oldest evidence of a sailing vessel in Egypt (about 3100 BC), sketch from the decoration on a vase (Casson, 1971). Note the extreme forward position of the mast

From the IIIrd millennium BC, Egypt also developed maritime routes, through the Red Sea to the legendary land of Punt (to the east of Sudan), and between the Nile and

30 Eridu, State Org. of Antiques and Herritage, Baghdad, 1982.

31 The Bahrain and Oman archaeological sites of the 3d millennium BC have been partially explored (Cleziou, 1987). It has been established that the copper used in lower Mesopotamia in this period came from the mines of the mountains of Oman. At the eastern extremity of the peninsula have been found remnants of tar used to waterproof seagoing vessels, as well as objects characteristic of the Indus civilization. The tasty dates of Dilmun are celebrated in Sumerian texts. Dilmun had numerous artesian wells and luxuriant palm groves.

the Phoenician ports of Byblos, and Sidon. The first evidence of a sailboat in Egypt, found on a vase (Figure 1.6), dates from about 3100 BC;[32] large seagoing sailing vessels appear in Egyptian engravings from 2400 BC.[33]

The first European civilization may be that of the Cyclades, in the Aegean Sea, in the Bronze Age during which commerce in metals plays an important economic role. Images of boats with tapered ends have been found on the island of Syros, engraved on objects of unknown usage, called "frying pans" by archaeologists, dating from about 2400 BC (Figure 1.8). These boats have many oars but appear to have neither mast nor sail; they are curiously similar to the boats with oars painted on Egyptian vases at the end of the IVth millennium BC. At the beginning of the IInd millennium BC, the Cretan navy operated shuttles connecting sites of the Aegean Sea, Syria (the port of Ugarit), Cyprus, Byblos, and Egypt. Cretan engravings and seals generally show boats with many oars, and a single mast carrying a sail. Starting in 1400 BC, the Greek navy, along with its Phoenician counterpart, protected and enabled long-distance commerce in the Mediterranean.

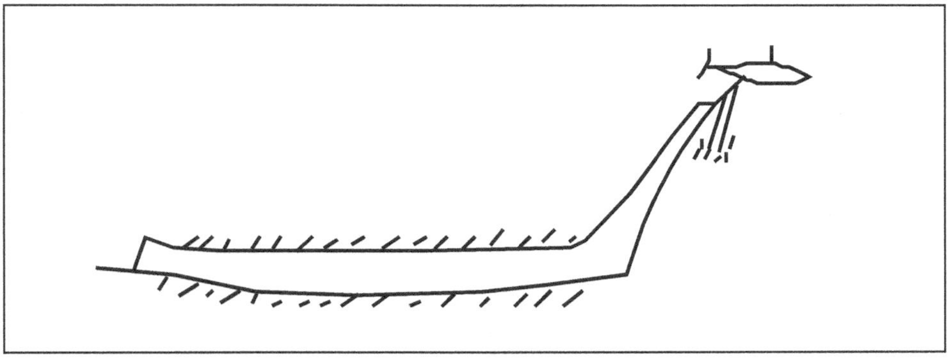

Figure 1.8. Image of a tapered boat with numerous oars, after one of the objects called "frying pans" found at Syros, in the Cylades (about 2400 BC). Note the similarity with the boat of Figure 1.7, in particular the bow figurehead, here in the form of a fish (Sketch after Casson, 1971).

Construction of hydraulic works thus was clearly driven by the need for irrigation and flood protection. But it was also driven by the need to maintain fluvial communication links, and to develop new maritime connections between the basins of the various major rivers. We will see several examples of this in later chapters, including on the Nile itself, between the Nile and the Red Sea, along the middle course of the Euphrates, between the Tigris and the Euphrates, and between the Yellow River and the Yangtze River – the famous Grand Canal.

32 Casson (1971).

33 The funeral chapel (mastaba) of the tomb of Akhethetep, on display at the Louvre museum, dates from 2400 BC and contains engravings of large sailing vessels with two masts, allowing a distribution of the wind force on the hull.

2. From Mesopotamia to the Syrian Shore: The land of the water pioneers

The triangle of land framed by the Tigris and Euphrates delta, Armenia, and the Syrian coast saw the development of the earliest large-scale techniques for water exploitation. From the IVth millennium BC through the conquest by Alexander the Great (in 331 AD), truly exceptional development occurred in this area.

The most important Sumerian city-states of lower Mesopotamia were Uruk and Larsa to the west; Umma, Lagash, and Girsu to the east; the large port of Ur to the south, and Nippur to the north. These cities imported wood and metals as raw material. The source was Bahrein *(Dilmun)* in the Persian Gulf, to which the following IIIrd millennium BC text attests :

> "Ur Nanshe, the king of Lagash (...) dug a canal (...) so that Nanshe could bring water into the canal. Boats from *Dilmun*, that far-distant country, brought wood to him."[1]

But these cities also traded with the upper valley of the Euphrates and Syria, and from this trade arose new cities on the Euphrates, like Habuba Kebira in the IVth millennium BC, then Mari from the IIIrd millennium BC, as well as a veritable explosion of Syrian cities like Ebla, Aleppo, and Qatna.

The first political unification in this vast area from the Persian Gulf to the Syrian coast was achieved by Sargon of Akkad in the Kish region. But his successors (2340 to 2200 BC) found it difficult to maintain this union. The fall of this first Empire ushered in a new era of autonomy of the Sumerian principalities, including the grand kingdoms of Lagash, Ur, and then Larsa and Mari under Semitic dynasties. Later came the establishment of the first empire of Babylon, which more or less included the domain of the conquests of Sargon of Akkad (1792 to 1594 BC).

A troubled period in the middle east began in the middle of the IInd millennium BC. This period saw competition among three great powers for Syria-Palestine: the Assyrian kingdom, the grand Hittite kingdom centered in Anatolia, and Egypt. At this time there was also rivalry for the plains of lower Mesoptotamia among the Assyrians, Babylonians, and Elamites.

Major migrations marked the transition from the Bronze to the Iron Age, around 1200 BC. In the near east, the *Sea People* (perhaps the Aegeans, themselves chased out by newcomers) left almost all the cities near the coast in ashes, and ended the Hittite Empire. Only Egypt, thanks to its power, successfully repulsed them. This troubled period does not end until the arrival of the Arameans from Arabia. They established a kingdom centered in Damascus about 1100 BC. Somewhat later, in about 1000 BC, David took Jerusalem from the Canaanites.

The great empires of Assyria, and then of Persia under the Achaemenids, were built on the ruins of this tumultuous period in the Ist millennium BC. The Assyrian Empire

---

1 Tablet of the Lagash Dynasty (2570 – 2340 BC), somewhat deteriorated. From Sollberger and Kupper (1971), p. 44.

reached its pinnacle between 890 and 606 BC, a period of delicate stability given the powerful rival Urartu to the north (Armenia), the revolts of Babylon, and the rise of the power of the Medes to the east. Assyria even extended its domination into Egypt, but only for a brief period. With the fall of the Assyrian Empire, Babylon again came to the forefront of the political scene in Mesopotomia, but not for long (604 to 539 BC). This period ends when the Persian, Cyrus the Great, and Cambyse, his successor, conquer the entire region, including Syria-Palestine, Anatolia, Egypt, and even Bactria.

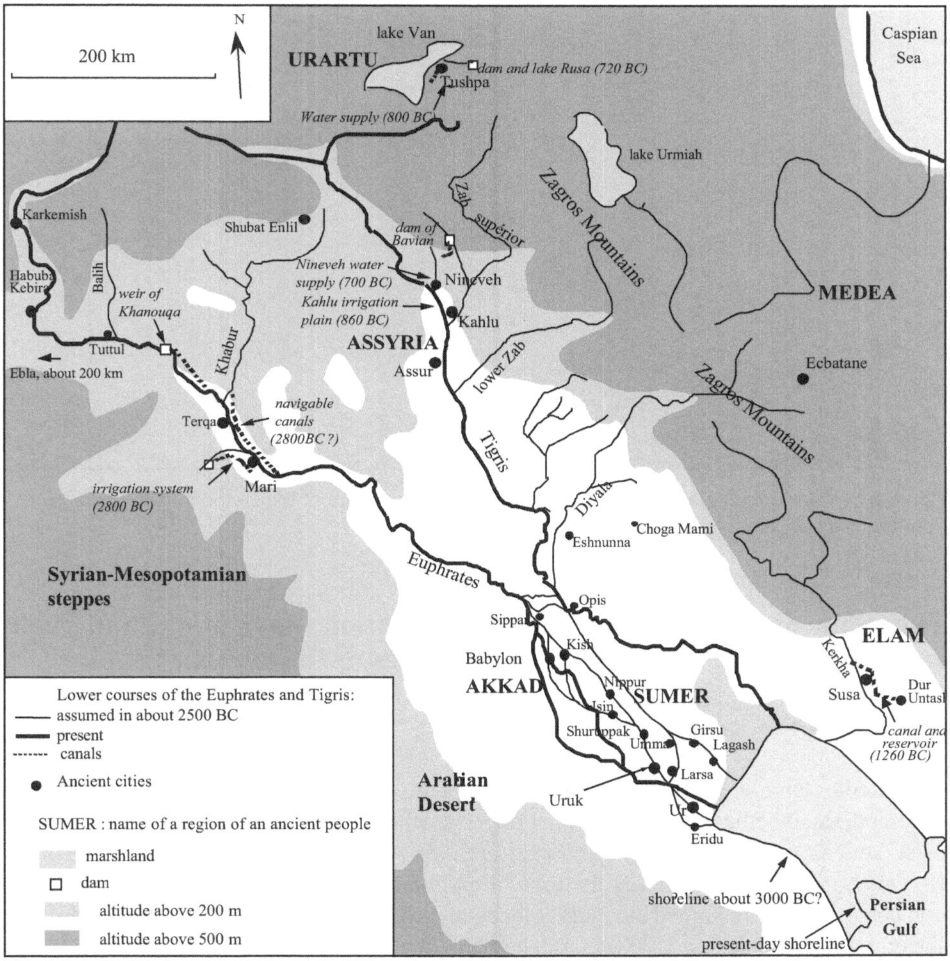

Figure 2.1 Principal sites of ancient Mesopotamia; overview of the major hydraulic works.

This region is topographically unbounded, without natural limits or constraints. Civilizations came and went, but all of them depended on the efficacy of the irrigation systems inherited from their predecessors. Hydraulic technology, including the first great canals and dams, are passed from one civilization to another and spread outwards from the region. Let us first look at the great alluvial plain of lower Mesopotamia, the ancient land of Sumer and of Akkad.

Figure 2.2 The Euphrates valley and the irrigated plain upstream of Mari – looking downstream from the cliffs of Doura-Europos (photo by the author).

The plain of lower Mesopotamia: irrigation, navigation, and river engineering from the Sumerian city-states to the Persian Empire

Irrigation practice in lower Mesopotamia

Field studies have shown that the urbanization of the IIIrd millennium BC developed along watercourses, whether they were natural river branches or artificial canals. Notable among these studies are those of the American archaeologist Robert McAdams (Figure 2.3). However, it seems likely that at the time of the independent city-states, secondary irrigation canal systems remained essentially local, their layout dictated by the nature of the soil surrounding each city: a band of gardens here, a band of irrigated cereals there. Canals effectively define the boundaries between cities. And of course conflicts over the shared use of water arose along such boundaries. For example, around 2460-2400 BC there was a long dispute between the neighboring city-states of Lagash and Umma. Land concessions in the plain of Gu-edinna, at the boundary between the two cities, were taxed through payments in grain. When water users defaulted on their payments, the irrigation water was simply cut off. It was then necessary to resort to force, as occurred on several occasions under Eannatum king of Lagash, then later during the reign of his nephew Entemena:

> "Eannatum, the *ishakku (prinice)* of Lagash, the uncle of Entemena, the *ishakku* of Lagash, marked off the boundary with Enakalli, the *ishakku* of Umma; led out its *(the boundary's)* ditch from the Idnun *(canal)* to the Guedinna; inscribed *(several)* steles along that ditch (....) He *(Ennatum)* levied a tax on them *(in compensation for conceded lands)* (....) Because this barley remained unpaid – *(besides)* Ur-Lumma, the *ishakku* of Umma deprived the boundary ditch of Ningirsu *(and)* the boundary ditch of Nanshe of water; ripped out its *(the boundary ditch's)* steles *(and)* put them to fire (....) and *(finally)* crossed the boundary ditch of Ningirsu-Enannatum fought with him in the Gana-ugigga *(where are)* the fields and farms of Ningirsu, *(and)* Entemena, Enannatum's beloved son, defeated him (....) At that time *(however)* Il, the temple-head of Zabalam, ravaged *(?) (the land)* from Girsu to Umma. Il took to himself the *ishakku*-ship of Umma; deprived of water the boundary ditch of Ningirsu, the boundary ditch of Nanshe, the Imdubba of Ningirsu, that tract *(or arable land)* of the girsu tracts which lies toward the Tigris (....) Entemena (...) made this *(boundary)* ditch from the Tigris to the Idnun in accordance with the straightforward word of Nanshe."[2]

Later on, development and use of the system of large canals saw coordination at a larger scale. Around 1800 BC, the great Babylonia king Hammurabi, who had just united the country, put forth a series of edicts, elements of a civil and penal code.

The prologue of this code notes that in the 33rd year of his reign the sovereign built a canal called "Hammurabi is the prosperity of the people", designed to supply the Sumerian cities of Nippur, Eridu, Ur, Larsa, Uruk, and Isin. This suggests that the branch of the Euphrates flowing between Nippur and Uruk (see Figure 2.3) had been channelized.

This is also suggested by the proclamation of the 33rd year of the reign:

> "Hammurabi has dug the canal "Hammurabi is the prosperity of the people" – the canal that is taken care of by (the gods) An and Enlil – and thus provided the cities of Nippur, Eridu, Ur, Larsa, Uruk, and Isin with a steady supply of water for their prosperity and made it possible for the inhabitants of (the lands of) Sumer and Akkad, who had been scattered (by war), to return to their settlements".

The code of Hammurabi includes edicts that regulate use of the irrigation system. It requires that riverside inhabitants maintain the dikes that protect the fertile lands near the rivercourses, and sets compensatory penalties for those who are remiss in this responsibility:

> "If a man has been slack in maintaining [the bank of] his [field] and has not maintained [his] bank and when a breach has occurred in his [bank] and so he has let the waters carry away (the soil on) the water-land, the man in whose bank the breach has occurred shall replace the corn which he has (caused to be) lost."
>
> "If a man has released the waters and so has let the waters carry away the works on his neighbor's field, he shall pay ten gur of corn for every bur of land."[3]

---

2 Kramer (1986), Chap 6. To commemorate his victory over Umma, Eannatum had the famous "stela (stone monument) of the Vultures" engraved; it can be seen in the Louvre museum.

3 Prologue adapted from Finet (1996). Laws 53 and 56 from the Translation of L.W. King (1910) Edited by Richard Hooker http://www.wsu.edu:8080/~dee/MESO/CODE.HTM . The proclamation of year 33 of the reign is from Renger (1990), from Driver and Miles, Babylonian Laws

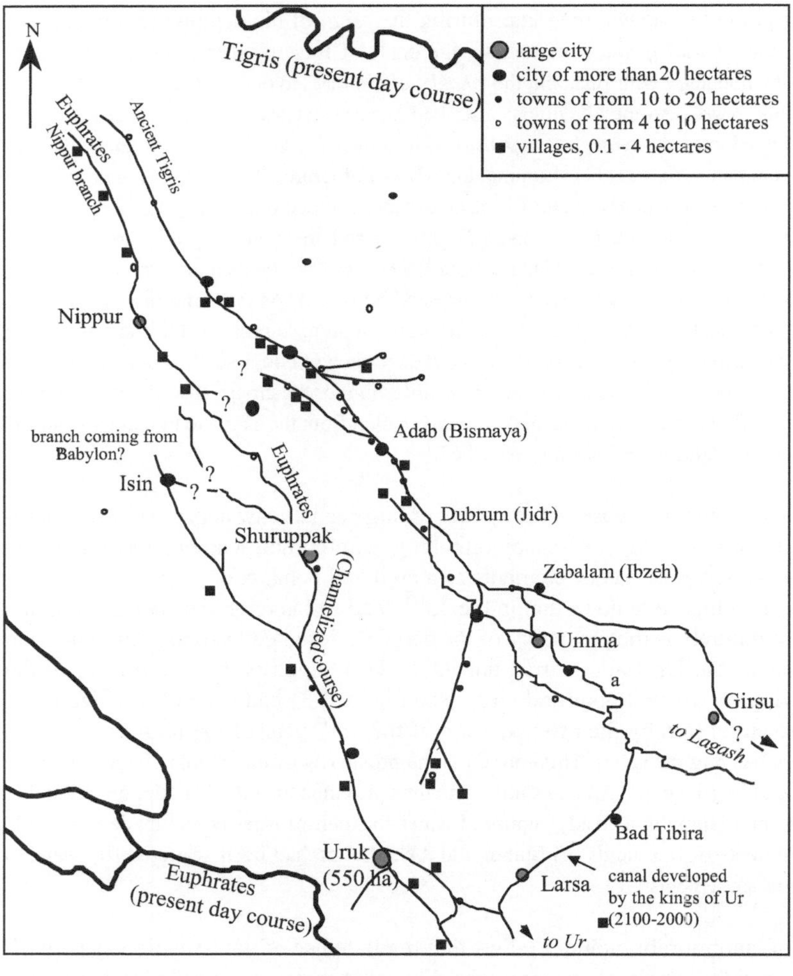

Figure 2.3 Urbanization of the hydrographic network reconstituted between Nippur and Uruk, around 2500-2000 BC. This map is based on the work of the American archaeologist Robert McAdams. The hydrographic network is reconstructed from (undated!) artifacts of meander fossils and from the juxtaposition of the sites. It was not until the second half of the IIIrd millennium BC (end of the ancient Sumerian dynasties and the empire of Akkad) that the rivercourses stabilized, undoubtedly under the influence of the artificial canals. The branch of the Euphrates between Shuruppak and Uruk, in particular, appears remarkably linear and regular from this period on. In the region of Umma, the principal course of the river shifts from (a) to (b) around 2400 BC. The large loop that flows toward the southwest and that supplies Bad Tibira and Larsa was most likely dug by the kings of Ur (2100-2000 BC).[4] Uncertain branch locations are shown by "?".

---

4 The ancient hydrographic network between Uruk, Larsa and Umma was analyzed by Hans Nissen (McAdams and Nissen, 1972). Regarding the confluence of the Tigris and Euphrates, see Gashe, Tanret, Cole, and Verhoeven (2002).

If gravity irrigation is to be used for large-scale cultivation of cereals, either the fields must be below grade compared to the river, whose bed is incised within the natural levees caused by progressive alluvial deposits, or the canal must be constructed on raised fill to bring the water above the level of the fields.

Other texts contain evidence of a hierarchy in canal structure. The river branches or large navigable canals are called, without distinction, *id* in Sumerian, or *naru* in Akkadian. Along these *narus*, intake works supply secondary canals that in turn deliver water to basins called *nag-kud* in Sumerian or *natbaktu* in Akkadian. These basins are essentially rectangular reservoirs, varying in length from 12 to 72 m, and 1 to 12 m wide. These *natbaktus* are built on a plain by means of earthen dikes reinforced with grass or brush, from 1 to 5 m high. These reservoirs, apparently fitted with outlet gates, first provided water storage, but also and importantly made it possible to redistribute water toward raised ditches along the top of small dirt ramparts that carried water into the fields.[5] The irrigation operations involved controlled flooding of fields to be cultivated. After sitting in the field for some time, the water is then drained, leaving the level and damp field ready to be plowed and planted. Once the new plants have germinated, the field is flooded once again, then twice more during the growth of the barley to improve the yield. This gravity irrigation was practiced from the very beginning, as shown in a collection of detailed advice from a Sumerian farmer to his son:

> "When you are about to cultivate your field, take care to open the irrigation works *(so that)* their water does not rise too high in it *(the field)*. When you have emptied it of water, watch the field's wet ground that it stays even...."[6]

We also have direct external testimony, albeit delayed, of these irrigation works from Greek travelers. Perhaps in trying to understand the origin of war, Herodotus of Halicarnassae, a Greek citizen of the Persian Empire, traveled the known world following the Median wars during which the Persians burned Athens. Around 460 BC he visited Egypt and Mesopotamia. Here is what he wrote about irrigation practices in the Babylonian region:

> "Very little rain falls in the land of Assyria, and this little is what nourishes the root of the crop; but it is in its watering from the river that the corn crop wins to its ripeness and the bread grain comes into being. It is not as in Egypt, where the river itself rises over the fields; in Babylon the watering is done by hand-operated swing beams."[7]

This text illustrates a second irrigation method - lifting of water to the crops to be irrigated. The technique was probably used only for small-scale agriculture on the marshes. The "machine" most often used in this period was the *shaduf*, a balance beam provided with counterweights (Figure 2.4). Its use appeared in Mesopotamia in the IVth millennium BC, and likely migrated into Egypt in the beginning of the IInd millennium BC.[8]

---

5 See Steinkeller (1988) for a description of the nag-kuds according to Sumerian texts, also Van Soldt (1988) for the equivalent natbaku from a later date. The concept seemed to last more than a thousand years.

6 after a Sumerian text called "The Farmer's Almanac" (Kramer, 1986, Chapter 11).

7 Herodotus, The History, Translation of David Grene, book I, 193.

8 after Bonneau (1993), p. 94.

The system relied on muscle power, but it was simple, efficient, and easy to maintain. It became a permanent feature of traditional irrigation techniques of the Near East.

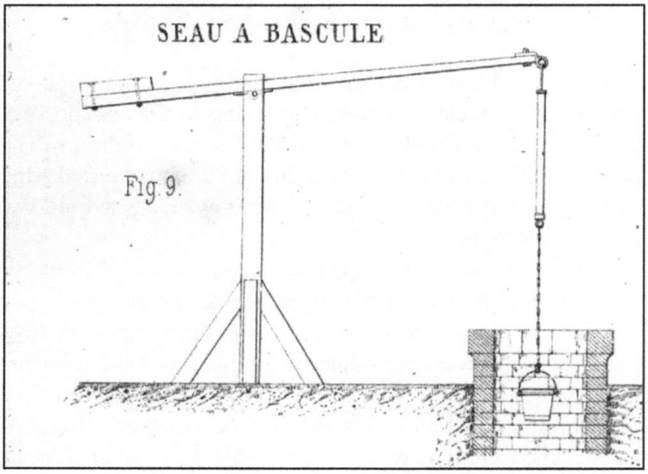

Figure 2.4 The balance beam, or shaduf. - reproduction of a bas relief from the palace of Sennacherib at Ninveva, Moussly, 1951

opposite, illustration from 1885 (Poillon, ancient archives of the Ecole Nationale des Ponts et Chausseés ENPC)

Another traveler, somewhat less of a passive observer than Herodotus, was Xenophon, a former student of Socrates. In about 400 BC, he joined in the adventure of an army of mercenary Greeks, the Ten Thousand, hired to support the revolt of a prince within the Achaemenids royal family. After the death of the prince and his generals, it was Xenophon who had to lead the difficult retreat. He crossed numerous canals between the Euphrates and the Tigris, a bit to the north of Babylon:

> "...crossing on their way two canals, one by a stationary bridge, and the other by a bridge made of seven boats. These canals issued from the Tigris river, and from them, again, ditches had been cut that ran into the country, at first large, then smaller, and finally little channels...."[9]

The irrigation principles of lower Mesopotamia were followed beyond the end of the Persian Empire of the Achaeminides. In the lowest areas of the plain, where the silts deposited by the two rivers caused the soil to be cohesive and sticky, constant work was necessary to keep the network in good operating condition. Later on, around the beginning of the modern (i.e. Christian) period, the Greco-Roman geographer Strabo writes:

> "Now this is the origin of the canals; but there is need of much labor to keep them up, for the soil is so deep and soft and yielding that it is easily swept out by the streams and the plains are laid bare, and the canals are easily filled, and their mouths choked by the silt; and thus it results again that the overflow of the waters, emptying into the plains near the sea, form lakes and marshes (....) And indeed there is also need of quick work in order to close the canals quickly and to prevent all the water from emptying out of them. For when they dry up in the summer, they dry up the river too; and when the river is lowered it cannot supply the sluices with water at the time needed, since the water is needed most in summer, when the country is fiery hot and scorched; and it makes no difference whether the crops are submerged by the abundance of water, or are destroyed by thirst for water."[10]

We will see in Chapter 7 that this delicate equilibrium is not sustainable, though this does not become apparent for several centuries to come. Even so, a large canal will be built parallel to the Tigris and on its left bank (the *nahr Awan*), departing from the river at a point situated 24 km upstream of Samarra (about a hundred kilometers upstream of the present-day Baghdad), and terminating about a hundred kilometers southeast of Baghdad, collecting the waters of the Diyala along its way. It will be completed in the sixth century AD under the reign of the Sassanide sovereign Khusraw I.[11]

Navigation between the Tigris and Euphrates

The importance of navigable waterways to the economy of Mesopotamia cannot be overestimated. The code of Hammurabi, from which we have already cited several

---

9 Xenophon, Anabasis, Book II, Chapter IV, 13, Translation of C. Brownson.
10 Strabon, Geography, (Translation of F. Lasserre), Les Belles Lettres, 1981, 16, 9-10, adapted.
On the other hand it seems now to have been established that contrary to certain legends, there is not a true dam constructed on the Tigris before the period of the Persian Empire.
11 See Schnitter (1994).

extracts, includes laws that regulate navigation on the rivers and canals. It sets compensatory payments for shipwrecks or breakdowns, and establishes right-of-way rules:

> "If a boat traveling upstream collides with and sinks a boat traveling downstream, the owner of the sunken vessel will officially declare, in the presence of God, all that was lost in his boat, and the boatman of the upstream-traveling vessel that caused the sinking will pay for the boat and everything that was lost."[12]

Several of the canals flow by gravity from the Tigris toward the Euphrates, reflecting the ancient confluence that existed in the IIIrd millennium BC in the region of Sippar (Figure 2.1). This did not go unnoticed by Greek observers, Herodotus in particular (this extract comes after the text cited earlier regarding irrigation):

> "Babylon is in its entirety, like Egypt, crisscrossed by canals; the largest one is navigable, oriented in the direction of the winter sunrise, and joins the Euphrates to the Tigris, the river on which Nineveh is situated."

According to Xenophon, four of these large canals are very respectable in size:

> "Here also are the canals, which flow from the Tigris river; they are four in number, each a plethrum wide *(about 30 m)* and exceedingly deep, and grain-carrying ships ply in them; they empty into the Euphrates and are a parsang *(about 5.5 km)* apart, and there are bridges over them."[13]

These canals also have a strategic function in the context of the Assyrian domination. The Assyrian capital Nineveh is indeed located on the Tigris as stated by Xenophon. At that time this river disappeared into the swampland and was not navigable to its mouth. Therefore the canals between the Tigris and the Euphrates enabled boats to get to the Persian Gulf from Nineveh. In about 700 BC a flotilla constructed at Nineveh for the Assyrian king Sennacherib descends the Tigris to Opis; it then transits across to the Euphrates in anticipation of a military operation in the Persian Gulf. It is this same Sennacherib, exasperated by numerous revolts against the Assyrian power, who destroys Babylon in 689 BC - not only by fire, but also by water. He floods the city using a branch of the Euphrates, called the *Ahratum*, that flows out from Sippar.

But later on, in the age of the Achaemenid Persians, it is by contrast from the Persian Gulf that invasions were dreaded. The Persians were not sailors; they decided to block navigation on the Tigris and Euphrates by constructing weirs across them. They brought traffic to a standstill on the two great arteries of the country. When Alexander the Great later becomes master of the country, he removes these weirs as a gesture of liberation.[14]

River engineering and flood protection

Hydraulic development involved not only the digging of canals, but also the restoration or maintenance of river courses. During the domination of Larsa in lower Mesopotamia

---

12 law No. 70, from the edition of André Finet (1996).

13 Xenophon, Anabasis, Book I, Chapter VII, 14, Translation of C. Brownson.

14 Strabon, Geography, 16, 10 (Translation of F. Lasserre), Les Belles Lettres, 1981, adapted.

(1932 to 1763 BC), it became necessary to rehabilitate the river system. The king Sin-Iddinam reestablishes the course of the Tigris, around 1845 BC, using paid labor:

> "When An, Enlil, Nanna and Utu (Sumerian gods) blessed me with a good reign of justice and long days (...), to obtain fresh water for the cities of my country, (...), I fervently prayed to An and Enlil. They answered my fervent prayers and, by their absolute orders, charged me with the mission to dredge out the Tigris, to restore it (into its previous state) and to give the days of a long life to my name. So (...) I grandly dug out the Tigris, the river of abundance of Utu. I raised the top of the slope, the old embankment (...); I transformed the Tigris into a freely flowing water; I established at Larsa, in my country, an eternal source of water, a never-ending abundance. When I dredged the Tigris, the great river, the wages of each worker were 3 *ban* (?) of Barley, 2 *sila* of bread, 4 *sila* of beer and 2 *sicles* of oil: this they received each day; nobody received more or less. With the strength of my country, I brought this project to a good end. With the reins and the decrees of the great gods, I restored the Tigris, the vast river, and I affirmed my good name for the far distant future."[15]

A succession of kings of Larsa, in particular Rim Sim who was the last, dredged and reestablished all the fluvial system from Lagash and Larsa down to the sea. This rehabilitation included the large branch flowing to the southwest on Figure 2.3 that supplies Bad Tibira and Larsa; this loop had been dredged earlier by the kings of Ur.

The floods of the Tigris and Euphrates, out of phase with the growing cycle of the grains, are more of a menace than a blessing. We know that the dikes or levees protect the cities and crops against floods, and stabilize the course of the rivers to the extent possible. At Babylon, major works were accomplished to this end. Herodotus attributes these works to two queens:

> "The first ruled five generations before the second; her name was Semiramis, and she built those dikes on the plain that are so remarkable to see; before that, the river used to run all over the plain and flood it."[16]

In the eyes of the Greeks, Semiramis is a veritable legend. According to Georges Roux, Semiramis is possibly Sammuramat, an Assyrian queen who occupied the throne around 800 BC. All of the works mentioned herein (and especially the works described below) can also, according to Georges Roux, be attributed to the queen Naqia, the widow of Sennacherib. It is in fact eleven years after the destruction of Babylon by Sennacherib, a destruction that was considered to be a sort of sacrilege, that the son of Sennacherib undertakes the reconstruction of the city (this then puts us in 678 BC). But the biggest astonishment remains to come; let us once again listen to Herodotus:

> "The second of these queens was called Nitocris (....) First, then, as to the Euphrates, which flows right through the middle of the city of Babylon. Formerly it was straight, but she made it so crooked, by digging canals above the city, that the river in its course comes three times to one of the Assyrian villages (....) This is what she did, she built an embankment along either shore of the river that is, in greatness and height, very wonderful in its dimensions. Far

---

15 From Sollberger and Kupper (1971), "Royal Sumerian and Akkadian Inscriptions". For the works of the kings of Larsa, see Renger (1990) and Charpin (2002).
16 Herodotus, The History, book I, 184-186, Translation of David Grene.

above Babylon she dug a basin for a lake, stretching it by the side of the river and a little away from it, and in depth she dug it always down to find water, and in breadth she made the circuit of the lake to be fifty-two and a half miles *(84 km)* (....) She did both of these things – the making the river crooked and turning the basin into a marsh – so that the river might be slower, as it was broken by the many bends, and that the courses into Babylon itself might be crooked, and that then, after this, should come the long circuit of the lake. These works were built at precisely the point of her country where were the passes of entry and the shortcuts from the road out of Media, so that the Medes might not get into contact with her people and learn of her affairs.

"With these defenses she surrounded her city, but she added another work that grew out of them (....) For when she had dug the basin of the lake (....) she had huge stones cut, and when the stones were ready and the basin had been dug, she turned the entire stream of the river into the place that was dug. While it was filling, the old riverbed dried out; and she bricked with baked bricks, in fashion like to the walls, the banks of the river in the city and the descents from the gates leading down to the river (....) When the dug part had become a lake, filled by the river (...) she turned the Euphrates into its old course...."

This is a fine example of river engineering. According to recent work by Charpin (2002), the artificial lakes and dikes might in fact have been built by Samsuilina 1749-1712 BC, the successor of Hamurabi. If the account is correct, cut stone blocks were used for the difficult operation of blocking the river to divert water into the artificial lake. It is interesting to consider the roles of the meanders and the lake. The meanders effectively reduce the slope of the river, since the same drop is attained over a greater length. If the width and discharge remain the same, one can estimate that reduction of the slope by a factor of two, over a long length, will reduce the current velocity a little less than 25%, and raise the water depth in about the same proportions. Therefore it becomes necessary to raise the dikes. The lake, located upstream of the city, very likely plays a role in attenuating floods, capturing excess water through overflow when the river rises above a certain level.

The accounts of Herodotus include other examples of the Mesopotamian know-how in river course modification. Cyrus, the founder of the Persian Empire, suffered the loss of one of the sacred white horses of his team when crossing the Diyala (Gyndes to the Greeks). Furious at this affront to his power, he put his entire army to work for a full year (according to tradition) to break down the Gyndes into 360 small canals. Herodotus also reports that Cyrus (in 539 BC) enters Babylon without resistance using the hydraulic works described above:

"When he came to the lake, Cyrus dealt with it, and with the river, just as the Babylonian queen had done; he directed the river by a canal into the lake, which had become a marsh, and so, when the river had sunk, its old stream became fordable."[17]

Once he had entered the city itself, Cyrus further extended the armoring of the banks. He didn't miss the chance to commemorate this feat in carving the following inscription:

"I added to (....) the banks protected by bricks, in the low-lying areas of the city, that a pre-

17 Ibid., book I, 191.

vious king had begun to build…"[18]

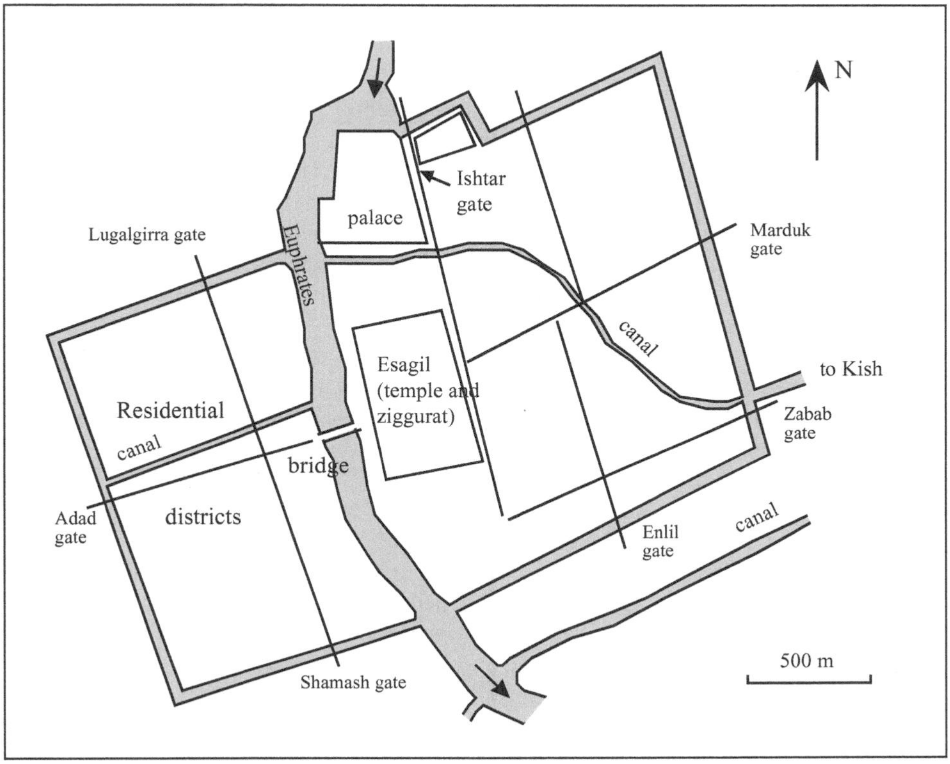

Figure 2.5 The Euphrates and the Babylonian canals (after Margueron, 1991).

Between the Middle Euphrates and the Syrian coast: dams and canals from the IVth to the IInd millennium BC

The mysteries of Jawa: the oldest known dams, on the slopes of Kjebel Druze (Djebel el Arab) – end of the IVth millennium BC

The site of Jawa, a hundred kilometers to the northeast of Amman in Jordan, is an enigma. It is an arid zone, in a desert of rough black basalt. The only source of water, apart from very infrequent rains, is the winter flood of a seasonal watercourse that comes down from Djebel Druze, the wadi Rajil. The site is somewhat off the track of communication routes, but on the other hand it is easily defendable. Jawa had some 2,000 to 3,000 inhabitants toward the middle or the end of the IVth millennium BC. In this region

18 Inscription on a clay cylinder called "the cylinder of Babylon"; Translation of Lecoq (1997), Chapter 3.

where there had never been any settlements before, and where there will not be any new ones for several hundred centuries to come, these people built a fortified walled city, a citadel in essence. A British expedition, led by Svend Helms, explored the site between 1973 and 1976, and postulated that these people were refugees or migrants from some urban culture.[19]

For their water supply, the inhabitants of Jawa built an elaborate system that fills reservoirs with the runoff of winter and spring rains, and also with water diverted from the wadi Rajil during the floods of November and May. Three weirs on the wadi Rajil divert water into stone-lined canals that are several kilometers long and convey flow to ten reservoirs (Figure 2.6). Some fifteen gates control the capture and distribution of water toward either the multiple reservoirs, or toward irrigated fields. Three of the reservoirs (Nos. II to IV, with a total volume of 42,000 m^3) supply the city itself. Several other smaller reservoirs, totaling some 10,000 m^3, supply animal pens. Cistern No.I, upstream of the city, is dug into a cavity of basalt. Open-air reservoirs Nos. VII, IX, and X are downstream of the city. One of the reservoirs (No. IV on the figure) is formed by a true dam, 4.5 m high and 80 m long. The dam comprises two stone walls confining a central impermeable earth core that is two meters thick (Figure 2.7). A hydrologic study has shown that the storage of water from the two combined sources (rainfall plus floodwaters of the wadi) could supply 3,000 to 5,000 inhabitants and their animals for an entire year.[20] The people of Jawa had even begun to raise the dam to a height of 5.5 m, and to build another similar reservoir along the course of the wadi Rajil itself. But these projects were destined to remain unfinished, as the city was abandoned after only fifty years of existence. The nature of the catastrophe that led these people to return to their desert wandering remains unknown to this day.

The hydraulic know-how seen at Jawa is not an isolated example. Somewhat later, around 3000 BC, semi-nomadic shepherds settle at the foot of the same mountain, 80 km to the north. Although they are not former city-dwellers like their earlier neighbors at Jawa, these shepherds are driven by the same preoccupation with their security, since they build a wall around their encampment. And like the inhabitants of Jawa, as soon as they arrive they master the same hydraulic techniques. Here, at Khirbet el-Umbashi (Figure 2.8), they form a reservoir in the bed of the wadi itself by building a dam right to the foot of the city wall. This earth dam, 30 – 40 m long, is later raised (as at Jawa) to reach a height of 6 to 8 meters. This reservoir then stores water brought from the two wadis to the north through a three-kilometer canal, as well as water from other direct runoff. An oblique weir, two kilometers upstream on the wadi Umbashi, diverts water toward another very large reservoir (some 30,000 m^3) developed in a natural depression bounded by massive levees that attain heights of 2 to 3 m and a base width of 25 m.

---

19 Helms (1987 a-b).
20 Helms (1987 b).

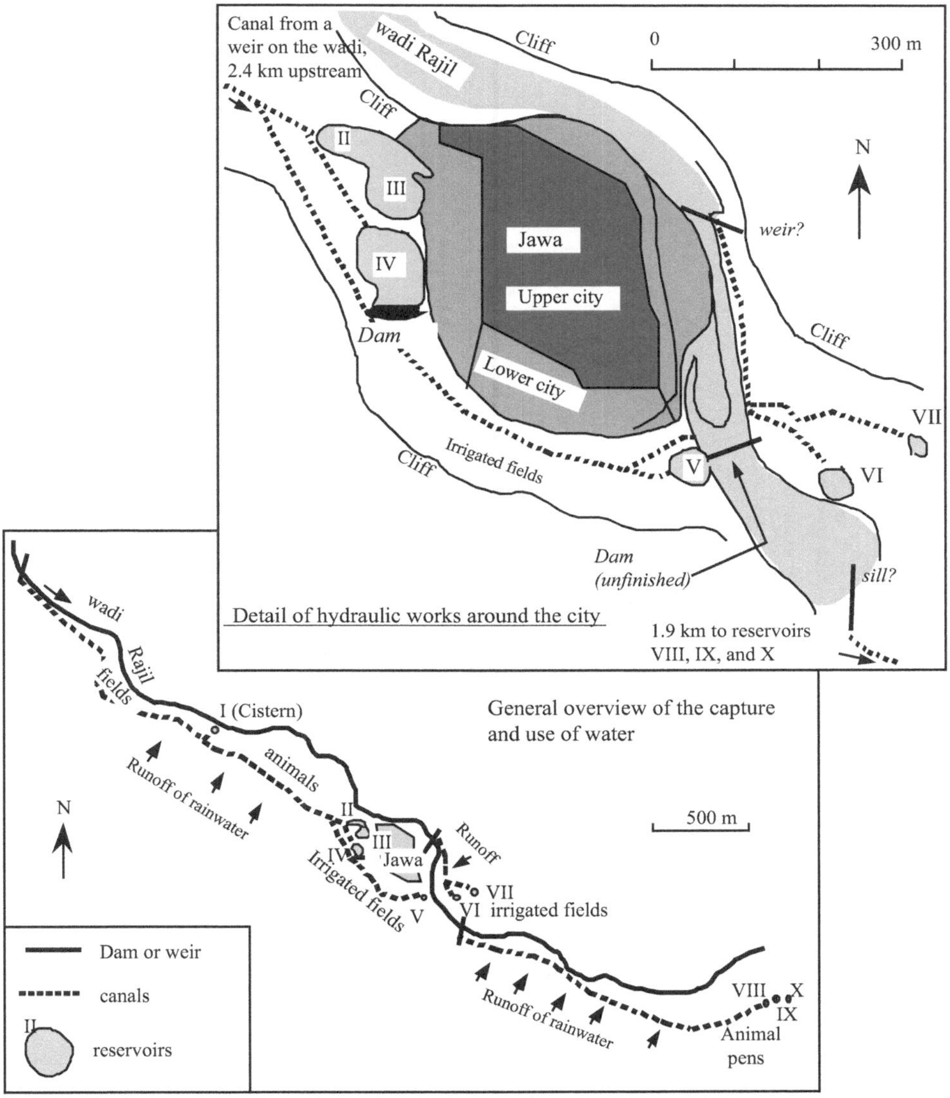

Figure 2.6 The hydraulic system of Jawa – end of the IVth millennium BC (after Svend Helms, 1987).

Figure 2.6b. The upstream wall of the dam forming reservoir IV in Jawa, looking toward the east. In the background, one can see the wall of the upper city of Jawa (photo by the author).

Figure 2.6c. The reservoir VI in Jawa, looking toward the west. In the background, one can see reservoir V, and on the right the upper city of Jawa (photo by the author).

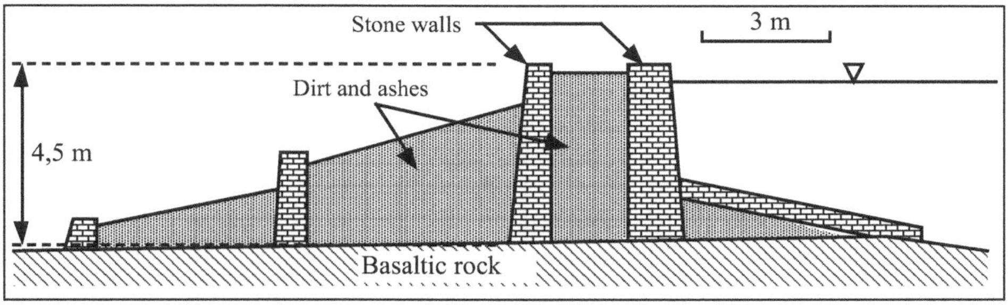

Figure 2.7 The dam for reservoir No. IV at Jawa; the oldest known dam (Helms, 1987b). Vogel (1991) gives a similar but slightly different reconstitution of this dam, with only two stone walls (the major ones), a downstream earthfill slope of 0.4 : 1, and a thinner upstream rock drainage layer.

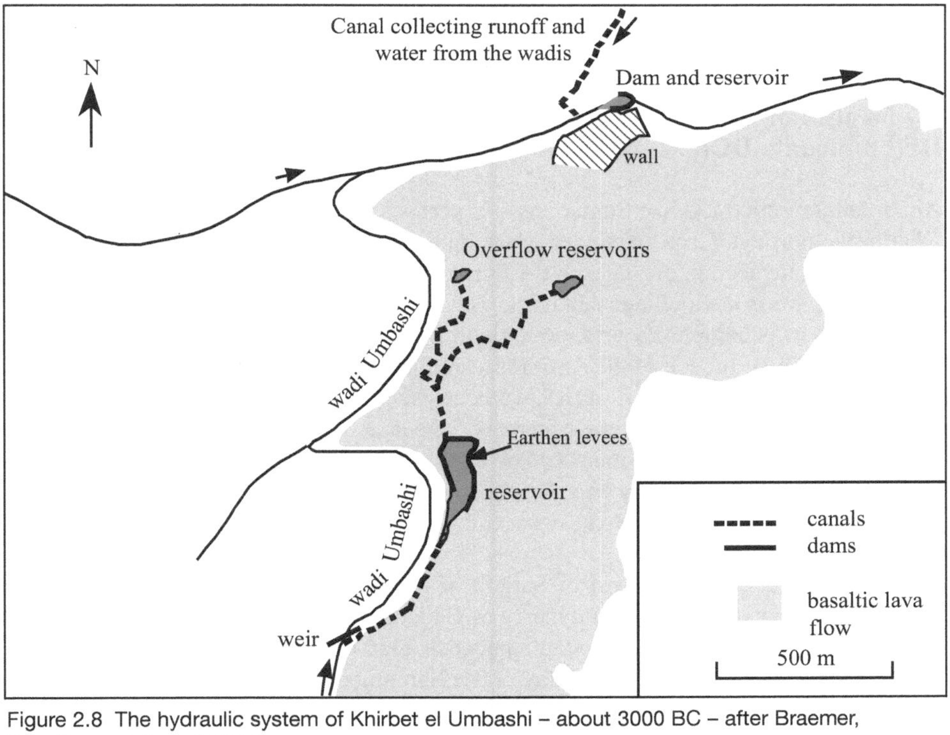

Figure 2.8 The hydraulic system of Khirbet el Umbashi – about 3000 BC – after Braemer, Echallier, and Taraqji (1996)

At a neighboring site, Hebariyeh (in today's Lebanon) one again finds the same kind of development, with a large watertight reservoir fed by diversions along a wadi and canals several kilometers in length. Unlike the ephemeral urban establishment of Jawa, these last two sites survive until about 1500 BC.[21] One can find additional remains of

21 Helms (1987b).

canals that collect rainfall along hundreds of meters of length[22] at other sites more to the west along the shores of the Dead Sea, from the beginning of the IIIrd millennium BC.

These hydraulic works represent *the oldest known dams*. It is impossible for them to represent merely local and isolated inventions. How could the new settlers at Jawa or at Khirbet el Umbashi have known how to construct, then and there, such relatively highly evolved hydraulic systems without having known of similar projects before? Therefore it seems clear that the construction of weirs and dam-reservoirs on intermittent watercourses had already become a technique known throughout the region from the end of the IVth millennium BC. The oasis of Damascus, occupied from Neolithic times, could have been at the origin of these techniques - though nothing is known of it during the period that interests us here. It should be remembered as well that the middle of the IVth millennium BC is a period of expansion of the Sumerian civilization. Habuba Kebira, with its highly perfected sewers, arises on the Middle Euphrates around 3500 BC, then is abandoned at some later date that could be close to the time of the founding of Jawa. With the beginning of the IIIrd millennium BC comes the period of the founding of Mari.

Hydraulics of the kingdom of Mari, on the middle Euphrates (IIIrd and IInd millennia BC)

About 2800 or 2900 BC, the Sumerians – or perhaps a people already established somewhat to the north at Terqa – founded Mari, on the middle course of the Euphrates. The site is at the intersection of routes to the Syrian coast, near the outlet of the fertile valley of Khabur. This is not a village that has grown and evolved, but rather a "new city". The region of Mari is completely arid, no agriculture is possible without irrigation. Yet 200 km to the north, at the toe of the Anti-Taurus mountains, one finds land that is naturally well watered. Therefore Mari must have been established where it is for reasons related to commerce and control of the waterway. The land of Sumer has to import raw material such as wood, stone, and metals. The early (and temporary) city of Habuba Kebira, further upstream, undoubtedly had the same needs. From the very beginning, Mari is a center of bronze metallurgy. Boats descending the Euphrates and the Khabur bring minerals and charcoal to the city.[23]

Mari falls under the control of Sargon of Akkad around 2300 BC. Its period of greatest grandeur occurs under a dynasty of Bedouin origin (called *Amorite*), capital of a great kingdom of upper Mesopotamia between 1850 and about 1761 BC. In 1761 BC it sees its final destruction by the Babylonian Hammurabi, even though Zimri Lim, last king of Mari, had helped Hammurabi in his wars against Larsa, Esnuma, and Elam. The palace fire buried, and preserved to the present time, the clay tablets that comprise the archives of the last thirty years of the kingdom.

The city of Mari is essentially circular in shape, surrounded by a dike and wall. The city is 1 to 2 km distant from the Euphrates, for protection from floods and erosion. It is crossed by a canal that links it to the Euphrates at each end. This canal is 30 m wide,

---

22 Ibid.
23 Margueron (2004).

and brings water to the city as well as providing access to the port of Mari.[24]

The city's water supply is from the Euphrates, lifted into it by manual labor; women carry the water and fill the palace's cistern (Figure 2.9). But in the palace of the IInd millennium BC there is also a network of brick conduits or pipes (Figure 2.10) that collects rainwater from the terraces to fill a reservoir.

Figure 2.9 The cistern of the Mari palace – beginning of the IInd millennium BC (photo by the author).

In Chapter 1 we mentioned that Mari also benefited from a sort of drainage system, with sumps or cesspools for the drainage of rainwater and wastewater. This is evident from certain remains, as well as from palace texts:

> "…On the subject of the sump, [….] according to the letter from my Lord, it is lined with asphalt from bottom to top. Over the layer of asphalt, there is a tar lining, and on top of that they put a coat of clay plaster…"
>
> "After two rainstorms in succession, the sump was filled with water to a depth of a cane. The next day, they investigated it: 4 cubits of water had flown out. There remained 2 cubits, but they have already drained out….."[25]

Development of the land relies on major hydraulic works. Around 1850 BC, the second king of the Amorite Dynasty, Yahdun Lim, founds a fortress-city about a hundred

---

24 Margueron (2003).

25 J.-M. Durant, Documents épistolaires du palais de Mari I, (1997), doc. 157-159, adapted.

Figure 2.10 Channels for capturing rainfall on terraces of the Mari palace, beginning of the IInd millennium BC (photo by the author).

kilometers to the north of Mari, on the right bank of the Euphrates upstream of its confluence with the Khabur. He gives his name to the city, and endows it with a canal:

> "Yahdun-Lim, the son of Yaggid-Lim, king of Mari, of Tuttul and the land of Hana, the strong king who rules the banks of the Euphrates; Dagan proclaimed my royalty (...). I opened canals, water-lifters were no longer needed in my country. I built the Mari wall and I dug its moat. I built the wall of Terqa and I dug its moat. Moreover, in the burning lands, in a place of thirst no king had been able to build a city, I, alone, had the vision and built a city. I dug its moat. I named it *Dur-Yahdun-Lim*. Then I opened a canal and named it *Isim-Yahdun-Lim*...."[26]

From texts cited further on, one can estimate that this canal is about 35 km long, a length necessary for the canal, with its flat slope relative to the Euphrates, to "raise" the water above the river to provide gravity irrigation of the crops. Traces of the canal remain visible today, but only in its upstream reaches.

Field studies[27] have uncovered traces of other important canals near Mari (Figure 2.11). There is a large irrigation canal, six to seven meters wide and 2.5 m deep, whose remains can be detected along a distance of 17 km. It is not entrenched, but for the most

26 Inscription on the head of a clay spike, found in the Mari palace, adapted from Sollberger and Kupper (1971).

27 Margueron (1988), Geyer (1990), Monchabert (1990), Margueron (1991), Calvet and Geyer (1992).

part is constructed of fill on the alluvial terrace by means of massive dikes, 2.5 m high and 50 m wide. This is probably the *Mari canal* mentioned in the archives of the IInd millennium BC. Extracts that we cite further on suggest that the canal ends some ten kilometers downstream of Mari, and that it is supplied by an intake on the Euphrates in the valley of wadi es-Souab. But it is also quite possible that the canal dates from a time preceding these texts, having its origin in a small dam whose traces have been identified on wadi es-Souab, to the west of Mari. This wadi, one of the most important seasonal tributaries of the Euphrates, has particularly abundant springtime floods. The dam, 19 km from the confluence of the wadi with the Euphrates, is 450 m long, 2.5 m high, and has two spillways.[28] There is yet another canal that could have provided drainage, between the city and the foot of the cliff. Mari could not survive without some degree of agriculture, and such agriculture was impossible without irrigation. Therefore it is nearly certain that all these canals date from the founding of Mari, around 2800 BC.

Written communications[29] found in the Mari palace include numerous descriptions of the work necessary to maintain the irrigation system. The water intakes are often blocked by silt deposits and the canals themselves encumbered with sediments and vegetation that must be cleaned out annually. "Barriers", apparently comprising tree trunks and branches, are placed to protect the intakes and limit deposits in their vicinity. Let's listen to Kibri-Dagan, the governor of Terqa, in extracts from two different messages:

> "As for the work on the canal of Isim-Yahdun-Lim that I had to undertake the fifth of the month of *Abum*, I undertook it. It is considerable. I am going to proceed with considerable dredging. At this canal, the barrier-*muballitum* that diverts the clayey silt toward the river is no longer there, and this has caused the canal to narrow near the river. (…) With my workers from the area, I am working on the interior of the canal. In ten days I will have cleaned out the reeds and brush down to Terqa, and there where the canal has narrowed, I will open it up again. I will be sure the work is done solidly so that the irrigation water will be blocked nowhere and so that the people can avoid famine."[30]
>
> "At the end of the month of *Abum*, I gathered the servile and common people of my district and together with districts of Mari and of Saggaratum, I set out to open the blockage of the canal of Mari. However, before working on the canal of Mari, I used up all the water of the canal Isim-Yahdun-Lim for the upstream district, saying to myself: before the fields of the countryside of Terqa are irrigated, once water is available, the *(upstream)* district must be able to drink so that later on there will be no basis for protest."[31]

---

28 Calvet and Geyer (1992), Chapter 9. This dam, built with limestone blocks cemented with a chalk-based mortar, could be a Hellenistic or Roman reconstruction, given the proximity of Doura Europos, of a structure from Mari or even from an earlier time, for there is a very ancient inhabited site here.

29 Akkadian texts engraved in cuneiform writing on tablets of baked clay; this evidence dates from the reign of Zimri-Lim, one of the Amorite successors of Yahdun-Lim, who reigned 14 years and was the last king of Mari.

30 Durand (1998), II, 793. This document suggests that the canal Isim-Yahdun Lim extends to Terqa, a hypothesis that we have adopted in Figure 2.11.

31 Ibid., II, 784.

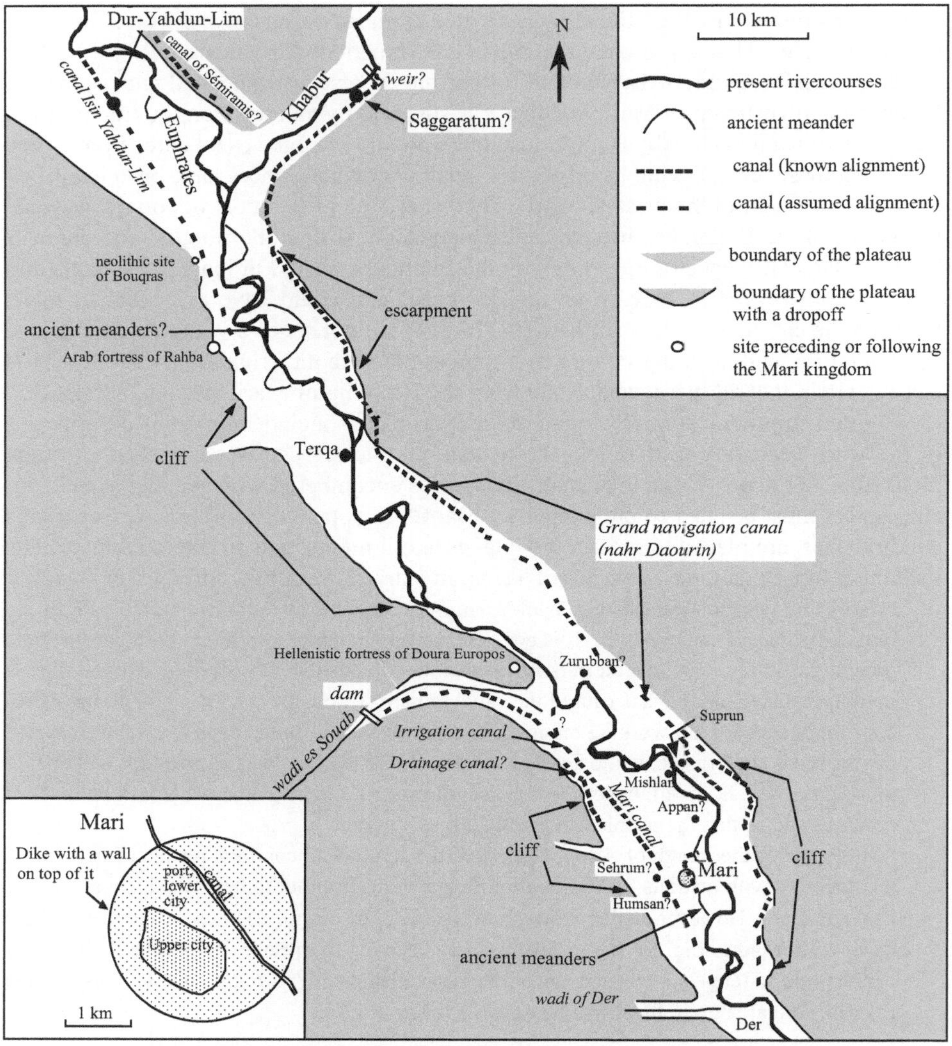

Figure 2.11 Hydraulic works of the Mari kingdom, synthesis of archaeological and epigraphical research.

Other correspondence gives indications of the manpower available for this work. At this time, Terqa can mobilize 400 workers. But for major undertakings, the order of 2,000 workers can be assembled, as the governor of Mari, Bahdi Lim, explains:

> "Tell my lord: thus speaks Bahdi-Lim, your servant. Concerning the wadi of Der (on the right bank, downstream), we got going on the dead (?) intake works and the diversion canal. The administrative scribes calculated the amount of work necessary. Beyond the dead intake works, there is the work needed for the diversion. A crew of 2,000 people, that's not much! After thinking about it, we go to work only on the diversion. The work that we have undertaken is good."[32]

The "barriers" mentioned above also play the role of raising the river's water level at the intake to facilitate the outflow into the canal, without completely blocking the river. At this time there are six similar "barriers" on the Khabur designed to supply the irrigation canals. A letter from Yaqqim-Addu, governor of Saggaratum (whose supposed location is shown on Figure 2.11), describes these barriers and mentions a canal that provides water to the left bank of the Euphrates from the Khabur as an important element of the region's irrigation system:

> "The Habur (*i.e. the Khabur*), like the canal of Isim-Yahdun-Lim and the canal of Hubur *(on the right bank, the north portion of the canal of Mari?),* is part of our irrigation system. The people who profit from this irrigation canal never undertake to maintain it, and they have not reinforced the weak spots. I had to take on six barriers *(muballittum)* — who else could assure they would be watched over? When one wants to take water for the ditches, right where the trunks form fences, it takes 3,000 bundles of brushwood to make a piled-up barrier. But this doesn't raise the Habur (*the Khabur*) a single finger! One must put in posts to form the fence: one makes brushwood during ten days: one then piles them up to form barriers. Today this system is damaged: alas! I am leaving for Habur; I am going to assess the damages." [33]

Later in this same letter, Yaqqim-Addu asks his colleague Dahdi-Lim, in charge of Mari, to loan him manpower, lacking which the Khabur flood will cause major damage:

> "At the present time, the Habur (*the Khabur*) is in flood at four cubits: it has covered all floodable places. The dike-*kisirtum* that is upstream of the breach, downstream of the *muballittu* that we built, Kibri-Dagan and myself, had slipped. I am going to rebuild it. The side, at present, slipped again. I undertook to rebuild it. Moreover, the breach that we had closed up has reopened: two arches, made of brushwood, were installed. These various efforts have been considerable and have exceeded my means; my Lord needs to give instructions to Bahdi-Lim so that he will send me 200 men so I can reinforce the weak points on the Habur. If a breach occurs in the aforementioned dike, no one will be able to close it."

One should not be surprised at how difficult it is to maintain these canals. As we have seen above, the slope of a canal must be less than that of the river, if the canal is to be used to irrigate terraces higher than the valley floor. The flow velocity must therefore be lower than that of the river, which favors the deposition of suspended silts in the canal.

The texts from Mari also show that fish farming is practiced at the beginning of the IInd millennium BC. The abandoned arms of the Euphrates are clearly exploited for this purpose, and it is again Kibri-Dagan who tells us about it:

> "Tell my Lord, thus speaks Kibri-Dagan, your servant. When the river flood returned, the pond of Zurubban swelled and became larger than normal. This made me fear for the fish: there is a risk that the fish will leave the pond toward the river. Now a hundred people must come to make the water of the pond go toward the river."[34]

Another text that we should cite tells us of a maneuver that makes the canal of Mari temporarily navigable, to carry boats loaded with grain from the harvest. All the second-

32 Durand (1998), II, 796.
33 Ibid., II, 804.
34 Ibid., II, 805.

ary canal intakes were closed to raise the level in the main canal. But this turned out to be a catastrophe, since the rising waters caused the dike to rupture, just as the governor of Mari, Sumu Hadu (predecessor of Bhadi Lim) was taken to bed, sick:

> "Tell my Lord, thus speaks Sumu-Hadu: one had retained the water in the direction of Der: because of the boats that must transport grain, one had blocked, from the upstream, *(all)* the irrigation ditches, and the water level thus rose (*in the canal*). But yesterday, at nightfall, in the end the water opened a breach upstream of the bridge that is the intake with the Balih (*here, the Wadi Der)*, there where there is a water conduit (*uncertain translation: a device allowing water diversion*). Immediately, despite my sickness, I got up, I harnessed my asses, and I went to turn aside the waters by a derivation system. Then I came back to stop the water in the Balih (*the wadi Der*). Early in the morning, I undertook to repair the damage: I am going to rebuild the water conduit (*?),* after which I will get to work compacting the soil. This breach caused an opening of two canes from top to bottom, on a width of four canes. By the first watch of the night, I will have finished blocking this breach and I will be able (again) to let the water pass. My lord should not worry! Moreover, I wrote to the various localities that I had turned aside the water during the night. At Appan, Humsan and Shehrum, the water was held in and there was not the least rise. As for me, I will be dealing with the sickness that I have contracted for a year!"[35]

All of these documents show that the leaders had strong personal engagements in the maintenance of the irrigation system. They called on specialists, likely trained from father to son, for positioning the gates, for the operational regulation of the network. The regions of Terqa and Mari are not the only ones in which such water management is practiced; further upstream the Balih is used to irrigate the region of the city of Tuttul during the Amorite period.

The question of the long navigation canal (the nahr Daourin)

Another project in the Mari region has left us with considerable evidence of its existence. This is the canal whose traces today are called *nahr Daourin*, with a width of 8 to 11 m, and a reconstituted length of some 120 km. Rising at Khabur, it joins the Euphrates downstream of Mari on the left bank, cutting into the cliffs in places (Figures 2.11 and 2.12). The canal has a fairly regular slope of about 0.2 m/km, though it is somewhat flatter in its upstream portions (0.12 m/km).[36]

The letter of Yaqqim-Addu, from which we have cited extracts earlier, mentions the existence of a canal issuing from Khabur at Saggaratum, at the beginning of the IInd millennium BC. This canal irrigates the Mari district on the left bank of the Euphrates, and is very likely the *nahr Daourin.* It was undoubtedly used for irrigation in particular given the context of the last years of Mari. But navigation was clearly the canal's primary purpose, as evidenced by its large cross section and length, and the fact that in

---

35 cit. after Lafont (1991); see also Durand (1998), II, 813. If indeed it is of the Mari canal that one speaks here, it must have extended to Der. The wadi Der is called Balih, from the name of one of the major tributaries of the Euphrates, which seems to be a common practice in usage.
36 Geyer (1990).

Figure 2.12 The nahr Daourin at one of the points in its course where it is deeply entrenched into the plateau on the left bank. (Photo: French archaeological mission of Tell Hariri – Mari (Syria)

places it is excavated into the high banks of the flood plain. It is entirely possible that the project is closely related to the very founding of Mari in the IIIrd millennium BC. Indeed, it was from this period on that bulk material, such as the charcoal necessary for metallurgy, had to be brought down to Mari from upper Khabur. It is much easier to imagine the empty boats being hauled back upstream on a canal then on the irregular course of a large river. This hypothesis would explain the particular site on which Mari was founded – on the right bank on a road that leads to Syria, but near the downstream end of the canal. Suprum, where the canal bends toward the river on the left bank a little to the north of Mari, could well be the port where merchandise was transferred between the Euphrates and the canal.[37] In connection with this navigation canal, there would have had to be a weir on the Khabur upstream at Saggaratum, to maintain the water level during low flow. But there are no longer any remains of such a weir.

The remains of another project of this type, including a weir across the river and a

37 Jean-Claude Margueron is a strong proponent of the hypothesis of an ancient navigation canal (see his various publications). Prior to the era of Zimri Lim, from which the texts cited herein are drawn, a message sent by a king of upper Mesopotamia, whose domains include Mari and its region, is of interest. The message asks that precious wood coming from Qatna in Syria, coming down the Euphhrates to Suprum, "be brought upstream by boat to Saggaratum, from there (again by boat) to Qattunan (further up the Khabur). From there, the wood can be transported on carts ..." (Durand, letter No. 187). J.C. Margueron brought this letter to my attention. The importance that it gives to Suprum as a transfer port, and to Saggaratum as a stopover, is quite consistent with the notion of a grand navigation canal.

lateral canal, have been found on the Euphrates somewhat further upstream. This is the *canal of Semiramis*.

The Khanouqa dam and the Semiramis canal

At the Khanouqa gap on the Euphrates some 80 km upstream of its confluence with the Khabur, are the remains of hydraulic works.[38] This includes a rock weir, built of loose natural basalt blocks, damming the Euphrates so as to provide all-season water to a canal whose offtake is immediately upstream of the dam (Figure 2.13). The left-bank canal is called the *canal of Semiramis*[39] in the writings of the Greek traveler Isidore of Charax, dating from the first century AD:

> "There is found the canal of Semiramis; the Euphrates is blocked by rocks so that, in its narrowed state, it floods the plain; but in the summer boats run aground there." [40]

Remnants of the canal are partially visible in the vicinity of Khabur, along about 80 km. Navigation is clearly the main purpose of the canal, but it could obviously have also been used for irrigation. It is not possible to assign a precise period to these remains. Yves Calvet and Bernard Geyer suggest that the most likely period is from the end of the IIIrd millennium BC until the beginning of the IInd millennium BC. This is a period of prosperity for Mari, but does not exclude the possibility that the works date from a later time, in the Ist millennium BC (but the construction does not appear to be of Assyrian workmanship, and therefore we are inclined to favor the earlier hypothesis.)

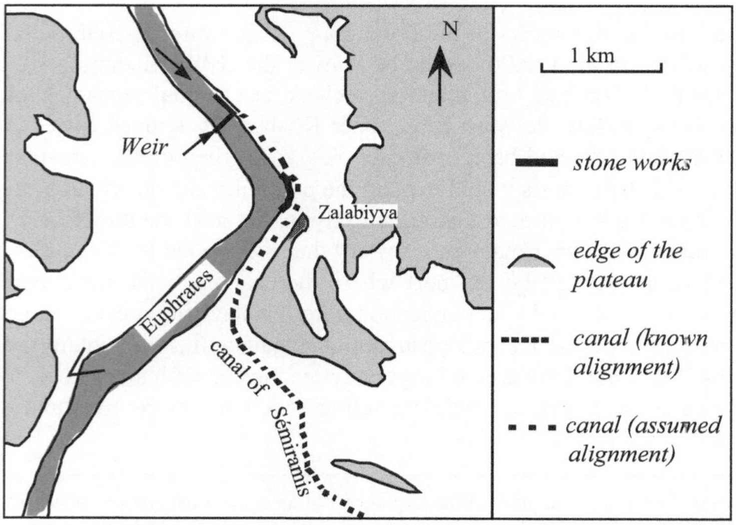

Figure 2.13
The weir at the Khanouqa pass and the canal of Semiramis (after Calvet and Geyer, 1992).

38 Calvet and Geyer (1992), chapt. 2.

39 This name does not have great significance; the Greeks tended to attribute numerous ancient works to the legendary Semiramis.

40 "Étapes de Parthie", cited from Calvet and Geyer.

Figure 2.14 The Euphrates at the Khanouqa gap. The path of the canal of Semiramis passes at the foot of the cliff in the background, on top of which one can distinguish the ruins of the Hellenistic city of Zalabiyya. The photo is taken at the location indicated by a V on Figure 2.13 (photo by the author).

Syria-Palestine in the IInd millennium BC

The city of Ugarit, on the northern Syrian coast (several kilometers to the north of the present-day Lattaquieh), has been occupied since very early times. It served as a maritime port for trade with Cyprus and Crete in the context of commerce among Mesopotamia, Mari, Aleppo, Ebla and the Mediterranean, and then as a port of the Hittite Empire of Anatolia. The city knew a period of great prosperity from the IInd millennium BC until its final destruction in 1200 BC by the *Sea People*. The city is built on high ground, using wells for its primary water supply. As at Mari, rainwater is captured on terraces and brought through gutters and vertical drops to the cisterns of houses.[41] The city is surrounded by two small temporary watercourses, the nahr ed-Delbe and the nahr Chbayye, on each of which there is a small dam; the dam on the nahr ed-Delbe is described in detail by Yves Calvet and Bernard Geyer (1994). The originality of this structure resides in its movable beams or stoplogs that can be removed to allow floods to pass. This is the first evidence of such technology that eventually became widespread.

More to the south, on the eastern slopes of the Anti-Lebanon mountains, is the city of Damascus. The city's water-resource infrastructure is developed toward the middle or end of the IInd millennium BC, under the control of the Arameans. Two canals flow out of the Barada, the perennial river on which Damascus depended from Neolithic times. The system is eventually completed by the Romans and the Arabs (Figure 7.6) and

41 Callot (1983).

remains operational to this day.

In discussing great cities and their infrastructures we must include Jerusalem. In the 12th or 13th centuries BC the Canaanites constructed a 537-m long tunnel to provide access, during sieges, to a reservoir on the flanks of a hill that is fed by an intermittent spring called Gihon (today called the "fountain of Marie"). Later, around 700 BC, Ezechias tapped this spring through an underground canal feeding a basin to the south of the city, known as the Pool of Siloam.[42]

The Zagros Mountains and their foothills from the 8th to 7th centuries BC: dams, aqueducts, and water supply for cities

We have seen that from the 13th century BC the political situation becomes very cloudy in all the Syro-Mesopotamian region. New powers rise to the east and north of the old lands of Sumer and Akkadia, at the foot of the Zagros mountains. The new power centers move from the region of Susa to the east, up the course of the Tigris (Assyria) before returning to the east with the Persians. The incubators of these new powers are valley, hill and mountain regions whose springs and streams can be developed to provide high-quality water for the settlements.

Water supply from Dur Untash, in the land of Elam

Elam, in the Susa region, is a very ancient civilization that developed expertise in hydraulic works starting in the IIIrd millennium BC. Three abundantly flowing rivers descend from the Zagros mountains and cross the region: the Kherka, the Ab-e Diz, and the Karun. Elam knew a brief period of glory when, between 1260 and 1160 BC, it took advantage of the weakness of the Babylonians to ravage lower Mesopotamia. Along with other spoils, the black stone on which is engraved the Code of Hammurabi passed through Susa at this time.

During this period Untash-Gal, sovereign of Elam between 1275 and 1240 BC, built a new city some forty kilometers to the southeast of Susa, next to the river Ab-e Diz; he gave the city his name in calling it Dur-Untash. But the course of the Ab-e Diz is below the elevation of the city, and the groundwater is brackish. So Untash-Gal dug a 100-km long canal to obtain water some twenty kilometers upstream of Susa from the Kherka, a river whose waters were known for their purity. The canal brings the water to a brick-lined reservoir just outside the wall of Dur-Untash. This reservoir, whose dimensions are 10.7 m long, 7.25 m wide, and 4.35 m deep, then feeds a basin just inside the city wall through nine openings. The city's inhabitants could take water directly from this basin.[43]

Development of Assyria. The waters of Nineveh

The sovereigns of Assyria begin development of their land on the upper course of

---

42 Contenau (1927), Volume 3, p. 1373.
43 Ghirshman, 1968.

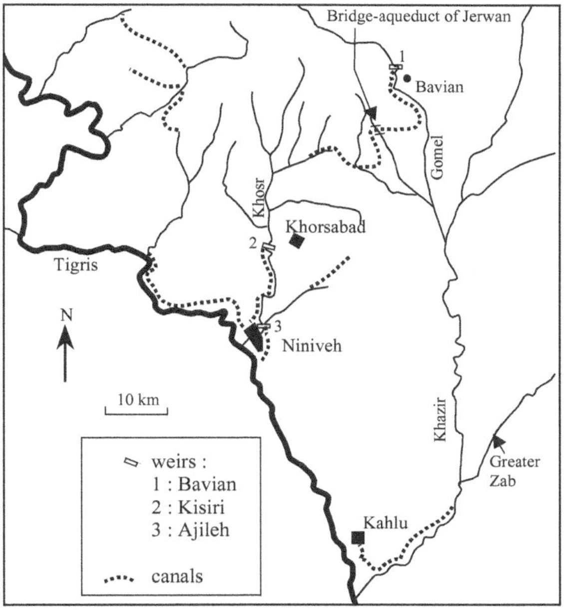

Figure 2.15 Irrigation and water supply works in Assyria in the 9th and 8th centuries BC (after Jacobsen and Lloyd, 1935; Roaf, 1990; Schnitter, 1994; Bagg, 2000)

Figure 2.16 Fluvial transport of wood beams. Bas-relief from the palace of Sargon II at Khorsabad, on display at the Louvre museum (photo by the author).

the Tigris in about 900 BC. This date marks the flowering of an Assyrian Empire destined to reign over all of Mesopotamia, and even as far as Egypt, for about three centuries. In 860 BC, Ashurnasirpal builds a new capital Kahlu (today Nimrud) on the left bank of the Tigris near its confluence with the upper Zab. A canal called *babilat nuhshi* ("bringer of abundance") is dug to irrigate the plain with waters of the Zab diverted into the canal by a dam or weir.

It is interesting to note also that a ventilation system comprising chimneys ("air doors") provides fresh air taken from the roofs of the royal palace to its grand rooms. But it is Sennacherib, the destroyer of Babylon, whom we are now going to see in a different light. He was a lover of gardens. Since Khorsabad, the ephemeral city created by his father Sargon II, was too austere, Sennacherib re-adopts Nineveh as his capital. Taking advantage of his unlimited supply of manpower, he immediately sets out to acquire the water necessary for his horticultural aspirations (Figure 2.15). His first project, in 703 BC, is a 16-km long canal, fed by the Khosr, that brings water to the plain to the west of Nineveh. The waters of this canal, diverted into it by means of an overflow weir at Kisiri, irrigate orchards on plots allocated to the inhabitants of the capital through a lottery system:

> "... from the environs of Kisiri to the plain of Nineveh, across mountains and valleys, using iron picks, I dug a canal. Along a distance of one *beru* and a half (*16 km*) I took water from the river and made it flow down to irrigate the orchards."[45]

Several years later, the area is in need of even more water. Sennacherib himself sets off into the mountains to see what springs existed. In 694 BC he had water tapped from the springs in the hills northeast of the city and brought to Ajileh, on the Khosr. The remains of two diversion weirs, constructed of large blocks of cut stone, are still visible there. But this new influx of water exacerbated the damaging floods in the Khosr. Therefore the king built a weir downstream of Ajileh to divert floodwaters into a canal going around the city to the south, and thence into artificial lakes. The king had plants and birds brought from the marshes of Babylonia, where he had admired them, to these lakes. The diversion weir was of serpentine shape, having a long crest length of 230 m that thus limited the rise of water during floods. The height of the weir itself was three meters.

But in the end even this additional water supply was insufficient, so its most spectacular feature was added to finally complete the system in 690 BC. Water was diverted from the Gomel, a tributary of the Zab to the north, and brought to the Khosr through a 55-km lined canal-aqueduct formed of lateral walls of cut stone crossing valleys on arches. At Jerwan (Figure 2.17) one can still see the remains of a magnificent bridge-aqueduct 275 m in length and 22m wide, crossing a valley by means of five arches each 4.75 m high. The diversion works on the Gomel at Bavian apparently included an oblique weir across the river. The canal, 6 m wide at its origin, passes through a short tunnel to cross a rocky spur.

Above the intake there are inscriptions that praise the hydraulic works of Sennacherib.[46] These inscriptions also describe an incident that occurred during the inauguration of the project, an incident that would surely have been very unfortunate for the engineers had the king not been in a good mood. The water pressure caused the closed gates to fail before they had been opened, allowing water to surge into the canal:

---

45 Inscription of Sennacherib, cited from Jacobsen and Lloyd (1935).

46 See Jacobsen and Lloyd (1935) for a report of field studies on the dam and the inscriptions of Bavian and the aqueduct of Jerman; and Schnitter (1994) for details of the weir of Ajileh.

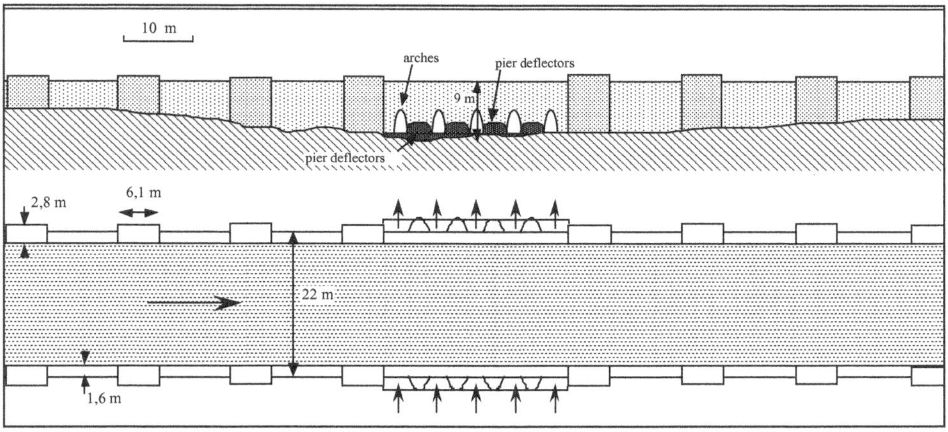

Figure 2.17 A central portion of the bridge-aqueduct of Sennacherib at Jerwan. It is made of essentially cubical 50-cm stone blocks, and is inscribed with the following text: "Thus says Sennacherib, king of the world, king of Assyria: over a great distance, bringing waters (...) of the river called Pulpullia (...), and from springs here and there along its course, I dug a canal to the edge of Nineveh. Across steep ravines, I threw a bridge of white stone blocks. And these waters, I made them cross over on this bridge." (sketch and citation from Jacobsen and Lloyd, 1935).

> "To inaugurate the canal, I summoned the priests (...) and made offerings of lapis lazuli (*a semiprecious stone)*, precious stones and gold to Ea, god of springs, fountains, and prairies, and to Enbilulu, god of rivers. I prayed to the great gods, and they heard my prayer. The gate yielded and let water enter in abundance. Even though the engineers had not opened the gate, the gods assured that the water found its way. After having inspected the canal and put things back in order, I offered sacrifices to the great gods (....) To the men who had dug the canal, I offered white linen cloth and colored woolens; I decorated them with rings and daggers of gold."

This system is perhaps the first example of cross-basin water transfer.
It is also likely during the era of the Assyrian Empire that irrigation canals were constructed on both banks of the Khabur, nearly continuously along the length of the river.

In the Mesopotamian north: the kingdum of Urartu. The oldest dams still in use

The kingdom of Urartu, in the south of Armenia, was a powerful rival of Assyria during the period from about 850 BC to 600 BC. Its capital was Tushpa, on the shores of the lake Van, whose water is too salty to be potable. Therefore this capital city's thirst led to the development of a vast water management program in this mountainous region, developed in successive stages (Figure 2.18).

In about 800 BC, the king Menua brought water to the capital from a perennial and abundant spring located some thirty kilometers to the southeast. The 56-km canal-aqueduct built for this purpose carried at least 1.5 m^3/sec, and was destined to be used

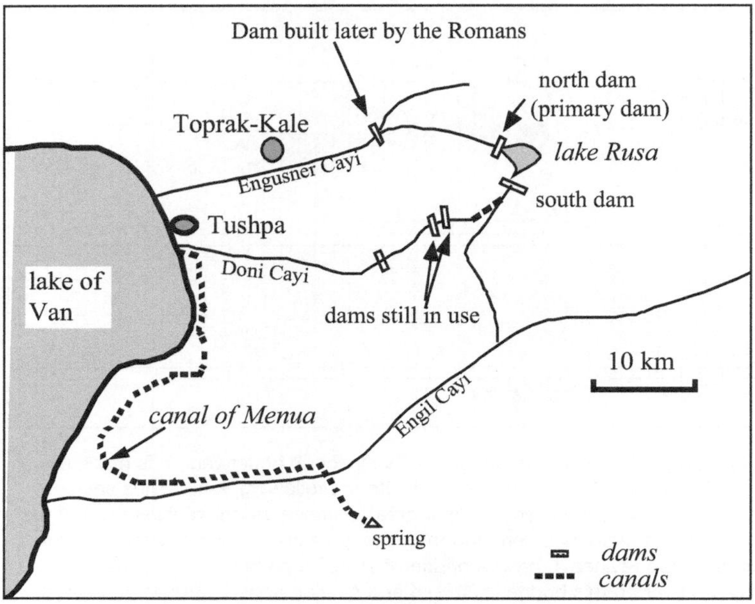

Figure 2.18 Water supply for the two successive capitals of Urartu (Tushpa and Toprak Kale) in the 8th century BC – after Garbrecht (1980, 1988).

for 2,500 years until it was partially renovated in 1950.
Later, probably about 670 BC, the king Rusa II[47] moved the capital some ten kilometers to the north (*Rusahinili*, today Toprak-Kale). The two rivers that supply the two cities, the Engusner and the Doni, are intermittent. Therefore the sovereign created an artificial reservoir, the Rusa lake (today this is the lake Keshish Gölü, whose water level is 10 m lower than at the time of king Rusa). The lake was created by damming two natural outlets of a mountain basin. The north dam has been measured at 15 m high and 75 m long, and the south dam at 7 m high and 62 m long.

This south dam has been well preserved (it is no longer in contact with water, for the lake is now 10 m lower). It is constructed of two walls of dry stone, each 7 m thick, containing an earth fill of thickness 13 m[48]. Water is conveyed from this south dam to a secondary dam-reservoir on the Doni Cayi, in part to provide for irrigation of the Tushpa region. Later, other dams are built downstream on the two rivers, to increase the storage capacity.

Thus we see that the water management system of lake Rusa includes several dams. Over the centuries, the north outlet of the lake has apparently been destroyed and rebuilt several times, which may explain the existence of a dam attributed to Roman times, downstream on the Engusner Cayi. Its most recent reconstruction, lower than its origi-

47 A stela mentions Rusa without further note. According to Paul Zimansky (1985), the king Rusa II (680 – 654 BC) deserves credit for this project.
48 Regarding supply of water from the Tushpa and Rusahinili, see Garbrecht (1980, 1988).

nal height, dates from 1950. The dam-reservoirs upstream on the Doni Cayi are still in service today. These are probably the oldest dams still in use[49] in the entire world.

The qanats: a new technique for obtaining water

When surface water cannot meet the needs of irrigation, one must tap groundwater. It was probably at the beginning of the Ist millennium BC, in Persia or in neighbouring lands, that a remarkable device for obtaining high quality water was invented: the *qanat.*[50]

This word means "reed" in Akkadian. The device comprises a gallery, or shaft, dug nearly horizontally from the flank of a natural slope back into the aquifer, but with a small slope (of the order of one or two per thousand) so that the water can flow out by gravity (Figure 2.19). In general, the construction of a *qanat* begins with the drilling of what will become its terminal well, called the "mother well", through which the nature and level of the water table can be determined. Then the excavation of the gallery begins from the downstream end, making it possible to work in the dry until the aquifer is reached; this excavation thus proceeds all the way to the mother well. Intermediate wells, normally spaced at from 50 to 100 meters, provide for removal of the spoil from

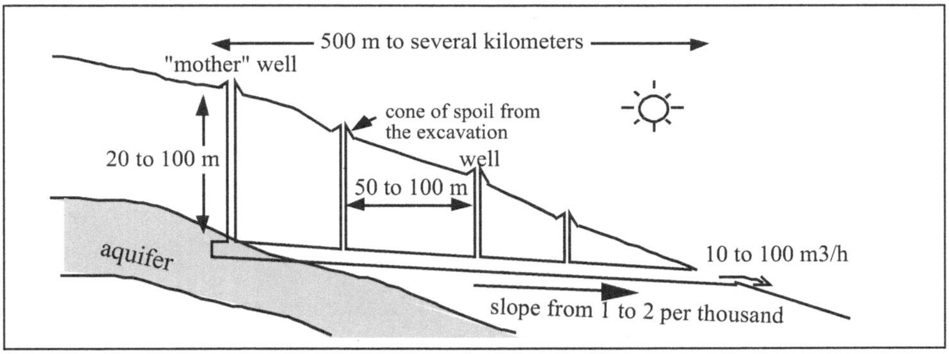

Figure 2.19 Principles of the qanat: a technique for mining groundwater, apparently first appeared in Urartu (Armenia) in the 8th century BC, and then spread throughout the Persian Empire (Goblot, 1979).

the gallery, and provide for ventilation. The gallery can be several kilometers long, even reaching ten or more; the mother well can be the order of twenty to a hundred meters deep. The flow delivered by the *qanat* is generally from several, to several hundred, liters per second.

In 714 BC, Sargon II, king of Assyria (and the father of Sennacherib), is at war with king Rusa I of Urartu. He destroys the outposts of Urartu in the region of the lake of

49 This title could be contested by the Homs dam in Syria that is attributed by some to the Egyptians, but that probably dates from the Roman occupation. We come back to this in Chapter 7.
50 This process is known to us through the quite complete study of Henri Goblot (1979). Modern studies tend to disagree with his belief that the qanats originated in Urartu.

Urmiah, as well as the *qanats* supplying water to the city of Ulhu, located to the east of this lake (60 km to the north of present-day Tabriz). The Assyrians, in a written account of this campaign, leave us an admirable description of the devices called "water outlets" that could comprise the first written evidence of the *qanats*:

> "Ursa (*i.e. Rusa*) the king and lord of this land, pushed by his intelligence, showed his people how one manages the water outlets and creates a flow of water as copious as that of the Euphrates".[51]

What is the genesis of this innovation? The Zagros mountains are a region of mines, especially in the area around the lake of Van. According to Henri Goblot, it was necessary to provide gravity drainage – to the surface - of certain mine shafts that were flooded after having pierced aquifers. In a country faced with the need to augment its water resources, the idea of making use of this drained water, and then to dig galleries expressly for this purpose, surely picked up speed rapidly. The idea had a grand future: the Persians adopted it to develop the Iranian plateau, and in particular to provide water for their capital, Ecbatane. This is reported by Ctesias of Cnidus, a doctor of Xenophon's expedition who was long held captive by the Persians:

> "Having arrived at Ecbatane, a city located in a plain, she (*again the legendary Semiramis! Here, it can only be a Mede or Persian sovereign)* built a luxurious palace and she watched over the entire region with great care. The city was without water and there were no springs in the vicinity; but Semiramis brought water to all the city, abundant and very pure water thanks to her heavy investments."[52]

Cyrus brought the technique of *qanats* to Oman, and Darius brought it to the oases of Egypt. As we will see in subsequent chapters, the Romans developed the technique in all of the Near East, and as far as Tunisia, and the Arabs took it to Spain and Morocco. Migrants coming from the East brought it to the benefit of Saharan oases.

On the steppes of central Asia: Irrigation in Bactria and Margiana before the arrival of Alexander the Great

Bactria, to the east of the Zagros mountains and the Iranian plateau, is connected to the Syro-Mesopotamian world through a continuous thread of ancient exchanges, and thus it also must be mentioned in this chapter. Bactria was a land of plenty and fertile valleys as noted by Strabo:

> "Man has only to take the trouble to irrigate, and all crops grown abundantly with the exception of the olive tree".[53]

In Chapter 1 we mentioned the appearance of the Bactria-Margiana civilization, or

---

51 Account of the eighth campaign of Sargon II against Rusa I, tablet conserved in the Louvre museum, citation after Goblot (1979).

52 Ctesias, History of the Persians, 13, adapted from the Translation of J. Auberger. Later, Polybius mentions even more clearly the "underground canals" of this region (extract cited in Chapter 7).

53 Strabon, Geography, (Translation of F. Lasserre), Les Belles Lettres, 1981, adapted.

the Oxus civilization, for which the mastery of irrigation is a fundamental pillar. It is doubtful that Bactria ever really fell under the Assyrian yoke, as is suggested in certain ancient sources. To the contrary, it was clearly part of the Achaemenid Persian Empire, and therefore belongs on the list of conquests of Alexander - but not without combat. Indeed, it was at Bactria that Alexander married Roxanne.

Successive hydraulic developments in the region became more and more generalized from the IVth millennium BC. In Margiana and western Bactria, rivers that used to disappear into the desert before reaching the Oxus River form deltas that are developed and irrigated in the IInd millennium BC. Among the oases on the deltas are those of Geoksiur (mentioned in Chapter 1), Merv (Marw) on the Murgab River (to be discussed further in Chapter 7), Sherabad, Ulambulak and Mirshada on the right bank of the Oxus in western Bactria, and Bactra (Balkh) to the south, capital of the region toward the end of the IInd millennium BC. Russian archaeological studies have shown that at Merv,

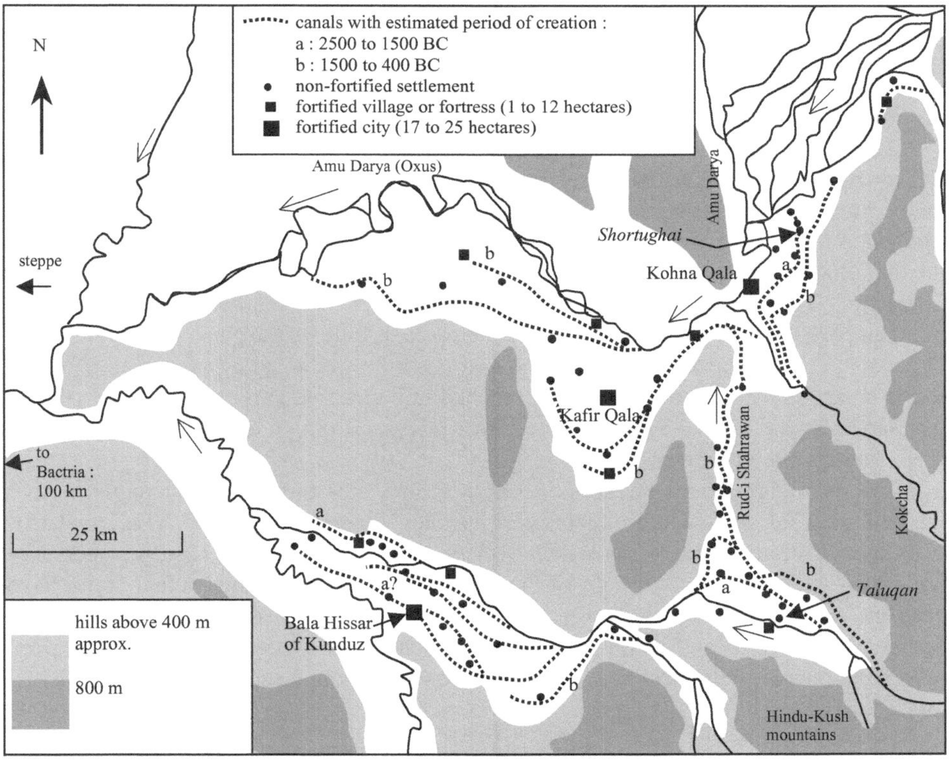

Figure 2.20 Irrigation canals in eastern Bactria on the eve of the arrival of Alexander the Great, near the confluence of the Oxus and the Kokcha (Gentelle, 1989; Gardiun, 1998). The irrigated fields are located between the canals and the rivers. The sites shown in italics (Shortughai and Taluqan) date from the IIIrd millennium BC.

54 Bader, Gaibov, Gubaev, and Koshelenko (1996);Gaibov and Kochelenko (2002): according to these authors, irrigation of the Merv oasis began as early as 2100 BC.

habitation progressively migrates upstream as aggressive irrigation starves the downstream extremities of the network.[54] In Sogdiana to the north, one can also find remnants of early irrigation at Sarazm, on the Zeravchan River (the watercourse on which Samarcand is later founded). One also finds such traces on the lower course of the Oxus, to the southeast of the Aral sea, and at Dehistan, to the southeast of the Caspian (see Figures 1.3 and 7.1 for location maps).

Irrigated agriculture develops naturally in the valleys of the Oxus and its tributaries in western Bactria, at the foot of the great mountains of Hindu Kush and Pamir. French explorations conducted between 1971 and 1977 led to restitution of ancient canals, and helped establish a chronology of their evolution based on the dates of inhabited sites.[55] As agriculture spread to terraces situated higher and higher above the rivers, it was necessary to extend the canals so their intakes would be higher than the irrigated land, and to route the canals from one terrace to another. Often it was necessary to supply a canal from a tributary of the main river, sometimes far away, to bring water to the irrigated terraces along the main river itself. For example, a plain that overlooks the Oxus at its confluence with the Kokcha was irrigated from the IIIrd millennium BC by a 25-km long canal fed by the Kokcha itself, and extending down to the Harappan settlement of Shortughaï (Figures 2.20 and 7.2).

But even larger water resource developments appeared during the period from 1500 BC to the arrival of Alexander the Great – a period that also saw the development of major water supply in Assyria and Urartu. An artificial branch of the Taluqan River was excavated to develop the left-bank region of the Oxus, proceeding to the north against the natural drainage (Figure 2.20). This 50-km branch is today called the Rud I Sharawan. That the canal is artificial can be seen from its often perpendicular orientation compared to the natural drainage. To the south, it follows the paths of old river arms or canals in the Taluqan plain; and to the north, it follows the valley of a small tributary to the Kokcha. The canal is deeply entrenched into unstable loess in its central portion that separates the drainage basins of the two rivers, attaining a depth of as much as 20 meters along one kilometer.

This is a significant operation of inter-basin water transfer, comparable to that implemented by Sennacherib to bring water to Khosr and Nineveh (Figure 2.15). Although it is impossible to say which of these developments came first, the necessary know-how clearly existed in both of these widely separated regions. What could have been the driving force for this project in Bactria: the decision of Persian leaders or the earlier work of a local Bactrian authority? Bactria had a strong cultural unity even before the arrival of the Persians. It may also have had strong political unity, but its history is unknown.

---

55 This exploration was led by Jean-Claude Gardin, with the participation of P. Gentelle and B. Lyonnet. For the layout of the canals, see Gentelle (1989); for a synthesis of the hydraulic works, with dating of the canals revised from analysis of ceramics, see Gardin (1998).

3. Ancient Egypt and the Arabia Felix, the rhythm of the flood seasons

Deserts border the two shorelines of the Red Sea. Along these shorelines are two countries whose verdant fringes have been struggling to resist the desert since the IVth millennium BC– two countries that are highly dependent on seasonal flood cycles. On the east there is Arabia Felix, present-day Yemen. On the west is Egypt, to which most of this chapter is devoted. The River Nile dominates and unifies the powerful and innovative Egyptian civilization. This civilization precedes the transition to the Hellenistic period and the flowering of Alexandria, described later in Part II of this book.[1]

Historical Points of Reference

The earliest Egyptian cultures evolve essentially in parallel with those of Mesopotamia, lagging only slightly. However in contrast to the agitated history of the Syro-Mesopotamian universe, the historical evolution in Egypt is relatively linear. The political unification of the twin lands (upper Egypt and the delta) occurs about 3100 BC, and clearly is the logical outcome of a common culture. This cultural and political union of the south and the north is a cherished aspiration of the Egyptians, and persists across the centuries despite several troubled periods. One of these periods is the separation of the ancient and middle Empires from 2180 to 2040 BC, and another is from 1730 to 1560 BC, a prelude to the establishment of the new Empire.

The middle and new Empires were marked by a commercial and military expansion to the south, up the Nile, and also toward the northeast. Egypt succeeded several times in extending its domination into Palestine, and even to the upper course of the Euphrates, under Thoutmosis III and Ramses II in the 15th and 13th centuries BC.

Around 1200 BC Egypt resists the land and river invasion of *Sea People*, but is weakened by the effort. Always capable of rising to new challenges, Egypt succeeds, for the most part, in preserving its unity. The Assyrian Ashurbanipal temporarily conquers Egypt about 660 BC; but the Assyrians are expulsed with the rebirth of the Saite Dynasty. The Persian Cambyse conquers Egypt in 525 BC, a date which marks the end of pharaonic Egypt and its integration into Achaemenid Persian Empire. In 331 BC, Egypt falls under the control of Alexander the Great, then of the heirs of Ptolemy, Alexander's general.

The ancestral principles of use of the Nile

An irrigation technique that is natural for the regimes of the Nile develops in the

---

1 We nevertheless discuss in this chapter certain projects of the Ptolemites, the successors of Alexander and sovereigns of Greek culture, when these projects are continuations of work of the Pharaonic era.

IIIrd millennium BC. The flood regime of the Nile is quite regular in time, from June to October, but is obviously of quite irregular magnitude. The earliest agriculture consisted quite simply in planting seeds in the moist soil fertilized by silty flood deposits. But a flood of small magnitude inevitably precedes a year of famine:

> "I was in mourning on my throne, Those of the palace were in grief....because Hapy had failed to come in time. In a period of seven years, Grain was scant, Kernels were dried up...Every man robbed his twin...Children cried...The hearts of the old were needy...Temples were shut, Shrines covered with dust, Everyone was in distress...."
>
> "My heart was greatly troubled for the Nile did not come soon enough during seven years. Grain was scarce, the grain was dried out, everything to eat was in very meager quantity, all were frustrated by the revenue."[2]

Although the Egyptians were never able to eliminate the effect of variable flooding on agriculture, they were nevertheless able to increase the amount of productive land through irrigation. Initially, the technique was to exploit natural basins on either side of the river. Water is stored in them long enough for the deposition of sediment (one or two months), and then drained to the Nile or to another lower basin, leaving the soil ready for cultivation. This practice was then extended to the development of artificial retention basins, and this required the construction of dikes and canals of increasing capacity. The *shaduf* (balance beam) appears during the IInd millennium BC; in a tomb of the Ramses period there is a depiction of an entire battery of *shadufs*.

The austere Greco-Roman Strabo, who visited Egypt about 25 BC at the beginning of the Roman domination, was not easily impressed, yet he wrote of the Egyptians:

> "Their practices concerning the river (*The Nile*) are so excellent that because of their diligence nature was conquered. For by natural order, one land will provide more yield than another, and more so if it has been flooded; and the greater the flood, the greater the extent of flooded lands. But often when nature falters, diligent activity can, even when the floods are weak, cause as much land to be flooded as during large floods, this by means of canals and dikes".[3]

Figure 3.1 provides an overview of the major sites of hydraulic engineering works in ancient Egypt and Nubia.

The Nilometers

One can readily see that in Egypt, measurement of the flood level has great importance. The management of the irrigation system is based on such measurements, as are the taxes, since the agricultural yield can be deduced almost automatically from the flood level. The level is quantified using graduated scales carved into stone; Strabo calls these

---

2 This account, called the "stela of famine", relates the difficult years that were said to have occurred under the reign of Djoser, 2nd Pharaoh of the 3rd Dynasty (about 2600 BC). The stela was engraved after the fact, under the Ptolemites, but it is likely the retranscription of a much older text or tradition. Translation by Lichtheim, http://www.touregypt.net/faminestele.htm.

3 Geography, book XVII, 1, 3, Translation of Pascal Charvet.

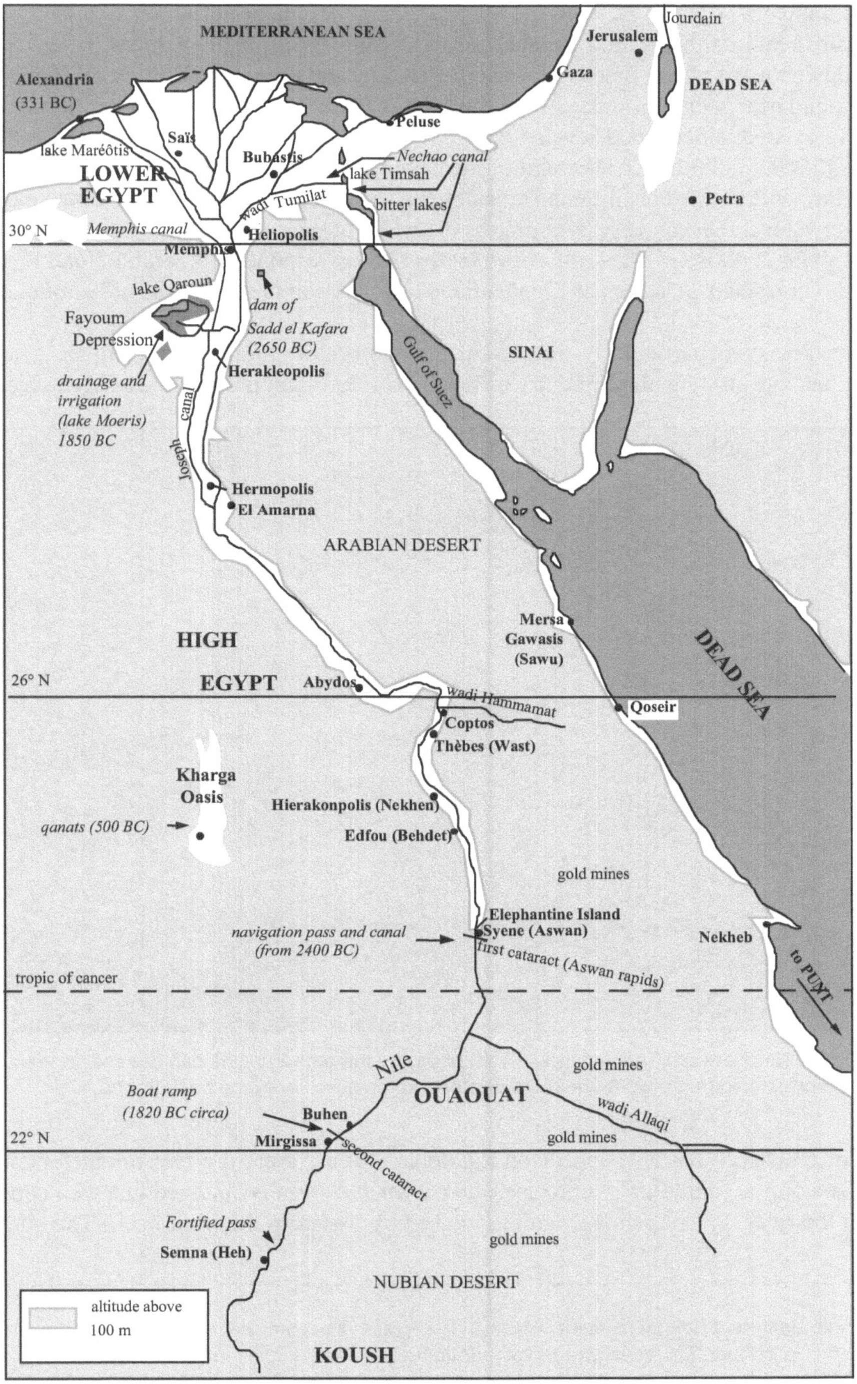

Figure 3.1 Major hydraulic works in ancient Egypt and Nubia.

scales "nilometers". The most well-known nilometers[4] are those at the fortified pass of Semma, upstream of the second cataract (around 1800 BC), on Elephantine Island at Aswan, downstream of the first cataract (1800 BC), at the temple of Karnak at Thebes (800 BC), and near Memphis upstream of the delta (Figure 3.1). But much older nilometers surely existed, since flood levels were reported in the annals of the IVth and Vth dynasties (2500 – 2000 BC).[5] The unit of measurement is the nilometric cubit, or 0.525 m. The zero, or datum, of the nilometric scales is quite probably set at the low-flow level of the river, a level that can vary over time as the width of the river varies (a scale change occurred about 2000 BC). The scales have marks that correspond to favorable flood levels: a little more than 21 cubits at Elephantine, 12 to 14 cubits at Memphis, 7 cubits in the delta.

There are two particularly important locations for flood measurement: at Elephantine (Aswan), the point of entry of the flood into Egypt proper, and at Memphis,

Figure 3.2 The Nile between Thebes and Aswan (photo by the author). One can see the contrast between the green irrigated plain (dark in the photo) and the arid hills in the background.

sentinel of the flood that will appear on the delta. In fact there are two nilometers at Aswan. According to tradition, a precise water level at Aswan is obtained in a well connected to the river, to dampen fluctuations caused by waves in the river itself. The date

4 See Danièle Bonneau (1986), Gunther Garbrecht (1987), and the commentaries of Jean Yoyotte and Pascal Charvet (1997) in "The Voyage to Egypt" of Strabo.

5 slabs of black stone, a fragment of which can be seen in the Cairo museum; the main slab, the "Palermo Stone", is in that city's museum (see for example Roccati, 1982).

of this concept is unknown. Let's again listen to Strabo:

> "The nilometer is a well, built of stone quarried from the banks of the Nile itself, in which there are marks indicating the greatest floods of the Nile, the smallest, and the average, for the water level in the wells rises and falls with that of the river. This is why there are marks on the walls of the wells, showing the peak flood levels and other levels. Inspectors examine the wells and communicate their observations to the rest of the population, for their information; they know well in advance, from these indications and their times, when the future inundation will occur, and can announce these forecasts. This information is useful not only to the farmers for the regulation of water distribution, for the dikes, the canals, and things of this nature, but also to the prefects for the estimation of public revenue, for these revenues increase with the strength of the flood."[6]

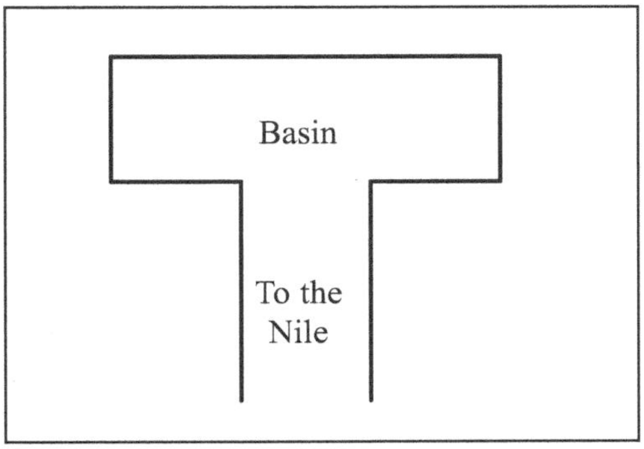

According to Danièle Bonneau, the measurements begin at the end of June, at the summer solstice, and continue through the period of inundation to the end of October, and are made known throughout the valley for general public use.

Of course the Nile is also the principal "highway" of the country. The paintings of boats of the Nile found on protohistorical pottery are among the first such known depictions. Each city, each temple, has its fluvial port, generally constructed in the form of a "T", with a basin connected to the Nile by a short canal.

The earliest developments in Egypt, IIIrd millennium BC

One of the first depictions of an Egyptian king is found on a macehead,[7] where the king is apparently opening a breach in a dike with a hoe, next to a man who is filling a basket with dirt. This king, called the Scorpion King from the ideogram on the macehead, is said to have lived around 3200 BC. This is the period during which writing first appeared in Egypt, and also a time of growing political unity between upper and lower Egypt.

Development of the capital Memphis

Herodotus visited Egypt in 460 BC when it was under Persian domination. He reports (on the basis of information from priests) that a king called Min (Menes) built the new capital of unified Egypt, called Memphis, at the border between the upper and lower

6 Geography, book XVII, 1, 48, Translation of Pascal Charvet.

7 Macehead said to be of Khashkemoui, Ashmolean Museum, Oxford.

portions. The construction of large dikes consummated the establishment of this city:

> "The Egyptian priests say that Min, who was the first king of Egypt, dammed off this place of Memphis from the Nile. For the whole river flowed close by the sandy mountain that is toward Libya, but Min, damming up the southern bend of it, about a hundred furlongs south of Memphis, dried up the ancient channel and channeled the river to flow through the middle of the mountains. Even to this day this bend of the Nile is most heedfully observed by the Persians, that it may flow in its confined course, and every year the barriers are built up again. For if the river should break out at this point, there would be a danger that all Memphis would go down in the flood. When Min, this first king, had made the cut-off part into dry land, he founded within it the city that is now called Memphis – for Memphis, too, lies in this narrow part of Egypt – and, outside it, he dug a lake away from the river to the north and west (for the Nile itself was the barrier toward the east), and he founded within the city the temple of Hephaestus, which is indeed a great one and exceedingly worth telling of." [8]

Who is this Min? He is very likely the first identifiable Pharaoh, Menes or Narmer, who probably reigned between 3150 and 3125 BC[9]. The dike he constructed at Memphis has been the subject of some speculation. Considering the difficulty of blocking a river like the Nile (and especially considering that this work would have been accomplished at the very beginning of the history of Egypt), it seems very unlikely that it was a true dam. Herodotus, who is considered to be a reliable witness, is careful to distinguish between what he sees and what he is told. He sees the dike, and he can see that it is maintained. It seems plausible that Memphis was founded by draining swampy land, perhaps in an abandoned branch of the Nile, and by then building the dike some twenty kilometers upstream to protect these drained lands from flooding. This dike had to be a dozen meters or so high, this being the height of the largest floods at Memphis.

The port of Memphis is without doubt the greatest in Egypt. It is thought to have been on the left bank, embedded in the valley of a wadi that flows along the edge of the plateau. The port is nearly a kilometer long and 200 to 300 m wide. It is connected to the Nile by a canal extending to the north along the edge of the plateau, providing access to the cultural and funereal sites of Saqqarah.[10]

Sadd el Kafara[11]: the first known large dam... and the story of its failure

In the IIIrd or IVth dynasty, about 2700 or 2600 BC, The Egyptians undertook the construction of a dam on an ephemeral tributary of the Nile, the wadi Garawi, some ten kilometers southeast of Memphis. This effort, coinciding with the period of construction of the great pyramids of Giza, is part of the development and improvement of the region around the capital. The remains of the dam are still visible today on both banks of the wadi, and were studied in 1982 by a Germano-Egyptian team.[12]

---

8 Ibid., Book 2, 99.
9 According to Nicolas Grimal (1992).
10 Kerisel (1999).
11 the Arabs have given this name to the remains of the works; it means "dam of the nonbelievers"
12 Gunther Garbrecht (1985) gives a detailed account of the conclusions of this exploration.

The dam comprises two rock faces with central core of random material. The upstream and downstream faces are protected by cut blocks forming steps of about 30 cm in height (Figure 3.3). It is the length (113 m) and especially the height (14 m) of this structure that earn its recognition as the oldest known large dam. Even more surprising is the dam's colossal thickness, nearly 100 m wide at the base and 66 m at the crest. This excessive thickness shows that dam techniques were still somewhat primitive in Egypt of the IIIrd millennium BC.

Figure 3.3 Remains of the Sadd el Kafara dam (about 2600 BC), on the right bank of the wadi Garawi . The oldest known large dam. View from upstream (photo of G. Garbrecht)

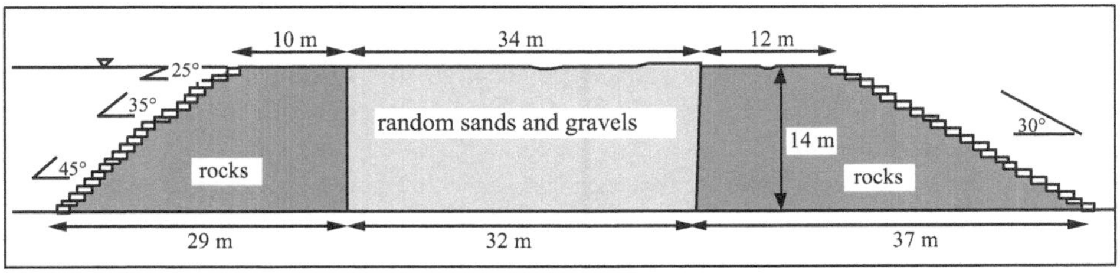

Figure 3.3a Cross section of the structure of the Sadd el Kafara dam on the wadi Garawi. The outside blocks form about 30 stairsteps. After Garbrecht (1985).

The most likely purpose of this dam was to protect downstream cultivated and inhabited areas riparian to the Nile from the violent floods of the wadi Garawi. Indeed, the structure's permeable core argues against the hypothesis of a reservoir to store water for agricultural use. In any case, irrigation in Egypt at this period remains entirely based on floods of the Nile. Curiously, examination of the reservoir area shows no trace whatsoever of sedimentation, suggesting that it was not the destiny of this important dam to survive very long.

We have already mentioned that the wadi Garawi is normally dry but can have violent floods. Its bed slope is fairly steep at about 10 m per kilometer, and the flood discharge can be estimated at between 50 and 250 m^3/sec. The storage capacity of the reservoir (620,000 m^3) is insufficient to capture all the water of a large flood. On the left bank, there is a terrace whose elevation is 1.5 m below that of the dam crest. This terrace therefore can serve as a natural spillway, whether or not this was intended by the dam's designers. The spillway has a capacity of the order of 85 m^3/sec, beyond which flow the dam can be overtopped. Given its broad dimensions and rock protection, the dam should have been able to resist partial overtopping. These analyses led Gunther Garbecht to hypothesize that the dam failed before it was completely finished, when the upstream rock face had been built to its nominal height but neither the downstream face nor the filling of the core were completely finished. Submerged by an exceptional flood, the dam could have been ruined by erosion of the central core and collapse of the upstream face. This failure would surely have resulted in downstream destruction, and the memory of this catastrophe could explain the absence of new dam construction in Egypt for many centuries to come.

On the river to Nubia, navigation works on the Nile (IIIrd and IInd millennia BC)

Navigation canal at the first cataract

Nubia is rich in quarries, and in gold and amethyst mines. A concerted effort to exploit these resources of the south began in the VIIth Dynasty, under the ancient Empire. But the Aswan rapids, comprising the first cataract of the Nile (Figure 3.4), present an obstacle to navigation. In about 2400 BC the Pharaoh Merenre I has his close lieutenant Ouni build a flume system to allow boat passage through this obstacle. Ouni, who later becomes governor of Upper Egypt, had his autobiography engraved in his tomb, where one can read the following:

> "Then His Majesty sent me to dig five canals *(flumes?)* in Upper Egypt and to construct three barges and four transport boats, from acacia wood of the land of Ouaouat. The chiefs of the lands of Ouaouat, Iam and Medla had the wood cut for this. I accomplished all of my task in a single year. When the boats were launched, they were also loaded with big wide blocks of granite for the pyramid *(of Merenre)*"[13]

This account tells us the main reason for Ouni's mission: the *descent* of boats com-

13 Funeral inscription of prince Ouni (2400 – 2350 BC), from Claire Lalouette (1984), I.

ing from the stone quarries. The current in the flumes was surely too strong for upstream passage. The flumes were rebuilt, or enlarged, under the reign of the Pharaoh Sesostris III (XIIth Dynasty) in about 1870 BC, height of the middle Empire. But now the reasons were clearly military, since the work enabled Egyptian expeditions to travel *upstream* to the second cataract (in years 8, 12, 16 and 19 of the reign).[14] But clearly these flumes either were undersized or filled with sand, which would be no surprise given the strength of the currents in this area and therefore the sand load carried by the flow. New work was conducted under Thoutmosis I and Thoutmosis III, between 1490 and 1425 BC, this time including the construction of a true canal 10 m wide and 7 m deep.[15]

Herodotus did not travel upstream of the first cataract during his voyage in Egypt. Here is how the navigation conditions in this zone were described to him:

> "From the city of Elephantine, going upcountry, the land is steep. There travelers must bind the boat on both sides, as one harnesses an ox, and so go on their way. If the rope were to break, the boat would be borne to its destruction by the strength of the current. This part of the country is four days' journey by boat, and the Nile here is as twisting as the Maeander; there is a length of twelve schoeni to pass through in this fashion."[16]

Strabo's much later account of his trip up the Nile (by land route) beyond the cataract does not mention these works, indicating that the canal was no longer in service. It is not hard to imagine that it was quite difficult to permanently maintain such a project, without locks, in an area of strong currents.

Figure 3.4 Rough terrain at the site of the first cataract, upstream of Aswan, at low water (photo by the author)

14 Grimal (1988), Chapter 6.
15 Goyon (1986).
16 Herodotus, The History, book II, 29, Translation of David Grene.

More to the south: hydraulic works at the second cataract

The pharaohs of the middle Empire tried to develop fluvial commerce with Nubia while at the same time protecting themselves from Nubia. Sesostris III extends the border of Egypt up to Semna, beyond the second cataract. A stela contains the following text:

> "Southern border established in year 8 of his majesty the king Khakaoure (Sesostris III), so as to prevent any Nubian from crossing it, by land or boat, neither any Nubian herds; except Nubians who would come to do commerce in *Iken* (Mirgissa), but not to the point that any Nubian boat travels to the north beyond *Heh (at the northern outlet of the pass of Semna)*, ever."[17]

The pass at Semna is fortified to enforce this proclamation, and to tightly control traffic on the Nile. The water surface is artificially raised and the Nile flumes are blocked, so that boats can pass only through the narrow passage between the two forts constructed on each bank.

At the second cataract, the site of Mirgissa is also fortified. A slideway is built for hauling boats around the natural obstacle formed by the cataract rapids. Boats enter this slideway from a harbor built in the calm upstream waters (Figure 3.5). The slideway, likely from the reign of Sesostris III (around 1870 BC)[18] has a useful width of about two meters, and is sloped to facilitate the hauling. It is lined with wood and silt kept damp.

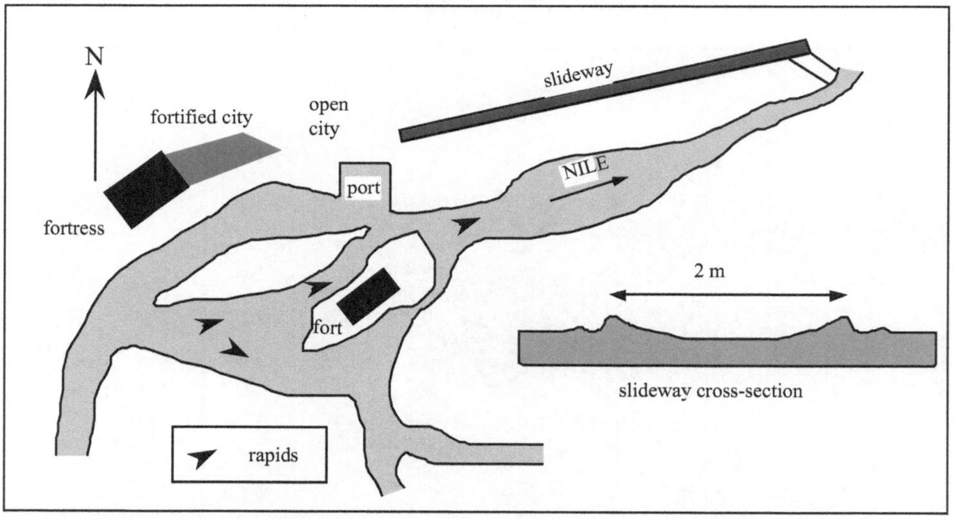

Figure 3.5 Hydraulic works at Mirgissa, at the second cataract (schematic reconstitution from Goyon, 1986; Vercoutter, 1991).

17 citation from Jean Vercoutter (1991).

18 Mirgissa was explored between 1962 and 1968 by a French expedition directed by Jean Vercoutter. For an overview, see Vercoutter (1991). For navigation between Aswan and Semna, see Goyon (1986).

The "marvelous" lake of Moeris. Fifteen centuries of work to develop Fayoum.

The Fayoum Depression, located 80 km southwest of Memphis (see Figure 3.1) in the "lake province" of the ancient Egyptians, was prized by the pharaohs and viewed as a marvel by Greco-Roman travelers. Strabo wrote:

> "(*This region*) contains also this admirable lake that is called the Lake of Moeris and has the dimensions and color of a sea."[19]

This region has a long history. It was first developed at the beginning of the IInd millennium BC by the pharaohs of the XIIth Dynasty. It was visited by Herodotus in 460 BC, and redeveloped by the Ptolemite successors of Alexander in the 3rd century BC. Strabo visits the region in 25 BC, at the dawn of the Roman domination during which Fayoum was one of the granaries. Today, the Qaroun lake sits in the depression, 70 m below the normal level of the Nile, with its rather barren shores. The observations of travelers, geologists, and archaeologists regarding the depression are often contradictory. This has lead to divergent interpretations of the developments in this region. But in forming such interpretations one must not forget that the history of the region spans more than fifteen centuries.

The pharaoh and the lake: the great hydraulic works of the IInd millennium BC

In prehistory the Joseph canal, or *Bahr Youssouf,* supplied water to Fayoum through an ancient arm of the Nile. At that time Fayoum comprised an immense body of water and marshes, with a water surface elevation somewhat below that of the Nile. Little by little, sedimentation raises the elevation of the plain.[20] In about 7500 BC, hydraulic connection of the region to the Nile is not continuous but episodic, causing periods of rising and falling lake levels. Throughout the IIIrd millennium BC, the lake level appears to have remained low, at an elevation that is thought by some to be around –2 m,[21] the level of the Nile being around +20 m at this time. At this time the large lake was natural, occupying roughly the area within the contour 0 on Figure 3.6. The lake supported fishing, and hunting along its shores.

About 2000 BC, possibly because of an exceptional flood, the water level in the Fayoun Depression abruptly rose to +22 m, and then fell again.[22] This surge of water inundated Kasr el Sagah (founded about 2400 – 2300 BC) where there was an embarka-

---

19 Geography, book XVII, 1, 35, Translations of Pascal Charvet, adapted.

20 10 cm per century, according to Butzer (1998). The normal level of the Nile at the latitude of Fayoum, today at an elevation of +24 m, was probably about +20 m at the beginning of the IInd millennium BC.

21 These data are from the geological studies synthesized and analyzed by Butzer (1997). The elevation of –2 m reached around 2000 BC is derived from the geological work of Gardener and Caton-Thomson (1929), cited by Garbrecht (1996). Annual evaporation in this region is estimated at 1.7 m; the lake level could therefore fall some 20 m in a dozen years.

22 Butzer (1998).

tion embankment. After this episode, men started to have a hand in the evolution of the lake, at the height of the Middle Empire. About 1890 BC, Sesostris II begins the extensive work that is eventually finished 50 years later by his grandson Amenemhat III,[23] the sovereign whom the Greeks later called Moeris.

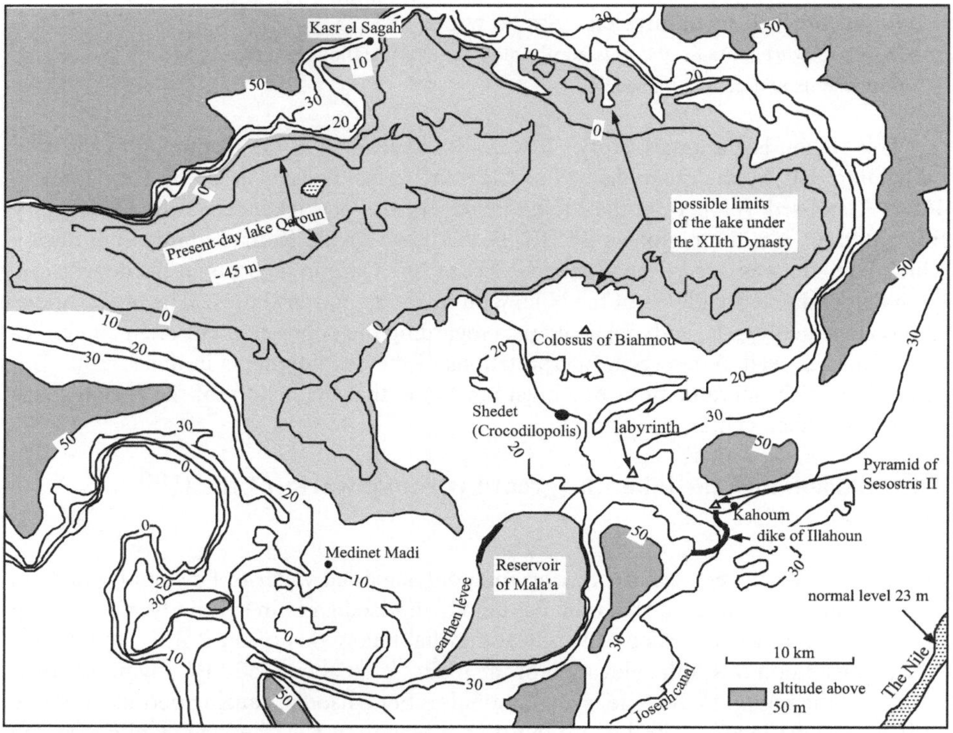

Figure 3.6 The Fayoum Depression and the "Moeris Lake". We have shown on this map the 0, 10, 20, 30, and 50 m contours, from Ball (1939), as well as monuments and establishments of the XIIth Dynasty. Also shown are the hydraulic works (the Illahoun dike and the Mala'a reservoir) which have been dated in the 3rd century BC (Garbrecht and Jarritz, 1992). The normal level of the Nile is today at an elevation of 23 m; the flood level can attain 30 m.

The work included widening the Joseph canal to 90 m, and building other irrigation and drainage canals. As a result, the now-cultivated region became an important economic center. The irrigation system undoubtedly used flood storage basins, as was the practice in Egypt. It is easy to suppose that the large artificial reservoir later recognized as Mala'a occupied the upper terraces to the south of the depression in one form or another, at an elevation sufficient to supply water to all of the region (there is no proof of this, however).[24] The capital of the *Lake Province* was then called *Shedet* (today Medinet el Fayoun), a city the Greeks will soon call *Crocodilopolis* since crocodiles are

23 Grimal (1988), Chapter 7.
24 see the synthesis of Günter Garbrecht (1996).

worshipped there. Amenemhat III had a tomb built so he could forever remain there, at Hawara. This is a pyramid associated with a vast temple called *the labyrinthe*, much admired by Greek and Roman tourists. This same king had a temple built near the present Medinet Madi, and another in the reconstructed colony of Kasr el Sagah. He also built two colossal statues at Biahmou, each twelve meters in height on a monumental pedestal, undoubtedly in his own image.

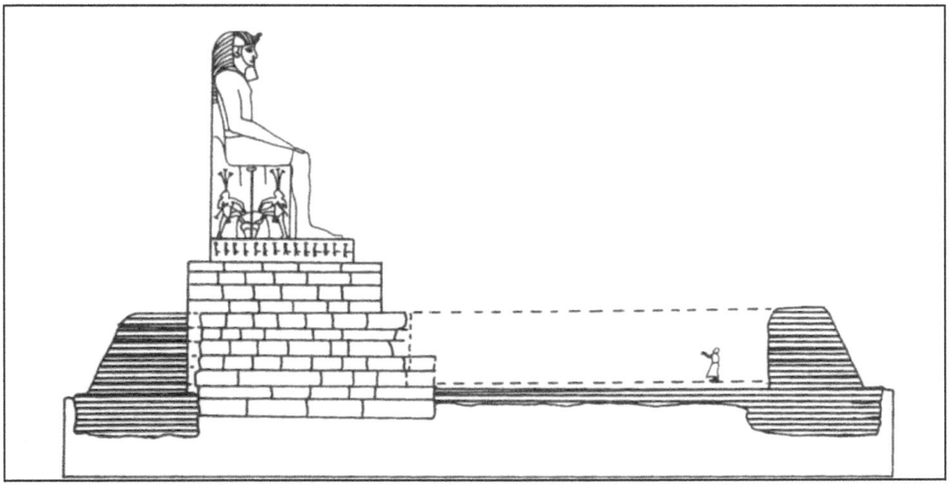

Figure 3.7 Amenemhat III looking over the Moeris lake – reconstitution of one of the colossi of Biahmou by the Egyptologist Sir Flanders Petrie (1899) (after Lane, 1985).

The Ptolemaic assertion that there were water control works at the entrance of Fayoun in this period (the Illahoun dike and its gates) has always been subject to some doubt. Yet one can see on Figure 3.6 that the monuments of the XIIth Dynasty are at an altitude of between +10 m and +20 m,[25] i.e. quite some distance below the flood level. Therefore it is easy to suppose that there must have been hydraulic works at Illahoun to isolate Fayoun from the rest of the valley and control the inflow of the Nile into the depression.

With the end of the XIIth Dynasty and of the middle Empire, in 1759 BC, came the troubles of the second intermediate period. During this period the maintenance of the Fayoun system was surely neglected. There is geological evidence of major inundations around 1700 BC, when the water level in the Fayoun Depression reached +22 m. Then again in 1500 BC it reached a record elevation of +24 m.[26] These floods were highly destructive, and the region never recovered its previous glory (at least not before the new developments of the 3rd century BC). Still, the region partially recovered in the period of the Ramses, when Fayoun hosted the great royal harem. A new series of troubles began in the 9th century BC, and with them surely came more uncontrolled incursions of the floods of the Nile into the Fayoun Depression. According to Butzer (1998), the

25 and even at +9.7 m for the temple of Medinet Madi, according to Butzer (1998).
26 Butzer (1998).

water level in the Fayoun remains around +20 m from the 9th to the 5th century BC. It is easy to see from Figure 3.6 that the quasi-totality of the depression would be flooded at this level.

The search for Lake Moeris

Among all the Greek travelers, Herodotus is the only one to have visited this region (in 460 BC) prior to the new hydraulic works implemented by Ptolemy II in the 3rd century BC. Having admired the Labyrinth, the funeral monument of Amenemhat III, he then describes a lake of very large dimensions, oriented approximately north-south:

> "Such was the labyrinth; but an even greater marvel is what is called the Lake of Moeris, beside which the labyrinth was built. The circuit of this lake is a distance of about four hundred and twenty miles *(670 km!)*, which is equal to the whole seaboard of Egypt. The length of the lake is north and south, and its depth at its deepest is fifty fathoms *(89 m)*. That it is handmade and dug, it itself is the best evidence. For in about the middle of the lake stand two pyramids that top the water *(these are the colossi of Biahmou)*, each one by fifty fathoms, and each is built as much again underwater; and on top of each there is a huge stone figure of a man sitting on a throne. The water that is in the lake is not fed with natural springs, for the country here is terribly waterless, but it enters the lake from the Nile by a channel; and for six months it flows into the lake, and then, another six, it flows again into the Nile."[27]

One would hope that this account reflects the work of the pharaohs of the XIIth Dynasty, but Herodotus' account is from more than a thousand years later. Herodotus is in fact describing the rather sad sight of the depression's complete inundation, consistent with the geological descriptions cited earlier. The dimension that Herodotus indicates (a perimeter of 640 km), and the fact that the colossi of Biahmou are "in the middle of the lake" leave no doubt that this is the case.[28] The alternating current in the Joseph canal was probably, at this period, a simple natural phenomenon caused by season variations in the level of the Nile, unregulated by man.

And yet, as we have said, there logically must have been one or more temporary reservoirs to store and distribute the flood waters in the era of Amenemhat III, distinct from the Qaroun lake that occupied the lowest portion of the depression. These reservoirs were situated above the irrigated lands, very likely near Shedet – Crocodilopolis and the mouth of the Joseph canal. Field studies conducted in 1988[29] made it possible to reconstitute the boundaries of a vast reservoir, located on the heights to the south of the depression as expected. The southern portion of the boundary approximately follows the contour +17 m, and to the north it is closed by a dam. This dam of Mala'a is 8,000 m long and four to five meters high. But the only remnants of its masonry construction that are visible today date from the Ptolemites (3rd century BC) – along with remnants

---

27 Herodotus, The History, Book 2, 149, Translation of David Grene.

28 Regarding the fact that the "lake of Moeris", that is to say the entire depression, was excavated artificially, it is clear that our author's sources were not straightforward with him. See for example the notes of Jean Yoyotte in his edition of Strabo's The voyage in Egypt, page 142.

29 Garbrecht and Jarritz (1992). See also Garbrecht (1996).

of repairs from the Roman and Islamic eras. Older vestiges have not been found. The visible traces of the Illahoun dike (remains dating also from the 3rd century BC) suggest that it was 5 km long and four meters high. It is unlikely that we will ever know the details of engineering developments from the period of Amenemhat III with any certainty. Since the Ptolemite engineers gave the ancient name *lake of Moeris* to their reservoir, it is possible that their work more or less replicated the preexisting system – but this is only speculation.

There remains another question: what has been the evolution of the "normal" level of lake Qaroun across the ages of its existence? As we have seen, the altitude of the monuments erected in the XIIth Dynasty argue for a lake whose surface is approximately at elevation +10 m. It rises to +20 m when the Fayoun Depression is not isolated from the valley, and fluctuates with the floods. In the Ptolemite period, as we will see later on (Figure 5.8), the new developments will be around elevation 0 (even –10 m), an altitude that is surely suggestive of the lake level at that time. It is likely not until the time of the Romans that the lake level was lowered to its present level, 45 m below sea level, to increase the amount of tillable land.

Fayoun owes its history as one of the most productive regions of Egypt to the hydraulic works of the successors of Alexander the Great. Strabo visits the region in 25 BC (the labyrinth remains one of the most attractive curiosities to travelers), long after these new works have been implemented:

> "It still remains that the lake of Moeris, by its dimensions and its depth, is capable of containing, during the floods of the Nile, the excess water, without overflowing onto inhabited places and their crops; and at the moment when the river waters recede, it is capable of returning this excess water by the same canal, in each of its two outlets, while keeping within itself and the canal, a reserve of water to feed the irrigation canals. Whatever be the acts of nature, they have placed locks *(ports or gates?)* by means of which the engineers regulated the flow of water that enters and leaves." (Strabo, Book XVII, 1-37)

Thus it is indeed the great reservoir of Mala'a that Strabo describes as the "*lake of Moeris*", rendering unwitting homage to the nearby remains of the old pharaoh.

The great accomplishments of Egypt in the first millennium BC: from the last pharaohs to the Persians

Who dug the first "Suez Canal"?

Egypt had a long tradition of maritime commerce with countries on the shores of the Red Sea (in particular with the country of Punt, situated approximately east of Sudan and north of Eritrea). The port of Mersa Gawasis was founded in about 1900 BC under Amenemhat II (Middle Empire), somewhat below the 27th parallel (therefore a bit to the north of Thebes). Merchandise is carried by land to or from Thebes through arid valleys, typified by the wadi Hammamat.

One can therefore readily appreciate the interest in a direct maritime link between the Nile and the Red Sea. Such a link was in fact realized though the valley of the wadi Tumilat with an east-west orientation. The remains of the canal were still visible in the 19th century, as described by Marice Linant de Bellefonds (1799-1883):

> "More than forty years ago one could see, in the northern portions of today's wadi Toumilat, the remains of an ancient canal that had rather small dimensions; it came from the west and flowed to the east along the desert and cultivated lands.... Near Tel-Retabee, this canal joined another much larger one, at a place called Ras el Wadi... this was the principal ancient canal.... It is there that the other canal from the northern portion joins the latter, which is much larger and looks to be a very old and well built canal."[30]

All of the classic authors mention the existence of this very ancient and large-scale link. Certain of them (Aristotle, Strabo, Pliny) attribute the paternity of the canal to a pharaoh whom they call "Sesostris."[31] But archaeology clearly rules out the existence of such a communication link in the middle or new Empire. At best, one may consider the possibility that in the new Empire a small canal was constructed to transport stones from the Nile to the monuments constructed by Ramses II at the site known today as Tell el Retaba. This is surely the small northern canal of the text cited.[32]

Modern studies[33] show that it is once again Herodotus who gives us the most precise information. He situates the real beginning of the construction of the large canal during the reign of the Pharaoh Necho II, of the Saite Dynasty, who reigned in about 600 BC. This pharaoh, like most of his predecessors, pursued a policy of expansion toward the east, taking advantage of the fall of the Assyrian Empire. He builds a fleet of boats and embarks on an African expedition. Necho II founds the city of *Tjekou*, on the site of today's Tell el Maskhouta, some fifteen kilometers to the west of the present Ismailia.[34] Most of the canal was therefore built in that period (at least as far as *Tjekou*).

Two independent sources, that we will cite below, indicate the digging of the canal was not effectively ended until a century later, about 5009 BC, by order of the Persian sovereign Darius I.

Here is what Herodotus says:

> "The son of Psammetichus was Necos, and he too became king of Egypt, and he was the first to attempt to dig a canal into the Red Sea; Darius the Persian was the second to dig it."[35]

According to this author, some 120,000 workers were employed for the task by Necho. In 1866 Ferdinand de Lesseps, during his preliminary reconnaissance for con-

---

30 citation after Redmount (1995).

31 The classical authors attribute many things to "Sesotris", as they do to Semiramis of Mesopotamia.

32 This zone is unoccupied in the Middle and New Empires, as attested to by Jean Yoyotte (see his note 266 in Strabo's *Travels in Egypt (Le voyage en Egypt)*; see also Carol Redmount, 1995). This author raises the possibility of a canal to Tell el-Rebatah built in the New Empire; the canal would be more to the north, and more modest, than the actual canal of the two seas. The text of Linant de Bellfonds nicely describes the remains of two distinct canals to the west of this site, one on the north flank of the wadi Tumilat valley, the other on its south flank.

33 See the synthesis of C.A. Redmount (1995).

34 Grimal (1998), Chapter 14.

35 Herodotus, The History, book II, 158, Translation of David Grene

struction of the Suez Canal, recognizes the traces of this ancient canal, well to the east of the vestiges found by Linant de Bellefonds.[36] At Kabret, 130 km from Suez on the isthmus of the same name, de Lesseps discovers a stela of pink granite engraved with the name of Darius, and with the following inscription in several languages:

> "the king Darius declares: I am a Persian. From Persia, I took Egypt. I ordered this canal to be dug, from a river of the name of Nile, that flows in Egypt, to the sea that comes from Persia. Therefore, this canal was dug, as I had ordered, and boats go from Egypt, by this canal, toward Persia, as it pleases me."[37]

The Persian's strategic interest in this canal is obvious. But it is clear that Darius wrongly attributes the conquest of Egypt to himself, since this conquest was in fact accomplished by his predecessor Cambyse. Therefore it is not surprising that he also exaggerates his own role in the digging of the canal. Necho's channel may have become clogged with silt or sand in the era of the sovereign Darius, who therefore would have re-excavated it. It is also possible that the canal project undertaken by Necho was simply not completed. In any case, three other stelas carrying the name of Darius will be discovered along the route of the canal, the most westerly being at Tell el-Mashkhuta (Figure 3.8).

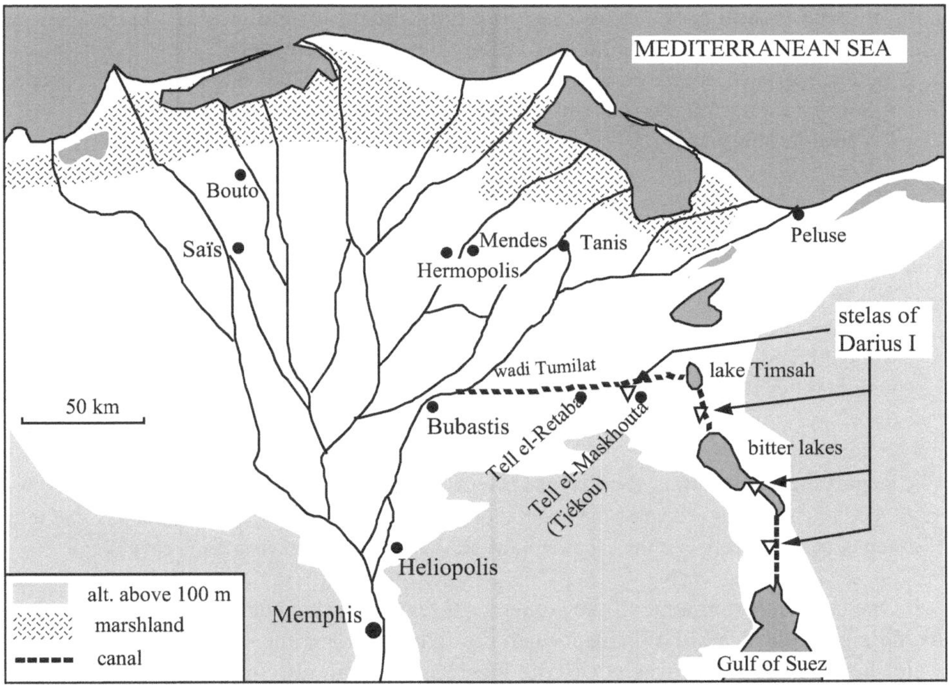

Figure 3.8 The Canal of Necho, or canal of the two seas.

36 Soulé (1997), II, 2.
37 Stela of Chalouf, after Pierre Lecoq.

A thousand years of traffic on the Necho canal

So what is the actual path of this ancient canal? We know it fairly well, since the remains observed in the twentieth century substantially agree with the descriptions of Greek and Roman travelers. The canal issues from the eastern branch of the Nile, follows the southern edges of the wadi Tumilat valley, passes by Tell er-Retaba and Tell el-Maskhouta (*Tjekou*), rejoins lake Timsah, then flows directly south toward the bitter lakes and the Gulf of Suez. Let us listen to a continuation of the account of Herodotus:

> "It takes four days to travel along it, and its width is such that two triremes could be rowed in it side by side. It is fed by the waters of the Nile, and begins a little above Bubastis (Tell Basta) by Pithon (Tell el-Maskhouta), an Arabian town. It ends in the Red Sea. The excavation was begun in the part of the Egyptian plain which is nearest to Arabia. The mountains where the stone quarries are and which are close to Memphis, are near this plain. The canal was dug along the foot of these mountains from west to east, passing through a gorge *(the wadi Tumilat valley?)*. It turns to the south out of the hill country toward the Arabian Gulf."

The outlet of the canal into the sea, at the end of the Gulf of Suez, is obviously an important and critical site. Apparently land transfer of goods, or boat portage (a common practice in Antiquity), is necessary up until the era of Ptolemy II. In 280 BC he built the terminal facilities. Diodore of Sicily describes them thus:

> "(Ptolemy II) conceived a barrier adroitly placed at the most favorable location. One opened it when one wanted to pass and immediately closed it, for it was well designed for this purpose."[38]

When Strabo visited the delta region, this installation was clearly in use. He certainly did not travel as far as the Red Sea, but his account confirms that the flow in the canal is from the Nile toward the lakes:

> "There is another canal that flows in to the Red Sea and the Arabian Gulf near the city of Arsinoe (...). It flows through the lakes that are called *bitter*. Originally, these lakes were, without doubt, bitter, but when the aforementioned canal entered them, their waters, mixed with those of the river, changed nature and are today full of fish and inhabited by aquatic birds."

Regarding the terminal facility of Ptolemy, Strabo's account is somewhat vague:

> "the Ptolemite kings finished the excavation and closed the passage, in such a way that they could, at will, freely exit the canal into the sea outside and reenter into the canal."[39]

One can get lost among all the conjectures regarding this "barrier". It was probably a single gate that could be opened when the tide equalized the water levels in the two water bodies, but surely was not a true lock[40]. One can only regret that Diodore and

---

38 Bibliothèque historique, book I, 33, 11.

39 Geography, XVII, 25, Translation of Pascal Charvet, adapted.

40 We will see in chapter 9 that it is not until the 10th century AD that a true gated lock appears, in China.

Strabo did not leave us a more precise description of this "barrier".

The canal will later be maintained and kept in operation by the Romans, who valued a direct pathway to the incense and myrrh of Arabia Felix. They named it *Trajan's Canal* after the work of this emperor, (who moved the point of entry toward Heliopolis, probably to improve the flow). The canal is also renovated and used by the Arabs, starting in 641 AD and up until its closure in 767 or 775 AD, to ship Egyptian wheat to Mecca and Medina. This closure was ordered by the Caliph Abu Jafr al-Mansur.[41] For twelve centuries, Necho's project assures communication between the Red Sea and Indian Ocean with the Mediterranean Sea.

Few technological achievements can lay claim to such long success.

The qanats in Egypt under the Achaemenid Empire

In Chapter 2 we described the invention of the *qanat* in Urartu and its development in Persia. After the conquest of Egypt by Cambyse, qanats are introduced in Egypt to irrigate the oases as well as the mountainous zones situated along communication routes (the wadi Hammamat, between Thebes and the ports of the Red Sea). Traces of three qanats constructed by order of Darius I (about 500 BC) have been found in the oasis of Kharga, some 300 km to the northwest of Asswan.[42] These qanats are as deep as 75 m, with a gallery several kilometers long and a rather flat slope compared to the usual practice, only about 0.5 per thousand.

Irrigation in the land of the Queen of Sheba

Irrigated oases at the threshold of the desert

To the south of the Arabian peninsula, the mountains of Yemen rise to 3,000 m and capture the seasonal monsoon rains. The high valleys are therefore well watered – but the most prosperous regions are not found here. The shrubs from which incense and myrrh can be harvested are located, rather, on the edges of the arid interior desert, at about 1,200 m of altitude, in the lands of Qataban and Hadramawt. These aromatic resins become quite the fashion in all the countries of the East, in Greece, and eventually in Rome, from the 8th century BC. The richness of this land, soon to be called Arabia Felix,[43] is built on incense and myrrh - their harvest, exchange, and associated control of the caravan routes that lead to the north of Arabia along the eastern edge of the great mountains.

However it is only through hard work that the kingdoms of Sheba, Qataban,

---

41 Redmount (1995); Mayerson (1996).

42 after Henri Goblot (1979).

43 Everyone knows the legend of the Queen of Sheba who is said to have visited king Solomon at Jerusalem. In reality, there is no historical evidence of this queen. Regarding Arabia Felix, the reader can consult, for example, the articles of Jaqueline Pirenne (1979) or the work of Jean-François Breton (1998).

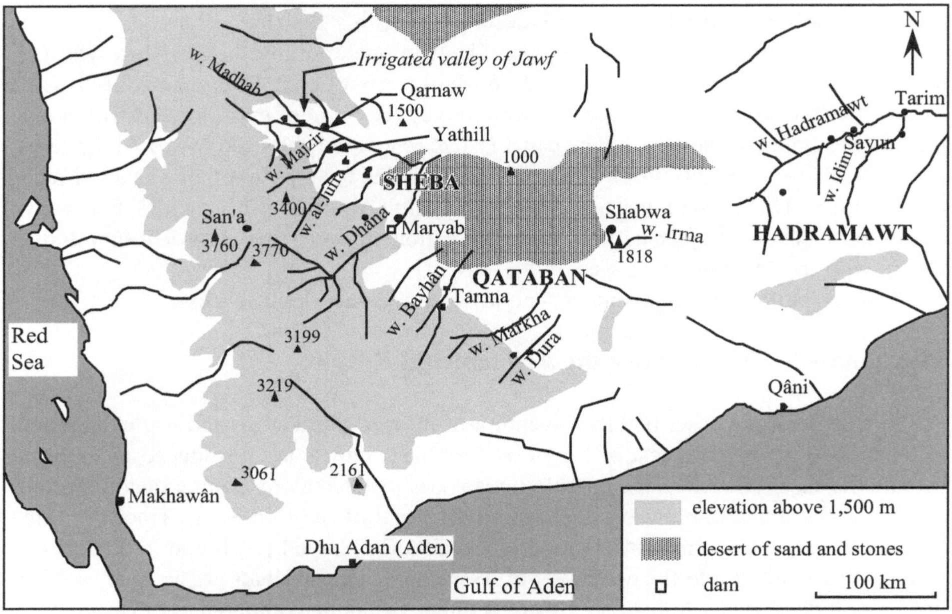

Figure 3.9 Arabia Felix, the land of Sheba, Qataban, and Hadramawt

Hadramawt, and Ma'in, whose capitals are *Maryab* (today Ma'arib), *Tamna* (Hajar Kulhan), *Shabwa*, *Yathill* (Baraqish), find prosperity. Nothing is possible without water, without crops. The land between the mountains and the desert is very arid; the only sources of water are the wadis. These are normally dry, but are subject to violent floods when heavy rains fall on the high mountains from which they issue, two or three times a year between March and August. People had learned how to make use of this water resource long before incense came into fashion – very likely from the IIIrd millennium BC along the wadis of Dura, Dhana and Markha; and from the IInd millennium BC in the basin of the wadi Hadramawt (along the wadi Irma upstream of Shabwa, in particular).[44]

The water engineering techniques remained about the same through the era of prosperity of Arabia Felix. Deflector walls, weirs and small dikes, and occasionally true dams, constructed in the beds of the wadis, direct some of the silt-laden floodwaters toward a system of branching earthen canals. These canals are provided with outtakes – stones with grooves into which beams could be slid to control the water flow. Since the canals have a much gentler slope than the wadi, the currents are slower and therefore only the finest of the sediments are diverted to the crops with the water. The fields are quadrilateral in shape and surrounded by earthen levees, the whole comprising a vast irrigated zone. Gates permit controlled flooding of the fields to a depth of several tens of centimeters. Thus two resources are being used simultaneously: the water itself, making it possible to plant as soon as the field flooding is over; and the sediments, whose

44 Breton (1998), p. 28. See also Breton, Arramond, Coque-Delhuile and Gentelle (1988) for the wadi Bayhan; also Coque-Delhuile (2001). The earliest settlements at Shabwa date from 1900 – 1800 BC.

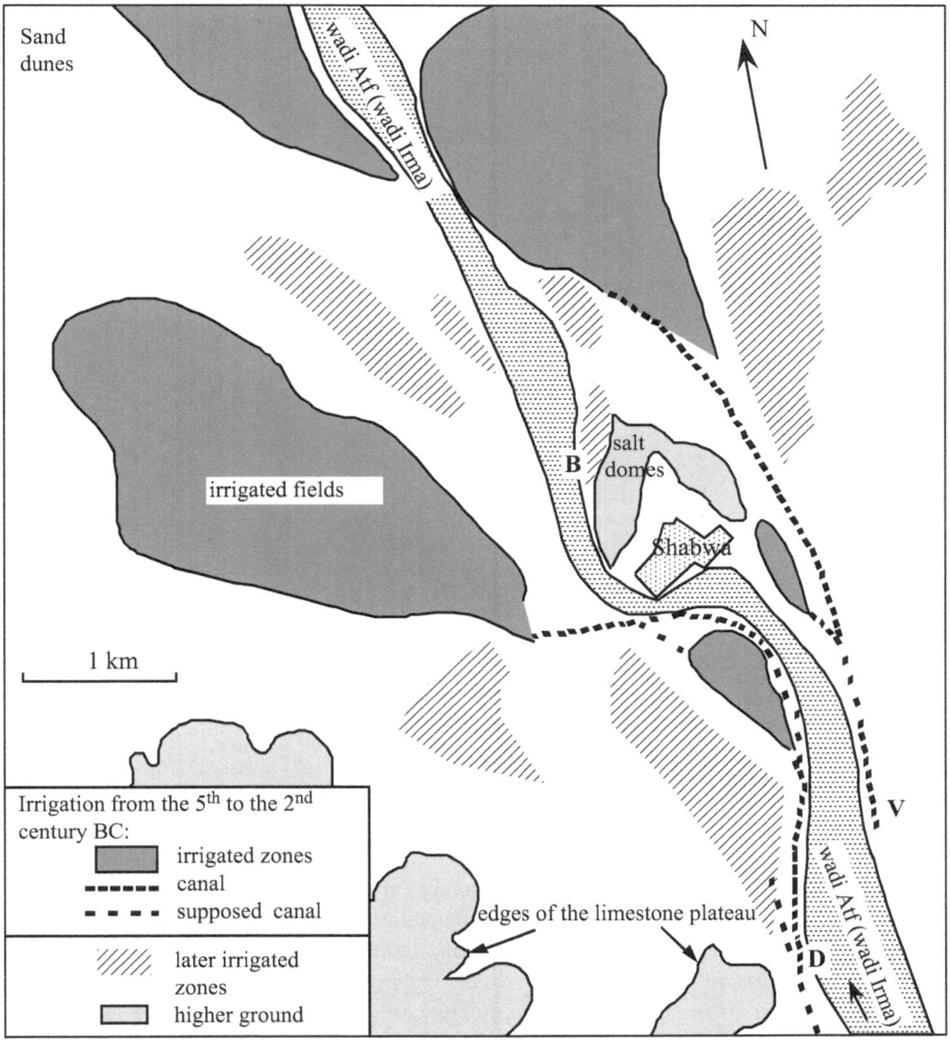

Figure 3.10. Reconstitution of the principal canals and irrigated zones around Shabwa, approximately between the 6th century (founding of the city) and the 2nd century BC – after Gentelle, 1992. The irrigated zones posterior to this period are also indicated. On the right bank, a large stone water gate (V) dates from the 2nd century BC; the deflector weirs (D and B) likely are from even later. On the two smallest ancient irrigated plots, just upstream of Shabwa, the thickness of the silt deposits reaches 15 m. The relatively steep slope of the wadi (5 m per km) makes it possible to cope with this downstream deposit, through modification of the locations of the intakes and canals. The desert is quite close, its first dunes appearing at the northwest extremity of this map.

deposition creates fertile layers of soil. Over the centuries, the thickness of topsoil increases, attaining some fifteen meters in certain locations (Figure 3.10), and even thirty meters at *Maryab*. This accumulation of soil requires that the canal system be reworked periodically, the intakes being raised or moved further upstream so that the water can still reach the fields.

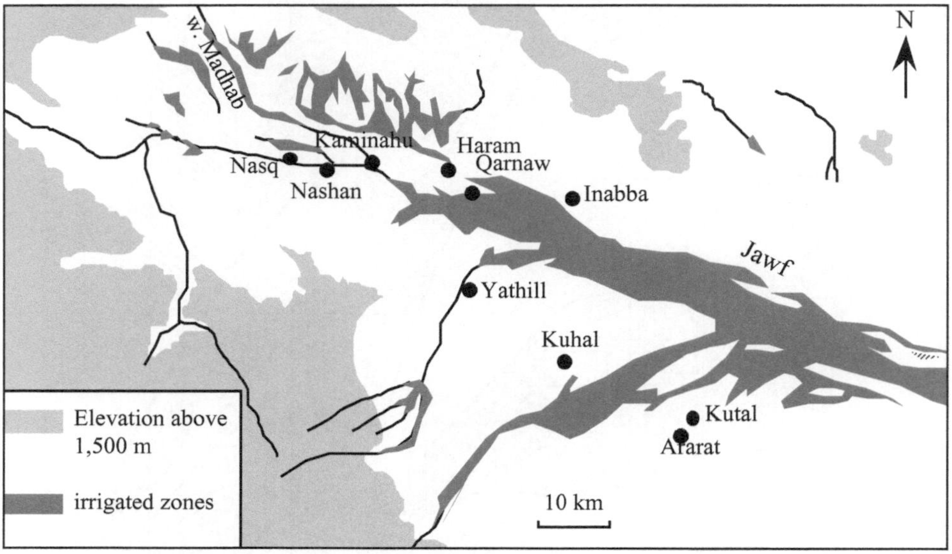

Figure 3.11. The irrigated valley of Jawf, and the city of the land of Sheba (Maryab is a little further to the south) - adapted from Robin (1991).

This mastery of complex hydraulic systems could not be achieved without a strong social organization. The kingdoms mentioned earlier are effectively organized and hierarchical. They are dominated by Sheba from the beginning of the 7th to the 1st century BC, under the hand of the great Sheban sovereign Karib'Il Watar, the "Karibilu king of Sheba" mentioned in texts of the Assyrian Sennacherib. The sudarabic writing is original, using an alphabet related to the Phoenician. Numerous inscriptions still exist, commemorating for example the redistribution of irrigation water following the destruction by Karib'Il Watar of the city of *Nashan,* at the convergence of the wadis Madhab and Jawf (Figure 3.11):

> "(Karib'Il Watar) took the rights of the king of Nashan and of Nashan to the waters of Madhab (...), took the water of dhu-Qafan at Sumhuyafa and at Nashan and granted it to Yadhmurmalik, king of Haram, took from Sumhuyaga and at Nashan the canals dhat-Malikwaqih and granted them to Nabat'ali, king of Kaminahu, and at Kaminahu the waters of the canals dhat-Malikwaqih (*that are)* beyond the boundaries of Karib'Il, built the enclosure of Nasq and peopled it with Shebans...."[45]

Another later text attests to the concession of irrigated land to a dignitary of the city of *Haram* (located several kilometers from *Qarnaw*):

45 Robin (1991).

"Wahab'il son of Ammidhara, of the clan of Thabaran (...), chief of the horse-mounted troops of Haram, honored Matabnatiyan, god of Thabaran, (...), when he dug, bore, and lined with stone his well, and that Yashhurmalik Nabat granted him a flooded public (?) land, of which he took possession, extending from the enclosure of Dhu-Arnab upstream to his property, and the concession that is upstream, between the canal of Dhat Batalan and the road of Ma'in, and he acquired and brought into cultivation a number of 300 *(parcels)*."[46]

Numerous texts also describe the construction of hydraulic works. The following extract, despite its gaps, describes the construction of a canal and dike, again in the region of *Haram*:

"*(the person in question)* bound and dug Yasim (?) with the upper canal of the plain of the city of Haram, next to (?) this canal and the tour, (...) and the land that is subject (?) to the king of Sheba (...) he granted his inscription and his dike."[47]

There are regulations to protect the waterways against thoughtless incursions. One of the earliest known is from the 5th century BC, after a certain Karib'Il Bayyin had built a runoff channel around the small city of *Nasq* (in Jawf, populated by Sheban colonists after the destruction of *Nashan*, as is referenced in one of the extracts cited above). His son is then led to issue a decree:

"Damar Alay Watar, son of Karib'Il, has decided and decreed, for Sheba and its colonists, the clearing of the water collector wall *(deflector?)* of the city of Nasq, that his father Karib'Il had opened according to the inscription of easement where his father delineated it in writing. Neither watered nor arid land shall be developed therein, and no irrigated produce or crop shall be harvested therefrom."[48]

Neither the regularity nor the amplitude of the beneficial floods is guaranteed, as is clear from the following text. Thus, if two clans have neglected the practice of a ritual hunt dedicated to the god Halfan, and there is a water shortage two times in a row, the clans must carve an inscription to do penance for their sin in the hopes that Halfan will again irrigate their lands:

"The clan Anir and the clan Athar have confessed and done penance to Halfan *(the principal god of Haram after the 2nd century BC)* because they did not render unto him the ceremonial (?) hunt in the month of dhu-Mawsab when they took refuge at Yathill during the war of Hadramawt, whereas they made the pilgrimage of dhu-Samawi at Yathill and put off the ceremonial hunt until the month of dhu-Anthar. Therefore he *(the god Halfan)* did not grant them water for their irrigation network from the spring to autumn, because of an extremely low quantity of water, and they will be careful not to do the same thing again."[49]

---

46 after C. Robin (1992), "Inventory of sudarabic inscriptions, I", Haram 2. The reference to the god Matabnatiyan indicates that this text dates from before the 2nd century BC.
47 Robin (1992), Haram 49.
48 citation from Jacqueline Pirenne (1982). The inscription of easement has also been found: "Karib'Il Bayyin reserved (for water flow) around the city of Nasq to its boundaries: 60 sawahit (?)"
49 Robin (1992), Haram 10. The date of this inscription is doubtless from between the 1st century BC and the 1st century AD.

Thanks to these irrigation systems, artificial oases flourish along the wadis Dhana (we will come back to it later) and Markha, in Hadramawt (see also Shabwa on Figure 3.10). Although these regions have the oldest irrigation infrastructure, other systems are found on the wadi Bayhan in Qataban, where a 45-km continuous ribbon of land is irrigated and cultivated .[50] Arabia Felix has some thirty cities between the 7th and 1st centuries BC, in the grand valley of Jawf and near the wadis Dhana, Ragwan, Juba, Harib, Markha, and Hadramawt. Most of these fortified cities have surface areas of from six to ten hectares; but they have neither sewers nor water supply infrastructure.

In the 1st century BC, merchants of Alexandria discover how to make an annual maritime voyage toward India, using the monsoon. This is a negative development for the economy of Arabia Felix, which thus loses its monopoly on the supply of spices. This century thus marks the beginning of a slow economic recession. Upland tribes (the Himyarites) begin their progressive domination beginning in the 1st century AD.

The oasis of Maryab; the great dike on the wadi Dhana

Maryab is the largest city of the region during the period of the sudarabic kingdoms. It occupies some 90 hectares enclosed by a 4.5 km wall. The site, apparently dating from the IInd millennium BC, is located some ten kilometers downstream of a gorge through which the wadi Dhana leaves the mountains. This wadi, typical of others in the region, has only occasional but particularly violent floods (twice a year, in April and July-August). Its flood flow can be as high as 600 m^3/sec.

At first, partial diversions are employed to make use of water in the wadi, as is done throughout the region. Inscriptions from the 6th or 5th century BC mention the construction or rehabilitation of water intakes. As completed, these are massive installations of cut stone, from which emanate two irrigation canals that branch out to deliver water to two vast irrigation zones on either side of the course of the wadi. At an unknown date (perhaps prior to, or perhaps after the 5th century BC), the bed of the wadi is blocked by a large earth dam, 15 m high and 650 m long, with rock protection on its upstream face. The reason for this dam, which is unique in the region, is probably found in the continuously rising elevation of cultivated lands due to the deposition of sediments. The thickness of such deposits can reach 30 meters. One can imagine that the initial response to this increase in level is to move the intakes further upstream (there are remains of even older works downstream of the dam). But once the intakes had been moved all the way upstream to the narrow gorge from which the wadi Dhana emanates, the only solution became to raise the water level for the intakes by completely blocking the gorge.

In its final form, the irrigation infrastructure of Maryab is truly impressive.[51] It includes a vast reservoir that, during floods, extends 4 km upstream of the dam. The intake at the north edge of the dam includes two passages, 3 and 3.5 m wide and 9 m high, that can be blocked by sliding beams into grooves carved into the massive stone

50 Breton (1998), p. 64.
51 Ryckmans (1979); Breton (1998), p. 34

blocks. These passages lead to a sort of stilling basin, whose outlet supplies a 14 m wide canal that is 1,100 m long. This principal canal ends at a stone distribution basin fitted with numerous channels leading toward secondary canals.

At the southern extremity of the dam, an 80-m long masonry revetment, 11 m high, protects the flank of the dike and borders the intake channel, 4.5 m wide. Here again, grooves are provided for controlling the water flow. This intake structure provides access to another narrower canal, as on the north side, but it is only 4 m wide.

On each of the two banks of the wadi, there is a side weir that permits overflow of excess water back into the channel (Figure 3.12). The two irrigated zones, covering some 5,700 hectares to the north and 3,750 to the south, comprise the "two gardens of Sheba", veritable oases that will later be celebrated by Arab writers.

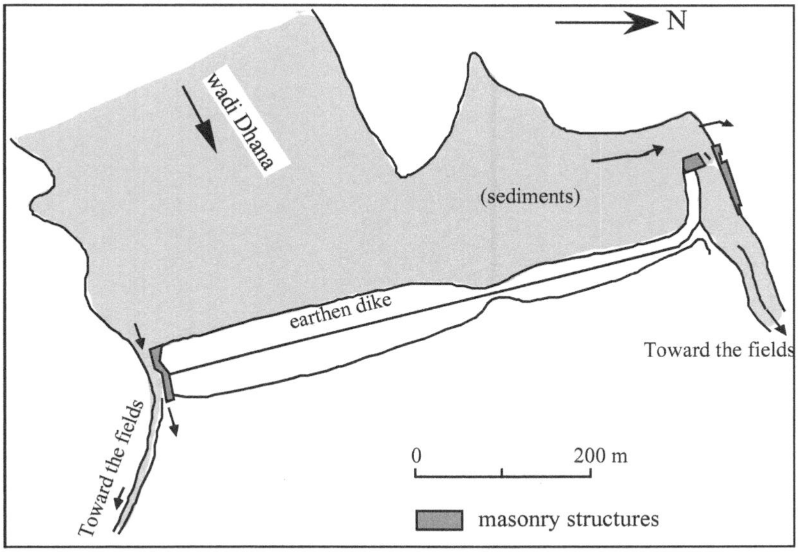

Figure 3.12 The dike of Maryab (Maarib).

It is quite likely that the capacity of the two outlets ends up being insufficient during large floods, given the continuous deposit of sediments in the reservoir. The dam is probably damaged, and perhaps fails, by overtopping on several occasions. According to inscriptions, there was a rupture that was repaired in the middle of the 4th century BC, and major reconstructions in 456, 462, 549, and 558 BC.[52] The definitive failure of the dam, occurring in the 6th century BC soon after the last reconstruction, is mentioned in the Koran:

52 Robin (1991).

"There was surely a sign in their country, for the people of Sheba: two gardens, to the right and left (*that is to say on each side of the wadi).*(…).
But they sidestepped this. Therefore we send against them the flooding of the Dam, and we changed their two gardens into two gardens of bitter fruits, tamarix and jujube trees."[53]

The city of Maryab cannot survive this catastrophe, and is abandoned.

53 sourate 34, verses 15 and 16.

4. The maritime civilizations of the Aegean Sea: urban and agricultural hydraulics

The first great European civilizations are found in and on the shores of the Aegean Sea – and thus in direct maritime contact with Egypt and Syrian ports. The earliest such civilizations are the Cyclades thalassocracy in the IIIrd millennium BC, the first maritime power of the Mediterranean; then Minoan Crete beginning at the end of the IIIrd millennium BC.[1] Civilization flourishes all around the Aegean Sea during the IInd millennium BC – especially in Crete, but also on the island of Cyprus, and in Asia Minor, with Troy to the north, and to the south Rhodes, Samos, and Kos. The first Hellenes came to continental Greece in several waves, and it is very likely around 1900 BC that Greek-speaking people appear - the Acheans. The warrior civilization that we call the Mycenaen developed from this time. Maritime trade is particularly active during this entire period – among all the European civilized lands, of course, but also with Syria, including the port of Ugarit to the north on the trade routes to Mari and to Hattusha, the capital of the Hittite Empire; and the port of Byblos to the south, threshold to Egypt.

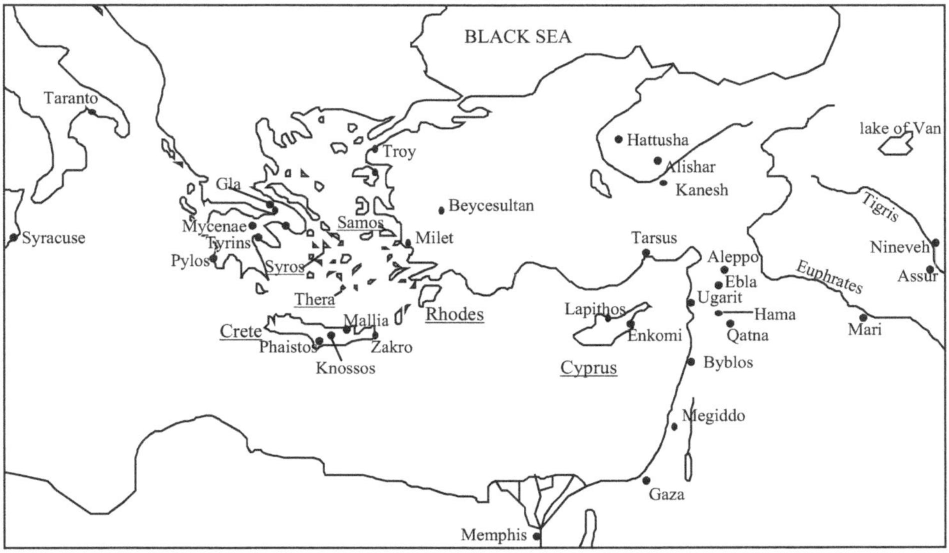

Figure 4.1 The Eastern Mediterranean in the IInd millennium BC.

New arrivals from the north appear between 1200 and 1100 BC: the Dorians. This begins the disappearance of the Mycenaen civilization, whose survivors desert the cities.

1 See for example Rachet (1993) for the prehellenistic period.

Some flee to the islands and coast of Asia Minor (Ionia). This is possibly the origin of the "*Sea People*" that we met in Chapter 2, responsible for the burning and sometimes the disappearance of the main cites of Levant. The destruction of Troy is likely one of the episodes of this drama. In the Ionian islands, the memory of the bellicose Acheans is preserved in the Homeric poems the *Iliad* and the *Odyssey*.

The Cretan cities and palaces: urban hydraulics brought to perfection

Crete was inhabited by migration from Anatolia, probably at the end of the VIIth millennium BC. Its more highly developed civilization, the Minoan, dates from about 2100 BC. This maritime empire was apparently a peaceful one, since neither palaces nor cities were fortified – and this despite the threatening face of the monster that the legendary king Minos kept in his labyrinth according to the Greek legends of Theseus and Minotaur.[2] A creative and sophisticated urban and palatial architecture characterizes this civilization. The Egyptians know the Cretans as the *Keftious*, and Mesopotamians know them as the *Kaptaras*. The Cretans could not have been ignorant of the Eastern developments in urbanism and urban hydraulics, because at the same time northern Syria is experiencing the "belle epoch" of Mari (destroyed in 1761 BC), Elba (destroyed around 1600 BC, and Ugarit (destroyed about 1200 BC).

The remains of numerous cities have been found in Crete, as well as the remnants of four great "palaces" (or temples?): Knossos, Phaïstos, Mallia, and Zakro. These cities and palaces have ample water supplies.[3] In the palace of Knossos, the largest in Crete, there are bathing rooms, latrines, and drains (Figure 4.4). These latrines have a "flushing" channel, dug into the floor, that is fed by an outside reservoir (Figure 4.3) and a sort of siphon, in an alcove of the back wall, connected to the drain. The bathing rooms have portable terra cotta bathtubs that can be arranged so that one can pour water over the bather from a separate room. Systems such as this, albeit of various degrees of sophisitication, can be found in the other palaces on the island. For example, the guest quarters of Knossos include a room where voyagers can wash their feet in running water.

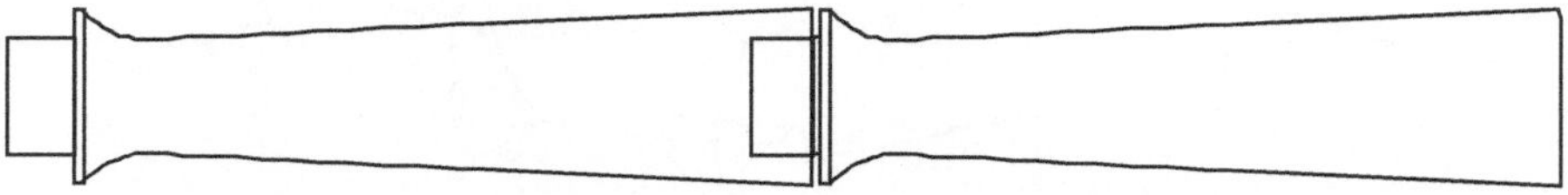

Figure 4.2 Elements of a Minoan terra cotta pipeline (after Graham, 1987).

2 Sporting games (or rituals) using the bull are practiced by the Minoans in many-roomed palaces that less civilized people might have taken to be frightening labyrinths. But in their frescoes, the Minoans seem rather to express the joy of living.

3 see J.W. Graham (1987), p. 219-221; N. Platon (1988), volume 1, pa. 350-410.

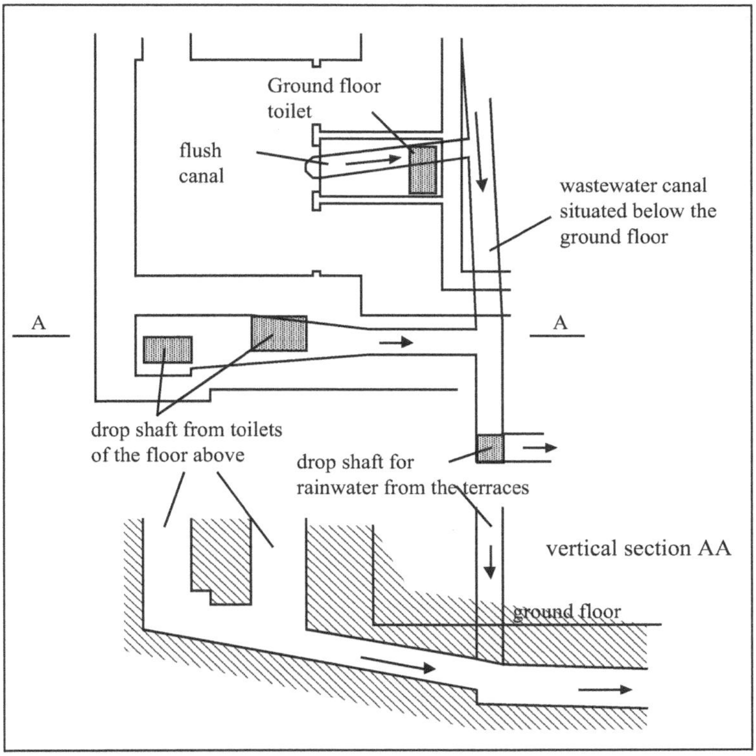

Figure 4.3 Drainage of rainwater and wastewater, discovered in the east wing of the Knossos palace (Castleden, 1990). This area is at the lower left in Figure 4.4, which shows the overall layout of the eastern wing (see "latrine").

But where does all this water come from? For Knossos, it is quite probably from a spring some 15 kilometers away. Terra cotta pipes following the natural slope of the land bring water first to the guest quarters, and then beyond to the palace. The pipeline crosses a small river on a little bridge that provides access to the palace from the guest quarters.[4]

These Cretan aqueducts are, as far as we know, the first true water supply works. They were surely made of terra cotta conduits, and laid along the natural slope of the land, as were the later Greek aqueducts. And actual remains of terra cotta pipelines assembled from conical sections (Figure 4.2), that can fit into each other using flanged joints, have been found in Minoan palaces.[5]

Cretan urbanism is also remarkable for the way it uses rainfall runoff for sanitary purposes. In the palace of Knossos (Figure 4.4), Arthur Evans has identified laundries located at the eastern extremity of the palace. These laundries are apparently supplied by a pipe that descends along a stairway, bringing water from a basin where runoff from

4 This is the building that Arthur Evans calls the caravanserai.
5 according to Rodney Castleden (1990), p. 73.

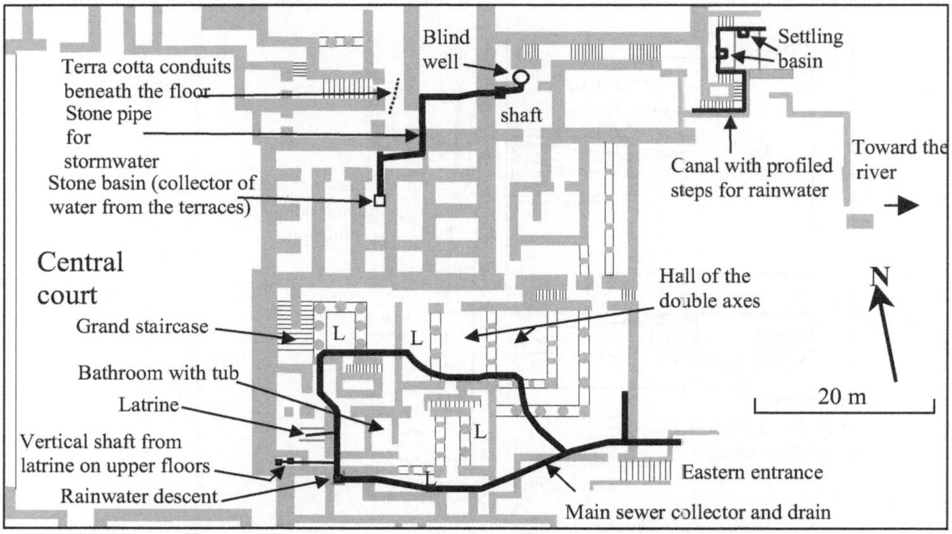

Figure 4.4 Hydraulic features identified in the eastern wing of the palace of Knossos. L = light well

the terraces is collected and stored.

In Knossos there are wastewater drainage systems, and the use of these same pipes for storm drainage further contributes to urban hygiene. Arthur Evans found two independent sewage networks in his excavations. One is made of terra cotta pipes, canals, and large collectors; the other uses smaller underground conduits. The system is serviced through manholes, and these manholes also capture rainfall runoff. Excavations directed by Nicolas Platon at Zakro have uncovered stone or terra cotta gutters under the floors, as well as masonry conduits covered with stone slabs. In the city of Palaikastro,[6] researchers have uncovered traces of stone gutters used to drain wastewater from houses, dumping the waste into the main sewer that flows under the central street. There is an analogous system of sewers in the Minoan city of Akrotiri on the island of Thera, buried around 1520 BC by the explosion of the volcano of Santorin.[7] Remnants of control gates have been found at several of these sites.

The elaborate drainage systems of Crete mirror similar developments at other sites of the IVth and IIIrd millennia BC, such as Habuba Kebira and Mari on the Euphrates, Harapa and Mohenjo Daro on the Indus. It seems reasonable to suppose that the Minoan engineers were inspired by oriental concepts of drainage networks. It is, however, only in Crete that these networks are developed and generalized to the point that we have seen above. But as for the capturing and distribution of water for domestic supply, we are not aware of other examples in the orient prior to the Ist millennium BC. Therefore this was very likely a Cretan innovation.

6 City situated near the northeast extremity of Crete, to the north of Zakro.
7 Akrotiri was uncovered in 1967 by Spiridon Marinatos. Thera is the ancient name of Santorin.

Hydraulic works were also developed in support of agriculture. A canal network was constructed on the plateau of Lassithi, south of Mallia, for the drainage or irrigation of cultivated land. Recent detailed exploration of the region of Mallia, between 1989 and 1995, shows that the coastal plain was very actively cultivated in this period, in sufficient measure to feed the inhabitants of the 60 hectares of urbanized land around the palace. Gutters deliver water diverted from mountain streams to the fields through temporary storage in cisterns that are dug into the tops of hills.[8]

Figure 4.5. Outlets of the principal collectors of two Minoan palaces (photos by the author): - left, near the east entry of the Knossos palace; right, south of the Phaestos palace (under the large central court).

The Cretan palaces are destroyed several times by earthquakes, and then reconstructed. But the palaces, and the brilliant civilization of the island, come to an abrupt end about 1400 BC. The causes are not precisely known. It is possible that the explosion of the volcanic island of Thera (Santorin) caused an ecologic catastrophe when it covered part of the island with cinders, and that a tidal wave destroyed the island's northern settlements. The bellicose Mycenaeans of continental Greece were also responsible for destruction of the palaces, symbols of the Minoan civilization. Only the Knossos palace survived for some time as the weakened seat of a power that seems now to have been Mycenaen.

8 after Sylvie Müller (1996, 1997).

These cataclysms, the explosion of Thera, the disappearance of Akrotiri, the abrupt end of Minoan commerce, are all probably the source of the legend of Atlantis, reported by Plato (427 – 347 BC) from Egyptian sources.[9] In any case, when he speaks of Atlantis, Plato gives tribute to the hydraulic developments of this legendary island, recognizing both urban hydraulics and irrigation as shown in the following significant extract:

"I will now describe the plain, as it was fashioned by nature and by the labours of many generations of kings through long ages. It was for the most part rectangular and oblong, and where falling out of the straight line followed the circular ditch. The depth, and width, and length of this ditch were incredible, and gave the impression that a work of such extent, in addition to so many others, could never have been artificial. Nevertheless I must say what I was told. It was excavated to the depth of a hundred, feet, and its breadth was a stadium everywhere; it was carried round the whole of the plain, and was ten thousand stadia in length. It received the streams which came down from the mountains, and winding round the plain and meeting at the city, was there let off into the sea. Further inland, likewise, straight canals of a hundred feet in width were cut from it through the plain, and again let off into the ditch leading to the sea: these canals were at intervals of a hundred stadia, and by them they brought down the wood from the mountains to the city, and conveyed the fruits of the earth in ships, cutting transverse passages from one canal into another, and to the city."[10]

Or again, this reference to the water supply:

"In the next place, they had fountains, one of cold and another of hot water, in gracious plenty flowing; and they were wonderfully adapted for use by reason of the pleasantness and excellence of their waters. Of the water which ran off they carried some to the grove of Poseidon (…) while the remainder was conveyed by aqueducts along the bridges to the outer circles (….)"

So here we may have direct tribute to the ancient know-how of the Minoans. But we must of course be careful to take these descriptions with a grain of salt, as they may just as easily reflect the ancient works of continental Greece.

Hydraulics in the Mycenaean civilization in the second millennium BC

At the beginning of agriculture in Greece, Vth millennium BC, the abundant precipitation of spring and autumn obviated the need for irrigation. But irrigation eventually does develop in the Greek world. The earliest written reference to it comes from the Iliad:

"As one who would water his garden leads a stream from some fountain over his plants, and all his ground-spade in hand he clears away the dams *(blockages)* to free the channels, and the little stones run rolling round and round with the water as it goes merrily down the bank faster than the man can follow (…)"[11]

---

9 Castleden (1998).

10 Plato, Critias, 118 Translated by Benjamin Jowett http://classics.mit.edu/Plato/critias.html

11 Homer, Iliad, XXI, 250, translated by Samuel Butler http://darkwing.uoregon.edu/~joelja/iliad.html#b21

The Mycenaen civilization that develops especially from 1600 BC is the cultural heir of the Cretan civilization. The Acheans are the heroes of the principal Greek myths, myths that are surely the distant echo of a measure of historical reality. Jason and his quest for the golden fleece are the story of a commercial maritime voyage – or of plunder. The adventures of Theseus and the Minotaur reflect the antagonism between the Acheans and the Minoans. The myth of Hercules, who changed the course of a river to clean the stables of Augias, surely echoes the mastery of river engineering that we illustrate further on.

Urban hydraulics in the Mycenaen palaces

The centers of Mycenaen power are in strongly fortified palaces (Figure 4.6), with an architecture that has been described as cyclopean due to its use of huge stone blocks. At Mycenae in the palace of Agamemnon, at Pylos homeland of Nestor, and at Tiryns, one finds bathing rooms equipped with permanent bathtubs of terra cotta that sometimes

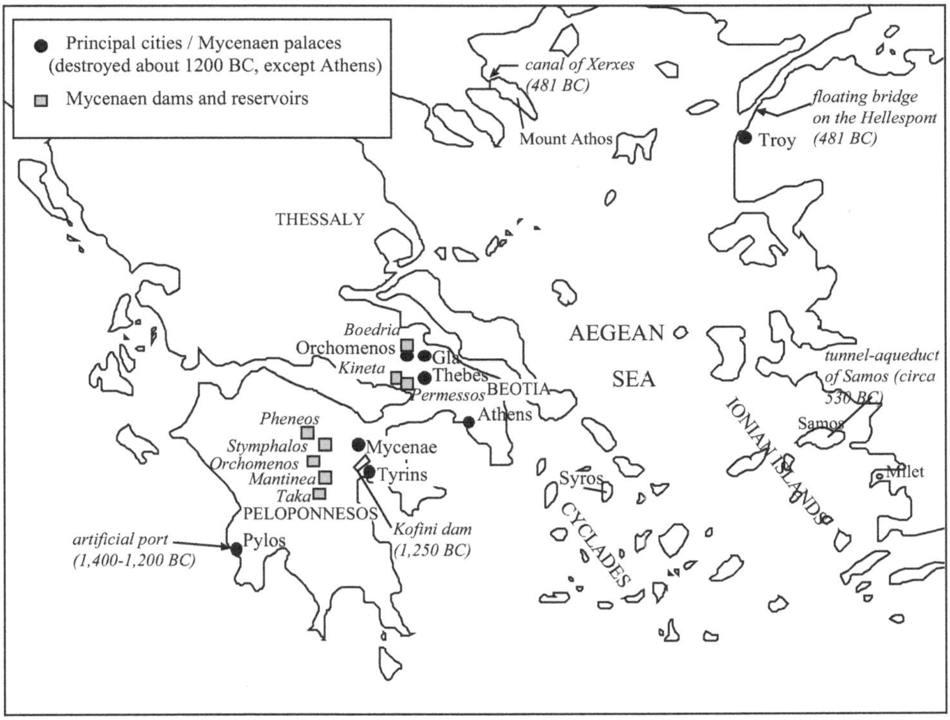

Figure 4.6 Sites of the great hydraulic developments in the Greek world before Alexander.

were equipped with a drain. Complex systems of drainage sewers in the floors of the palace drained both rainwater and wastewater, including water from the bathing rooms. At Pylos, stone drains collect wastewater and dump it into collectors large enough for a man to stand in.[12]

Like the Cretans, the Mycenaens bring water to their palaces using aqueducts. At Pylos, it is thought that a two-kilometer aqueduct brings spring water to the palace; part of the aqueduct is made of terra cotta in the shape of a U; another part is made of wood; and yet another part is incised into the rock. At Mycenae, an underground cistern is accessed from the citadel by going down three flights of stairs that pass under the outer enclosure wall. This cistern is fed by a rock tunnel that brings distant spring water to it (Figure 4.7).[13]

Figure 4.7 The underground cistern of the palace of Mycenae (photo by the author).

Port development at Pylos

Between 1400 and 1200 BC, Achean shipping dominates long-distance commerce in the eastern Mediterranean, extending to Sicily, and perhaps even as far as to Spain. The ports of Antiquity are often developed in natural bays (that do not always provide good shelter), or in river mouths.

Recently, a detailed study of the Pylos region has made it possible to reconstitute the development of an artificial port.[14] This port was created by excavating a closed basin into marine sediments, and linking it to the sea through a channel (Figure 4.8, 4.9). The sinuous path of the channel keeps ocean swells from entering the port itself.

The port would rapidly have become unusable without additional engineering efforts, either due to silting-in of the basin itself, or by the blocking of the entrance chan-

12 Taylour (1983), Chapter 5.

13 Ibid.; Nicolas Platon (1988), volume 2, pp. 277-292.

14 This reconstitution is due to E. Zangger. Our source is the article of C.W. Shelmerdine (1997).

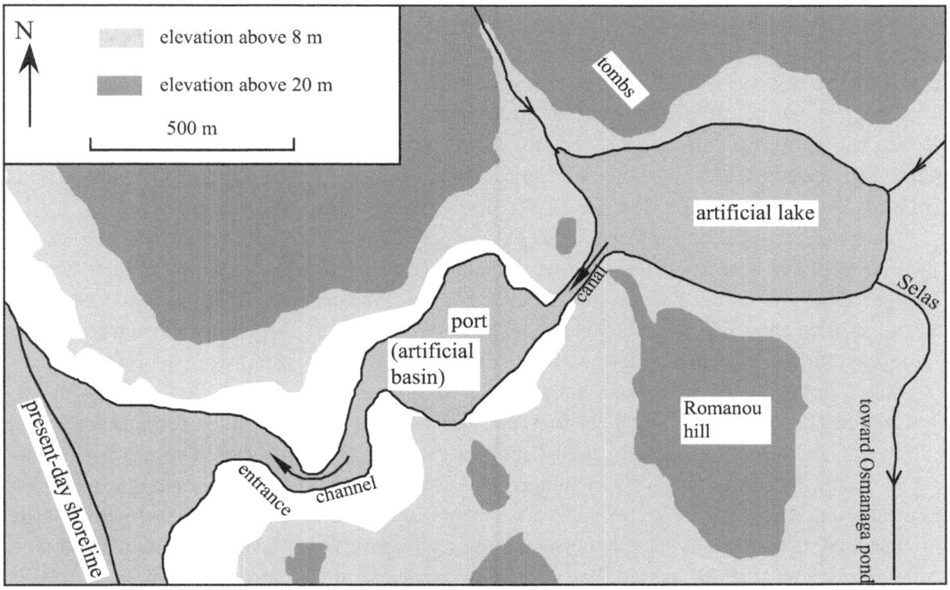

Figure 4.8. The artificial port of Pylos (1400 to 1200 BC) (after Shelmerdine, 1997).

nel by wave-transported sand. A through-flow was necessary to keep the port open, as would be the case in the natural estuary of a river. Therefore a river called the Selas, naturally flowing into the Osmanaga pond to the south, was relocated so that part of its flow, very likely controlled, passed through the port on its way to the sea. An artificial lake, linked to the port by a canal dug through a rocky barrier, served as an intermediate storage and settling basin for the river flow.

Study of the history of sedimentation in the Osmanaga pond enables us to date this wonderful project to about 1200 BC. Such analysis also suggests that after the destruction of Pylos, in about 1200 BC, the totality of the river flow takes the direct path to the sea through the port.

Figure 4.9 The site of the ancient port of Pylos from the heights to the east of the entrance canal, looking to the north; the flat area with the orchard is the ancient basin of the port (photo by the author).

Drainage and land improvement in the Mycenaen civilization

Many of the regions of Peloponnese or Attica are karsitic. Entire rivers disappear into abysses or caverns (called *catavothres* in Greek), only to reappear at some distant point. When the subterranean cavities fill up or are blocked, for example after earthquakes, water can accumulate in marshes and lakes, the water level varying from season to season and from one period to the next. Strabo describes these phenomena:

> "Some of these plains are marshy, since rivers spread out over them, though other rivers fall into them and later find a way out; other plains are dried up, and on account of their fertility are tilled in all kinds of ways. But since the depths of the earth are full of caverns and holes, it has often happened that violent earthquakes have blocked up some of the passages, and also opened up others, some up to the surface of the earth and others through underground channels. The result for the waters, therefore, is that some of the streams flow through underground channels, whereas others flow on the surface of the earth, thus forming lakes and rivers. And when the channels in the depths of the earth are stopped up, it comes to pass that the lakes expand as far as the inhabited places, so that they swallow up both cities and districts, and that when the same channels, or others, are opened up, these cities and districts are uncovered; and that the same regions at one time are traversed in boats and at another on foot, and the same cities at one time are situated on the lake and at another far away from it."[15]

To be arable, these valleys had to be drained and the lake levels stabilized. The most important of such efforts were developed in Beotia by the Mynians (subjects of the king Mynias), near Orchomenos, their capital. The memory of Mynian power and management of the lake Copais was still fresh at the time of Strabo:

> "They say that the place now occupied by lake Copais used to be dry, that it then belonged to the Orochomenians, their close neighbors, and that all sorts of crops were grown there. Here one sees an additional confirmation of the wealth of this city."[16]

Lake Copais is fed by runoff from rainfall and by the Kephissos river. The natural grottos already mentioned, i.e. *catavothres*, were used to drain it to the sea.[17] The waters of the Kephissos, previously flowing into the lake, were detoured to an underground passage through a 25-km long canal. The canal may have also been used for navigation. The land reclaimed by emptying of the lake was itself drained, surrounded by protective dikes, and brought under cultivation. Water from what remained of the lake provided irrigation for these lands. The Mynians built the palace of Gla (Homer's *Arne*) in one area reclaimed from the lake, in the middle of the depression. The lake, now fed only by runoff from rainfall, was nearly dry in the summer, so that part of its bed could

---

15 Strabo, Geography, IX, 2, 16, Translation of H.C. Hamilton, Esq., W. Falconer, M.A. http://www.perseus.tufts.edu.

16 Strabon, IX, 2, 40, Translation of R. Baladié, adapted.

17 see Guy Rachet (1993), pages 71 and 470, and Nicholas J. Schnitter (1994), pages 11-13. The reports of both these authors are based on the work directed in the 1980's by J. Knauss, University of Munich. The studies show that, contrary to what had been thought beforehand, lake Copais had never been completely drained by the Mynians.

also be cultivated. Figure 4.10 is a map of developments around the lake that were very likely in operation about 1300 BC. With the end of the Mycenaen civilization, the lake's systems were no longer maintained, and the marshy lake reestablished itself. Between 334 and 331 BC one of Alexander the Great's engineers, a certain Crates of Chalchis, again set out to drain and dry the lake. But these efforts were not completed, either because of technical difficulties in the region or some other troubles.

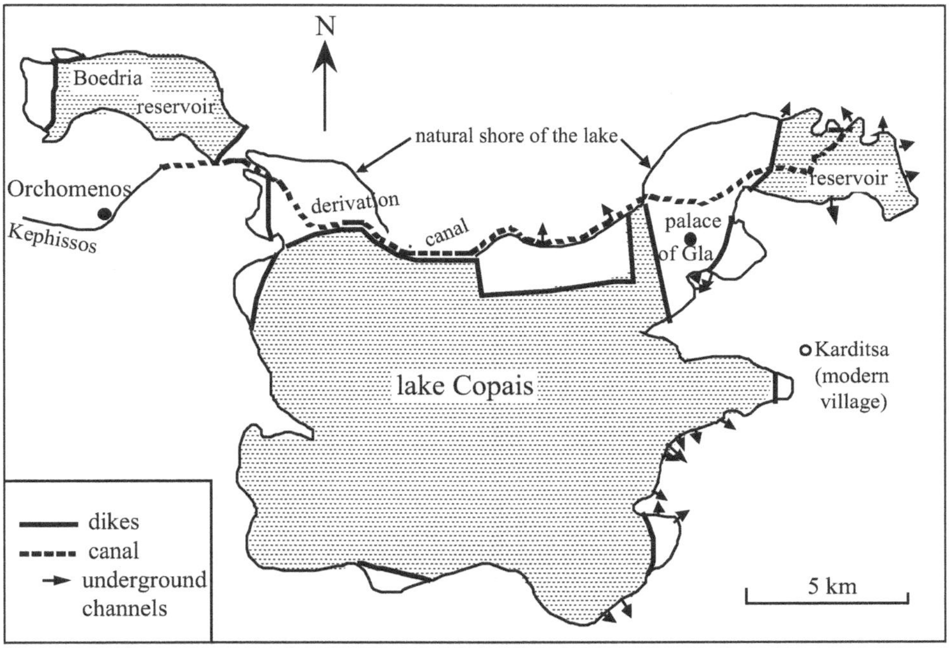

Figure 4.10 Hydraulic developments around lake Copais from the Mycenaen era (after Schnitter, 1994).

These same techniques are used to reclaim marshy valleys in many other locations. Typically there are dikes surrounding a reservoir-lake fed by rainwater and snowmelt, the dikes also serving to protect dry areas that are drained into grottos or caverns. Eight such reservoirs, including the Boedria reservoir shown in Figure 4.10, are listed in table 4.1. Five of these sites are in Peloponnese (Figure 4.6). For all of these projects, the dikes are between two and four meters high, and of variable length from 200 to 2,500 m. They most often are built of earth fill between two exterior walls. Some of these dikes are still visible today (Figure 4.11).

Table 4.1 Mycenaen dam-reservoirs (after Knauss, 1991; Schnitter, 1994)

Name (see Figure 4.6)	**Height** (m)	**Length** (m)	**Reservoir Volume** (million m^3)
Boedria	2	1,250	24
Kineta (Thisbe)	2.5	1,200	4
Mantinea	3	300	15
Orchomenos	2	2,100	16
Permessos	4	200	2
Pheneos	2.5	2,500	19
Stymphalos	2.5	1,900	9
Takka	2	900	9

Figure 4.11 Remains of the Kineta dam (Thisbe), 1,200 m long. The modern road was built on top of the ancient dam; at the left, remains of the wall of large cut stone (photo by the author).

A dam to protect Tiryns

A catastrophe strikes the city of Tiryns toward the end of the Mycenaen civilization. The city is located on an alluvial plain about one kilometer from the sea; the palace is 24 m above sea level, on a limestone hill. The city itself is at the foot of the palace, to the east and south. The watercourse along which the city is located, the Lakissa, leaves the mountains on a steep slope of nearly 15 m/km as dictated by the local topography.

Levees normally protect the city from the caprices of the river. But in about 1200 BC, at essentially the same time that the nearby city of Mycenae and its palaces were destroyed and burned, an exceptional event occurred. The Lakissa River left its bed, flowed to the north across the lower city, and covered most of the city with a thick layer of sediments, as much as 4 m deep in places. Only the palace on its hill in the southern part of the city was spared the effects of the deluge. This event could have been caused simply by an exceptional flood, or more likely by a major earthquake, the same one that destroyed Mycenae. Such a quake could have caused the collapse of the riverbanks causing a wave of debris-laden water rushing down the steep slope of the riverbed.[18]

But this event did not mark the end of the Mycenaens. To protect the city from the Lakissa, they built a dam 3.5 km upstream to divert waters of the Lakissa into a 1.5 km canal for conveyance to another watercourse, the Manessi, that flows more to the south (Figures 4.12, 4.13 and 4.14). This 10-m high dam, extending 57 m and 103 m across the left and right banks, respectively, is built of earth fill between two walls of the same type of cyclopean masonry[19] that is used in other Mycenaen works. The lower city is rebuilt on top of the sediments that buried the old city, eventually occupying about the same overall area as before (see inset in Figure 4.12).

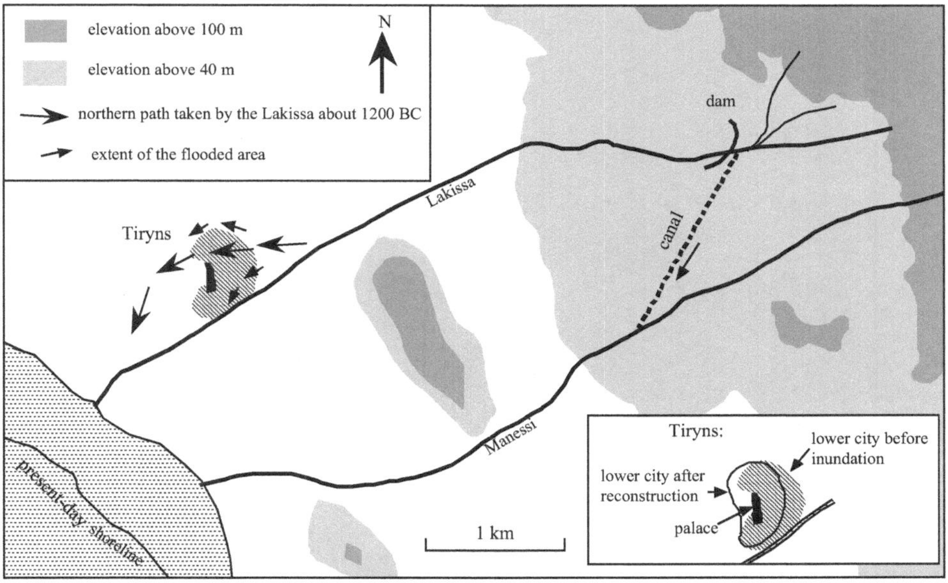

Figure 4.12 Diversion canal for the city of Tiryns (after Zangger, 1994)

18 All these details are known thanks to the work of a German team, directed by Eberhard Zangger, who studied the site between 1984 and 1988 (Zangger, 1994).
19 The structural details of the dam are provided by Schnitter (1994), from an early study of the dam by Balcer.

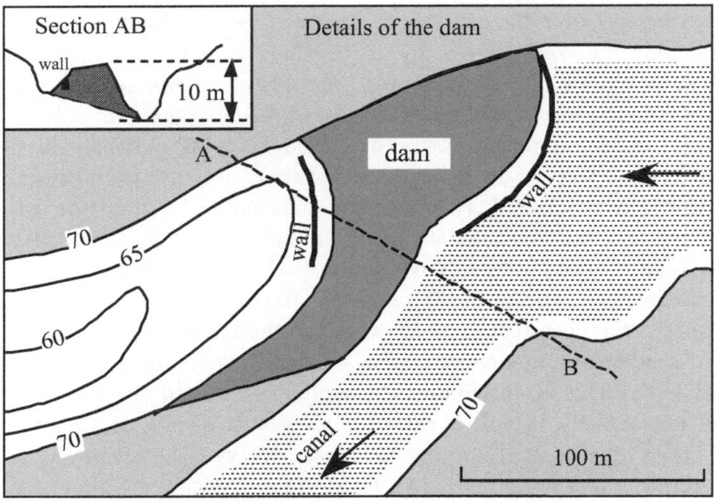

Figure 4.13 The Kofini dam for protection of the city of Tiryns against floods of the Lakissa (present-day situation) - Zangger, 1994.

Figure 4.14 Lakissa diversion canal, looking from the top of the Kofini dam (photo by the author).

This account illustrates the kinds of natural catastrophes that, in the troubled climate of the years around 1200 BC, may have contributed as much to the destruction of the palaces as did the Dorian invaders. Nonetheless, the Mycenaen civilization was still strong enough to implement the hydraulic works that returned the city of Tiryns to safety. A century later this civilization of Nestor, Menelaus and Agamemnon, warrior-builder descendents of the Minoans, does finally succumb to the overwhelming pressure of the Dorians, leaving only legends behind.

The Greek world in the classical age

With the disappearance of the Mycenaen civilization, Greece enters its dark age. For no less than four centuries writing is forgotten, not to be rediscovered until the 7th century BC with the adoption of the Phoenician alphabet of the 10th century BC. In the 5th century BC, Greek history is punctuated by the revolts of the Ionian cities against the yoke of the Persian Achaeminides; then by the Median wars with two Persian invasions in continental Greece; then by internal struggles (the Peloponnesian war). Travelers, sometimes imbued with a sense of Hellenic superiority, but also sometimes remarkably open and observing like Herodotus, open the eyes of the Greeks to Egypt and the Orient.

These "tourists" were not all peaceful; Xenophon and the successful of the Ten Thousand may have inspired the subsequent invasion of Alexander. These centuries also witness the expansion of the Greeks toward the west, with the founding of Marseilles (about 600 BC) and with the founding of Greek colonies in Sicily and the south of Italy toward the end of the 5th century BC. This expansion formed a cultural ensemble today called greater Greece.

Water supply for Greek cities

Greek cities develop their water supply using local springs and aqueducts of terra-cotta conduits, following the centuries-old Cretan and Mycenaen traditions. These conduits are set underground, both for their protection and to accommodate irregular topography.

They are assembled from interlocking pre-fabricated elements from 60 cm to 1 m long, and between 11 and 22 cm in diameter.[20] Some of the individual elements have a hole in their crown, normally plugged with clay, very likely intended to provide access for inspection and cleaning of the pipes. The presence of these inspection holes, as well as the thinness of the walls (2 to 4 cm), clearly suggest that these pipes conveyed water through free-surface gravity flow, not under pressure.

The hydraulic works of Samos: record achievements in the Greek world

The city of Samos in Ionia is located near the coast of Asia Minor on an island of the same name. A spectacular tunnel more than 1,000 m long was dug for its water supply (Figure 4.15). The tunnel was bored in two sections starting from its extremities (the meeting point of the two bores is shown by an arrow on Figure 4.15). The water supply conduit was laid at the bottom of a trench dug into the floor of the tunnel. Because of this trench, it was possible to dig a horizontal tunnel (a relatively straightforward task); the depth of the trench is zero at the entrance to the tunnel and progressively increases

---

20 from Praxitelis Argyropoulos (1979). See also Trevor Hodge (1995), Chapter 1.

to reach nearly 8.5 m at the tunnel's exist, assuring the slope necessary to convey the flow.

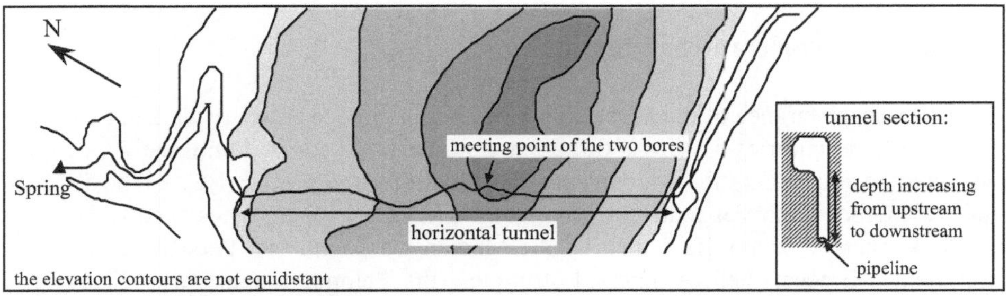

Figure 4.15. Plan view of the aqueduct of Samos and the tunnel of Eupalinos.

Herodotus considered this project, which is still partially visible today,[21] as one of the marvels of the Hellenic world:

> "I have dwelt the longer on the affairs of the Samians, because three of the greatest works in all Greece were made by them. One is a tunnel, under a hill one hundred and fifty fathoms *(265 m)* high, carried entirely through the base of the hill, with a mouth at either end. The length of the cutting is seven furlongs *(1,240 m)* - the height and width are each eight feet *(2.4 m)*. Along the whole course there is a second cutting, twenty cubits deep *(8.5 m!)* and three feet broad *(0.9 m)*, whereby water is brought, through pipes, from an abundant source into the city. The architect of this tunnel was Eupalinus, son of Naustrophus, a Megarian. Such is the first of their great works; (....)"[22]

Herodotus' dimensions approximately coincide with those that can be deduced from the remains of the works. The project was very likely undertaken under the reign of the tyrant Polycrates, around 530 BC, before Samos fell under Persian domination.

Herodotus also mentions a mole (breakwater) at the port of Samos, once again a record achievement for the period. Here is the continuation of the passage above:

> "the second is a mole in the sea, which goes all round the harbor, near twenty fathoms *(35 m)* deep, and in length above two furlongs *(355 m)*. The third is a temple (....)"

Engineering projects developed in Greece by the Persians during the Median wars

Herodotus' history of the Median wars is interesting for its representation of the first confrontation between the classical Greek world and the Orient. Certain elements of Herodotus' writings show the technical and cultural abyss that separated the two civilizations, both in their relationships to the sea and their practice of fluvial engineering.

---

21 Numerous works mention the tunnel of Samos; among them, see Jacques Bonnin (1984), Chapter 9; Trevor Hodge (1995), Chapter 1.

22 Herodotus, The History of Herodotus, III, 60, Translation of George Rawlinson http://classics.mit.edu/Herodotus/history.html

During the first Median war, the Persian fleet suffered major losses during its passage around Mount Athos, a cape that extends quite far into the northern portion of the Aegean Sea (see Figure 4.6). Anticipating the second war, king Xerxes spent three years, according to Herodotus, digging a canal to get around the mountain on the land side of the isthmus. Here is how the historian describes the organization of the project:

> "(...) a line was drawn across by the city of Sand; and along this the various nations parceled out among themselves the work to be done. When the trench grew deep, the workmen at the bottom continued to dig, while others handed the earth, as it was dug out, to labourers placed higher up upon ladders, and these taking it, passed it on farther, till it came at last to those at the top, who carried it off and emptied it away. All the other nations, therefore, except the Phoenicians, had double labour; for the sides of the trench fell in continually, as could not but happen, since they made the width no greater at the top than it was required to be at the bottom. But the Phoenicians showed in this the skill which they are wont to exhibit in all their undertakings. For in the portion of the work which was allotted to them they began by making the trench at the top twice as wide as the prescribed measure, and then as they dug downwards approached the sides nearer and nearer together, so that when they reached the bottom their part of the work was of the same width as the rest."[23]

The Greeks, given their familiarity with the sea, would never have built such a canal to avoid sailing around a dangerous cape.

During the military campaign that ensued, king Xerxes had reason to be astonished at the Greeks' lack of experience in large hydraulic works. To Xerxes, relocation of a river for military purposes is a classic maneuver, as seen in his analysis of the vulnerability of Thessaly, in the north of Greece:

> "When Xerxes came and saw the mouth of the Peneus, he was in great amazement, and, summoning his guides, he asked them whether it was possible to turn the river aside from its course and lead it into the sea somewhere else.
>
> > "'They are clever men, the Thessalians *(said the King)*. This is why they took their precautions long ago and conceded victory to me; it was especially because they have a country that is easy and quick to capture. It would only be a matter of letting the river in upon their country by shifting it out of that channel and turning it from the course in which it travels with a dam, and all of Thessaly, except the mountains, would be beneath the waves'." [24]

The floating bridge that Xerxes cast across the Hellespont for passage of his immense army must also be included among the many great hydraulic works of the Persians during the second Median war.

---

23 Ibid., VII, 23.
24 Ibid., VII, 128-130.

Greek philosophers and thinkers of the classical era

The hydraulic works of Greece during the classical era do not measure up to those of the Mycenaen era. The few extracts cited above also show that in Greece, hydraulic know-how lags behind that of the Orient. And yet this classical Greece is known to us as the privileged cradle of development of philosophical and mathematical thought. Two facts are important in this regard. First, Greek science at this period took it as a point of honor to be disconnected from engineering, i.e. not to be associated with practical applications. Second, and it is here that one can effectively and for perhaps the first time refer to science, the "sages" sought to apply reasoning to the explanation of great natural phenomena.

Such explanations of natural phenomena are often incomplete, and sometimes false. Regarding the hydrologic cycle, the brilliant mathematician and astronomer Thalès of Miletus (610 – 545 BC?) – who had established that the world is round – postulated that water is the primordial substance and that the land masses float upon it. This is a theory perhaps inspired by the ancient Mesopotamian and Egyptian cosmologists. Later, Anaximandre of Miletus (610 – 545 BC) identifies the origin of precipitation as water torn from the earth's surface by solar action. Xenophane of Colophon (570 – 475 BC?) writes that the sea, source of all water, is also the origin of clouds and winds, that fresh water comes from evaporation of the sea, that rivers result from rain, and that it is by rinsing the ground that watercourses carry salt to the sea. The hydrologic cycle is therefore fairly well understood in Greek civilization of the 6th century BC, except for the role of groundwater. Indeed, during this period it was thought that the water in springs and wells comes from the sea.

Plato (427 – 347 BC?) adopts the idea, articulated before him by Empedocle of Agrigente, of the four primordial elements: water, air, fire, and earth. His student Aristotle (385 – 322 BC), born in Macedonia, frequents Plato's Academy until the death of the master, and becomes preceptor of Alexander at the court of Philippe of Macedonia. At Athens he creates the institution called the Lyceum. Aristotle in his turn adopts the theory of the four elements, a theory that will remain a standard reference until the Middle Ages. He believed that each of these elements tends to return to its natural place, e.g. water below the ground, air above water, etc. Thus, for Aristotle, wood floats because it contains air, and it is in the natural order of things that air be above water. To each element is attributed two of the four fundamental "qualities": cold, moistness, heat, and dryness (Figure 4.16).

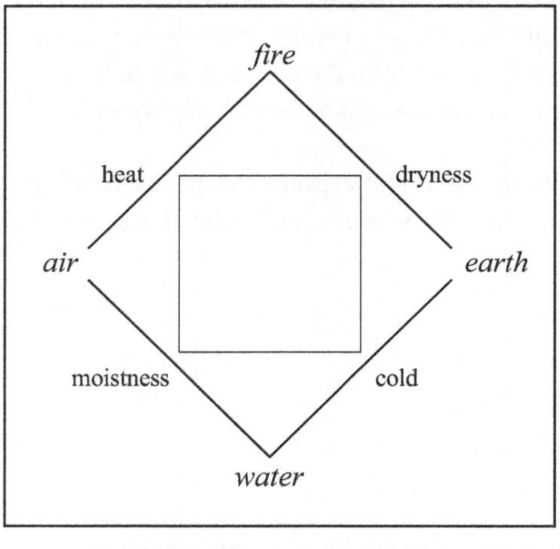

Figure 4.16 The Aristotelian vision of the four primordial elements.

Attempts to explain nature from such a postulate can obviously lead to absurd conclusions. Here is how Aristotle, reflecting his ignorance of what we now call inertia, shows in his *Physics* that a vacuum cannot exist, and in so doing sets down an absurd theory of wind resistance. First, in virtue of the natural "position" of each element, as we have seen above, if an object (for example a stone thrown by a man) moves through the air, "outside" of its natural position, there is an action, a "motor" that sustains its movement:

> "...earth and all other bodies should remain in their proper places and be moved from them only by violence."[25]

Then, since "it seems that everything in motion is impelled by something", the stone that continues its movement through the air must be sustained, in its movement, by the air itself:

> "(...) things that are thrown move *(al)*though that which gave them their impulse is not touching them, either by reason of mutual replacement, as some maintain, or because the air that has been pushed pushes them with a movement quicker than the natural locomotion of the projectile wherewith it moves to its proper place."[26]

Aristotle continues his demonstration in postulating that in a vacuum, all movement would be impossible – i.e. in the absence of this air that "maintains" movement. Therefore "there is no vacuum separate from things".

At about this same time Pytheas (380 - ? BC), of the Phoenician colony of *Massalia* (Marseilles), sets out on a voyage that leads him to the great north, probably as far as Iceland. During this voyage he establishes a correlation between the periodic phenomenon of the tides and lunar cycles.

Overall, one can only admire this quest for truth as representing progress and promise for the future. Yet it is also clear that during the Greek classical era, hydraulic theories are, at best, rudimentary. It is not until the Hellenistic period and the intellectual movement coming out of the school of Alexandria that mathematics and the faculty of reasoning lead to practical innovation.

---

25 Aristotle, Physics, VIII, III,Translated by R. P. Hardie and R. K. Gaye http://classics.mit.edu/Aristotle/physics.8.viii.html.
26 Ibid., IV, VIII

Part II

THE EMPIRES OF THE BUILDERS

For thousands of years, the hydraulic know-how of the East was no more than an oral tradition. But with the conquests of Alexander the Great, this oral tradition comes into contact with the Greek spirit of observation and analysis. The city of Alexandria, at the maritime front door of Egypt, for several centuries serves as the scientific center of the known world. The understanding and know-how of the scientists of the Alexandrian school remain unequalled during the following millenium, both in the East and the West, until the advent of the Renaissance.

Roman engineers, the inheritors of Etruscan and eastern techniques, influenced by the Greeks and endowed with a strong practical sense, leave the influence of their hydraulic achievements around the entire Mediterranean perimeter.

The fall of the Roman Empire ushers in the intellectual decadence of the West, but the East continues to develop up until the Mongol invasions of the 12th century AD. Then the East, in its turn, enters a period of profound reversal.

In China, whose hydraulic development began later, technical developments transcend the millennia up through modern times. The scale of these developments reflect the vast expanse of the country itself. From the 1st century BC through the 15th century AD, China serves as mankind's principal nursery of technical innovations.

Meanwhile, during the Middle Ages the West even forgets that the earth is round. Nevertheless during the West's demographic expansion of the 12th and 13th centuries, hydraulic development blossoms as lands are drained and mills are constructed.

5. Mathematicians and inventors of Alexandria and the Hellenistic world

In 335 – 331 BC, Alexander the Great conquered the totality of Greece and the Persian Empire, including Egypt and Mesopotamia. The spirits of analysis and hydraulic know-how now were brought together in the same crucible, fueled by the need to innovate – in order to ensure survival in a world whose boundaries were suddenly pushed enlarged, and to increase agricultural productivity to meet the needs of the new ruling classes. This crucible has a name: Alexandria.

Figure 5.1 The Hellenistic World.

Brief history of the Hellenistic kingdoms and their successors[1]

Setting out from Macedonia, with contingents of Macedonian and Greek soldiers, Alexander achieved a first victory over the Achaemenid King Darius II, who ruled the Persian Empire, at the battle of Issus. Egypt falls without resistance into the clutches of Alexander, who makes a long sojourn there from 332 – 331 BC. During this sojourn, he founds Alexandria on the seafront, an ideal site for the development of commerce, lying

1 See Dickinson (1994).

between the sea and a lagoon. But the site is poorly supplied with fresh water, being some distance from the Nile. Considerable engineering efforts are undertaken to support the burgeoning activity of the city (Figure 5.2). These include a kilometer-long dike (hepstastade) to link the island of Pharos to the coast, a 30-km canal to bring fresh water from the Nile, and numerous cisterns to store this water. Later a lighthouse is built to guide maritime navigators along this low-lying and dangerous coast.

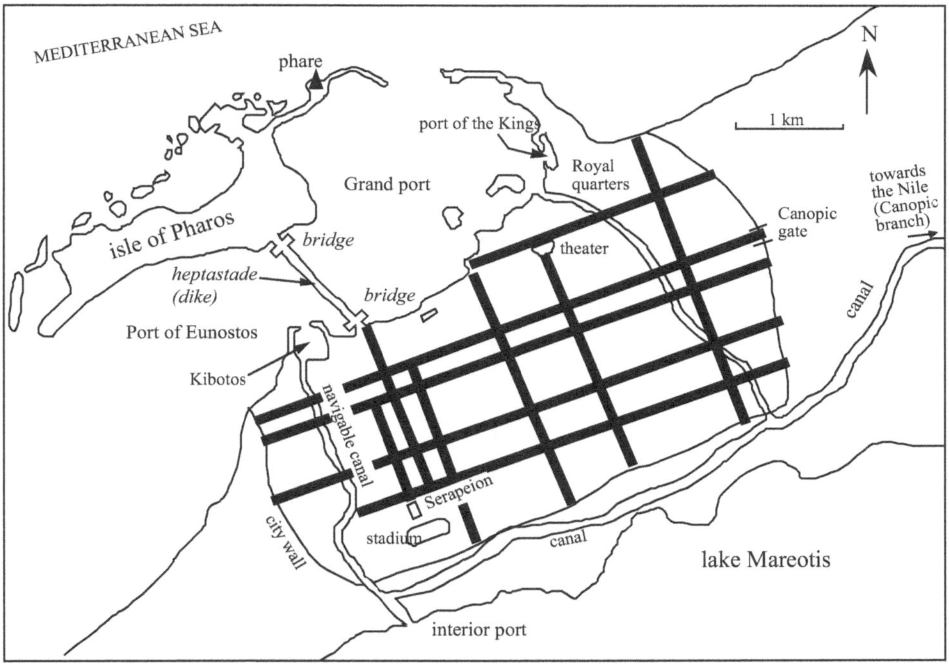

Figure 5.2 The city of Alexandria and the harbor works described by Strabo who visited it around 25 AD. "The point of the island is a rock assailed by the waves on all sides and supporting a tower made of white stone, admirably constructed, having several levels, and having the same name as the island. Sostrate of Cnide, a friend of the king's, dedicated it as a salute to navigators, as shown on its inscription. [....] The opening to the West [.] forms a second port, that of Eunostos, which sits opposite the closed and artificial port. The port whose entry is next to the tower of Pharos, mentioned above, is the major harbor; two others are contiguous to it at the back of the bay, separated from the main port by the Heptastade dike. The dike forms a kind of bridge [...] with two open passages. [....] This dike actually served as both a bridge and an aqueduct, during the period when Pharos was inhabited. [...] The city has many advantages. First, the location touches two seas, to the north the Egyptian sea, as it is called, and to the south by the Mareia Lake, also called lake Mareotis. This lake is fed on its upper boundary and sides by numerous canals coming from the Nile. [...]. When leaving by the Canopic gate, one sees to the right the Canal, which connects the Lake and Canope." (Geography, XVII, 1 6-7 and 16).[2]

2 From the Translation of Pascal Charvet, after Strabo, The journey to Egypt commented by Jean Yoyotte (1997).

Leaving Alexandria, the conqueror succeeds in completely destroying the Persian Empire through a decisive victory, and pushes on to the Indus. He dies an early death at Babylon in 323 BC, before having solidified his Empire. Indeed, his generals divide up – and fight over – the elements of that Empire. The Macedonian Ptolemy, son of Lagos, founds the dynasty of the Lagides in Egypt. Lysimachus receives Thrace. Seleucos obtains Babylonia several years later, and then Syria. But his family line, the Seleucids, only intermittently govern lower Mesopotamia, as it is fought over by other powers such as the Parthes, who progressively push the Seleucids toward Syria. New kingdoms of Hellinistic culture appear to the north of the empire of the Seleucids, along the Black Sea: Pontus, Bithynia, and Cappadocia. The power of the city of Pergamon, initially just a simple fortress, increases in Asia Minor and it becomes a dependency of the Seleucids from about 282 to 260 BC, capital of an increasingly powerful empire. From 260 BC, the influence of Pergamon becomes comparable to that of Alexandria.

The Seleucids are conquered by the Romans, allied with Pergamon, in 189 BC. In 133 BC the last king of Pergamon bequeaths his kingdom to Rome. From the middle of the 2nd century BC, a certain decadence of the Lagide Dynasty sets in, including economic difficulties and revolts. With the death of Cleopatra, in 31 BC, Egypt in its turn falls under the control of the Roman Empire.

The influence of Alexandria in the 3rd and 2nd centuries BC

Mathematicians and inventors[3]

Is it to increase their prestige that the first Ptolemites set themselves up as protectors of the sciences, techniques and arts? Ptolemy I created the Library of Alexandria – more of a personal collection than a true institution. The Library is then completed by the Museum, either by Ptolemy I himself, or by his successor Ptolemy II Philadelph, who reigned from 285 to 246 BC. The Library is dedicated to the acquisition[4] and conservation of books, whereas the Museum was what one would call today a research institute. The director of the Library and the members of the Museum are supported by the Ptolemites and in general, the financing of the two institutions is entirely assured by the state. It is believed that the Museum and the royal Library – the latter thought to have contained 500,000 rolls of papyrus[5] - were housed in the same building, in the interior of the royal palace grounds. A library annex, open to the public, was set up outside the palace grounds in a temple called the Serapeum.

Scholars from the entire Hellenistic world flocked to Alexandria, either to live there, or to study or make extended stays. This is why one can confidently use the term school

---

3 Our principal general sources concerning the Library and Museum of Alexandria are the works of Geoffrey Lloyd (1973), Luciano Canfora (1986) and the article of Anita Measson (1994).

4 Sometimes acquisition by force: on transiting boats stories were told of books that were confiscated, only copies to be returned to the owners.

5 But many fewer books: only the shortest of books could be contained in a single roll. This estimate is discussed in detail by Luciano Canfora (1986).

of Alexandria to describe the vast intellectual movement associated with it.

First there was mathematics. The foundations had been laid by the Greeks during the classical period, and the easterners, Babylonians and Egyptians, were equally known for their mastery of geometry. In about 300 BC Euclid set forth the foundations of our modern geometry, and through his students the mathematical school of Alexandria becomes a reference for Antiquity. Geometry underpins and enables many lines of thought and activity; we will see this later in the context of Archimedes of Syracuse. Geometry enabled Eratosthene of Cyrene, director of the Library during the era of Ptolemy III Evergete (who reigned from 246 to 221 BC), to determine the circumference of the earth with remarkable precision.[6] Eratosthene deduced this circumference from measurement of the height of the sun at noon at Alexandria when the sun is at its zenith at Syene (Aswan) – a city which is near the Tropic of Cancer - taking into account the measured distance between Alexandria and Aswan.

The new element here, compared to Greek science, is that the applied sciences are for the honor of Alexandria, even if their function is often only to "please our senses in charming our eyes and ears."[7] Ctesibios of Alexandria (around 270 BC) and Philon of Byzantium (around 200 BC) invented "marvelous machines" – water clocks, pumps, automata, diverse mechanical devices. The invention of the hydraulic organ is credited to Ctesibios. The invention of the hydraulic screw (called *cochlea (snail)* by greco-roman historians, with reference to the snail's spiral shell) as a mechanism for lifting water, is attributed by ancient authors to Archimedes at the time of his sojourn at Alexandria. During the Hellenistic period it is these Archimedes screws (Figure 5.3) which make it possible to create the famous *hanging gardens* of Babylon.[8] Other lifting devices, always muscle-powered,[9] appear during this period: the waterwheel and the bucket chain, or *saqqya* (Figure 5.4). Strabo, at the beginning of the roman occupation, notes the use of the lifting waterwheel in Egypt, a device that will be widespread in Asia during the roman period.

---

6 We recall that Greek science had established that the earth was round in approximately the 6th century BC. The circumference obtained by Eratosthene was 250,000 stadia or 39,690 km, taking the stadium to be 157.5 m, compared to the modern accepted Figure of 40,009 km (Lloyd, 1973)

7 Vitruvius, X, 7, 4.

8 These hanging gardens are described by Strabo (Geography, XVI, 1, 5), but not by Herodotus who wrote during the period of the Achaemenid Persians. Thus it seems probable that they were not developed until the Hellenistic period.

9 A passage in the work of Philon may suggest hydraulic force to turn a lifting waterwheel (noria). But we know this work only through an Arabic transcription, and there is some doubt as to the authenticity of the passage (it could be an addition from the Islamic era, during which the use of hydraulic energy is widespread). We share this doubt. There is no evidence of the use of hydraulic force before the 1st century BC (see further on), and even during the era of Strabo's voyage in Egypt (25 BC), there are no hydraulic machines on the Nile. His account is perfectly clear on this point (we cite an extract further on).

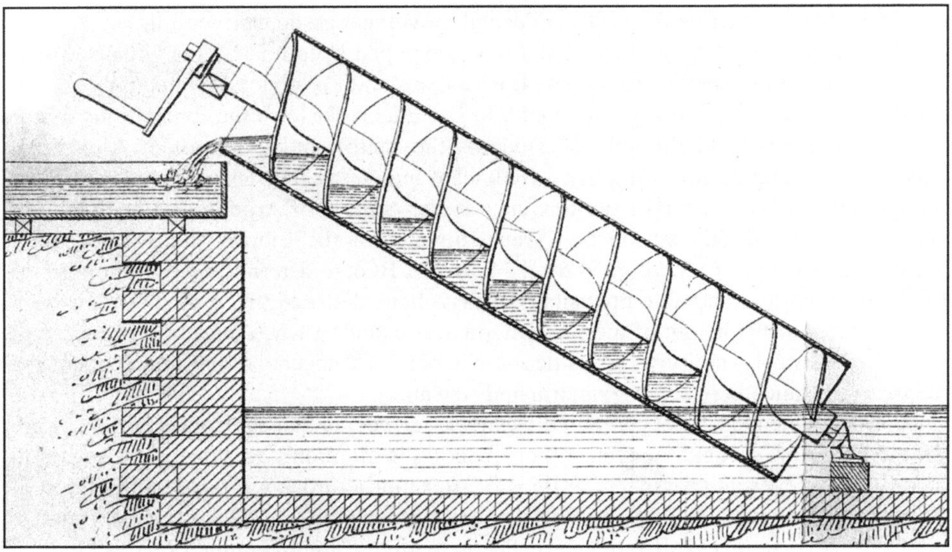

Figure 5.3 The Archimedes Screw (Poillon, 1885 – ancient archives ENPC)
In Antiquity, the Archimedes screw was used on a relatively small slope, and it was operated by foot, rather than with a crank handle.

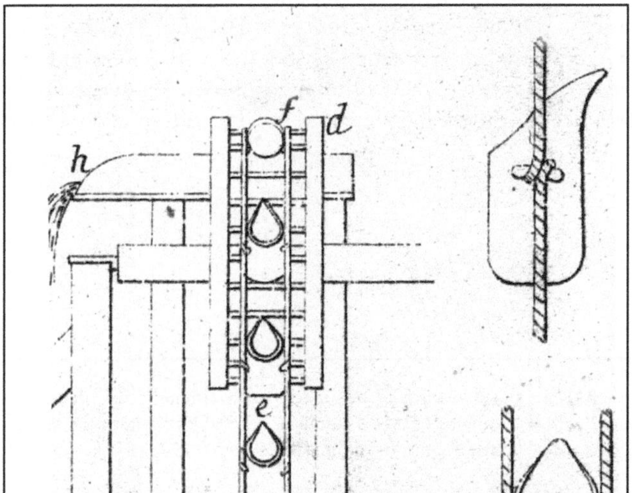

Figure 5.4 The Bucket Chain, to be called saqqya by the Arabs (Poillon, 1885 – ancient archives ENPC).

Vacuum, pressure, and the first hydrodynamic devices

The writings of our 3rd century BC authors have unfortunately been lost for the most part. We know them primarily through citations and references contained in the writings of the subsequent period, in particular those of the Roman Vitruvius (about 25 BC) and of Heron of Alexandria (around 60 BC). It is thought that Straton of Lampasaque conducted the first studies of a vacuum and may be at the origin of the concept reported by Heron that "an absolute vacuum does not exist, but one can artificially produce vacuum in opposition to nature." Straton was the private tutor of the future Ptolemy II Philadelph in about 290 BC, and, after 286 BC, he was the successor of Theophraste as the head of the Academy of Athens. One of the experiments reported by Heron, but very likely inspired by Straton, consisted in blowing air into a hermetically sealed metallic sphere through a small tube: "This shows clearly that the compression of bodies contained in the sphere enables them to reside in the dispersed pockets of the vacuum." Inversely, one can suck out the air contained in the sphere through the same tube, "a considerable quantity can be withdrawn, without any substance taking its place inside the sphere."[10]

Vitruvius credits Ctesibios of Alexandria with the invention of the "fire pump" (Figure 5.5). This would appear to be the first hydrodynamic device in which the flow of water is not driven by gravity, but by the action of an artificially induced pressure. Philon of Byzantium appears to have pursued this interest in flow devices "under pressure" through his interest in connected vases and siphons, and perhaps also under the influence of the work of Archimedes that we will now discuss.

Archimedes, and the first theoretical formulations of fluid mechanics

Archimedes (287 – 212 BC) was born at Syracuse, in Sicily. In all probability, he spent time in Alexandria where he studied geometry with the followers of Euclid. Though Archimedes belonged to the mathematical School, it would have been quite natural and possible for him to see the inventions of Ctesibios during his stay. Upon his return to Syracuse, he continued to correspond with the scholars of Alexandria, in particular with the mathematician Conon of Samos, and with the director of the Library, Eratosthene.[11] This justifies the association of Archimedes' work with the school of Alexandria.

First and foremost a mathematician, Archimedes was interested in the problem of buoyancy of bodies of arbitrary shape. Extreme rigor distinguishes his work - he first proposes axioms, then demonstrates their consequences. The initial postulate of his work on "floating bodies" introduces the notion of pressure:

> "We take as a principle that liquid is of such a nature that, its parts being arranged in an equal and contiguous manner, the part that is the least compressed is displaced from its position by

---

10 Heron, "Pneumatics", I, 16, cit. after Lloyd (1973).

11 Note the dedication of Archimede's treatise "The Method": "Archimedes to Eratosthene, prosperity! I earlier sent you certain theorems that I had discovered, giving you only the statements and inviting you to discover the proofs…" (Vol. III of Works of Archimedes, Edited in French by Les Belles Lettres, 1971, adapted).

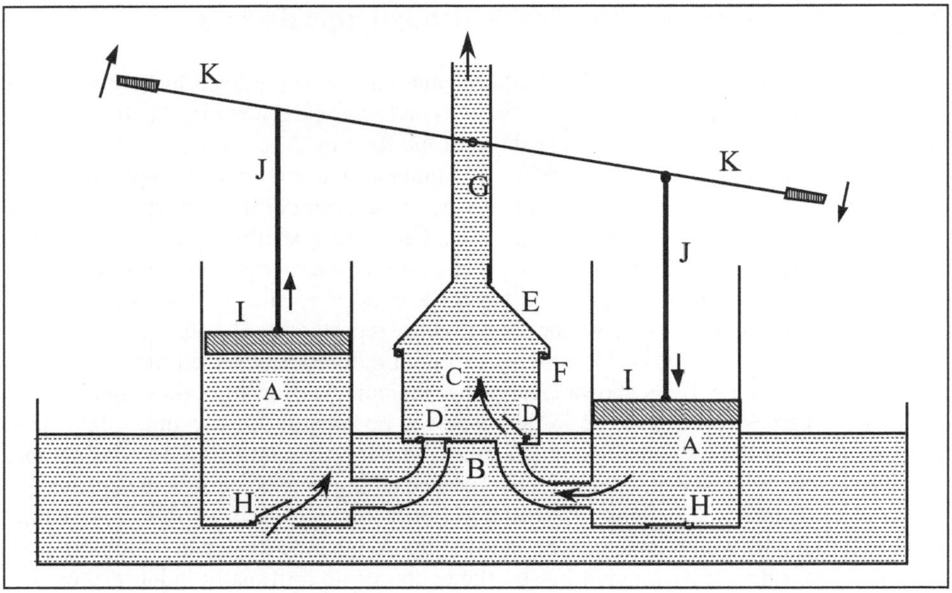

Figure 5.5 The two-body pump of Ctesibios of Alexandria, as it is described by Vitruvius: "It is appropriate now to describe the machine of Ctesibios, which lifts water. This machine must be made of bronze. At its base, and quite near to each other, are twin cylinders (A), to which are connected tubes which, forming a fork (B), converge symmetrically to an intermediate chamber (C). In this chamber are two flaps (D), precisely adjusted to cover the upper openings of the tubes; blocking the openings of these orifices, they prevent the return of fluid that air *(it is not air, but the water pressure!)* has driven from the chamber. Above this chamber is a cover in the shape of an inverted funnel (E) that is held down by a slotted key to resist the raising effect of the water pressure. Another tube, called a trump, is vertically mounted above the cover. Under the lower openings of the tubes, the cylinders have additional flaps (H) covering openings in their bases. The pistons (I), nested in the upper portions of the cylinders, smoothed and lubricated with oil, are activated by rods (J) and levers (K); [....] thus, from a reservoir at a low location, water can be made to gush forth."[12]

a more compressed part, and that each of these parts is compressed by the liquid above it, unless the liquid is in a closed receptacle and is compressed by something else."[13]

As early as on the second page of his treatise, he reaches a remarkable conclusion - the surface of water at rest is not horizontal, but spherical:

"The surface of any liquid at rest will have the form of a sphere having the same center as that of the earth."[14]

Later on, the pragmatic Roman engineers were perplexed by this proposition: was it

12 On Architecture, book X, 7, 1 – 3, adapted from the Translation of Louis Callebat.

13 Of floating bodies, Translation of Charles Mugler.

14 Here, in summary, is how this proposition is deduced from the initial postulate: if this were not the case for a liquid at rest, two points inside the liquid, situated on a sphere centered at the center of the earth would be compressed by different water heights above them, thus "the part that is the least compressed is displaced from its position by a more compressed part; it follows that the liquid cannot remain at rest."

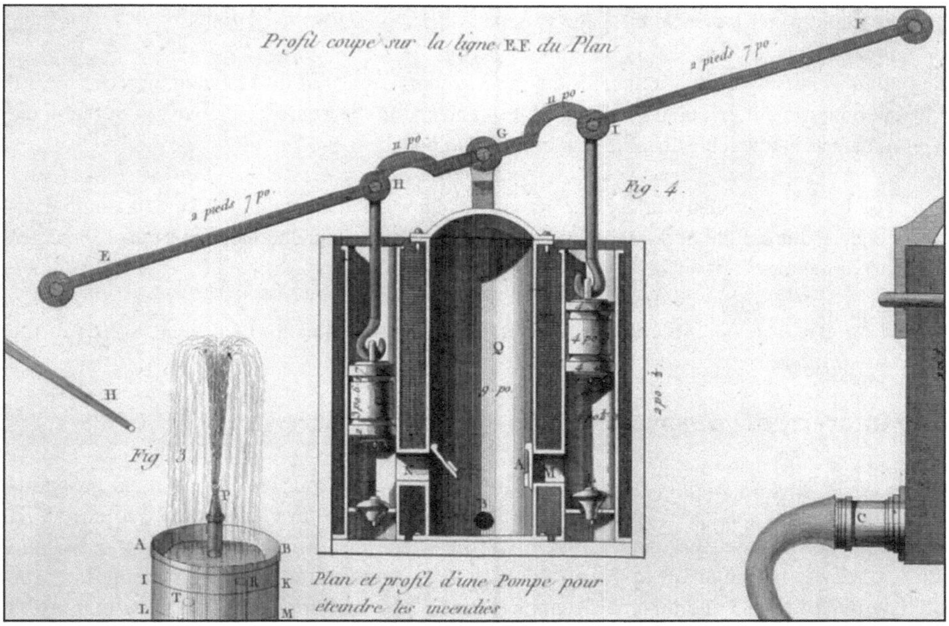

Figure 5.6 A variation of the pump of Ctesibios: a fire pump from the 18th century (Bélidor, 1737) ancient archives of ENPC

then not possible to rely on a water surface to represent a horizontal plane?[15]

Further on in this same treatise, laid out with the same rigor, are the various propositions that comprise the theory of floating bodies and the well-known formulation of Archimedes' Principle:

> "Solid objects which have (for the same volume) the same weight as the liquid in which they are immersed, when released remain submerged in such a way as not to rise to the surface of the liquid or descend further down within it.
>
> "Any solid object lighter than a liquid (for the same volume), released in this liquid, will be submerged to a level such that the liquid which occupies the volume of the submerged portion has the same weight as the entire solid body.
>
> "Solid bodies lighter than a liquid (for the same volume), plunged into the liquid by force, are pushed upwards by a force equal to the excess of the weight of the body over the weight of

15 The Roman engineer Vitruvius writes later, in book VIII of his treatise On Architecture (V, 3): "Perhaps those who have read the works of Archimedes will say that one cannot establish an exact level using water, since Archimedes teaches that a water surface is not a level plane, but a sphere having as its center the precise center of the terrestrial globe. But whether the water surface is planar or spherical, if a horizontal straightedge is laid upon the surface of water in a gutter, then this straightedge, at its left and right extremities, necessarily is at the same distance above the water surface; if, on the other hand, the straightedge is laid on a slope, one end of it will be above the water while the other touches it."

the liquid which occupies the same volume as the solid body.

"Bodies heavier than a liquid (for the same volume), released in the liquid, descend toward the bottom until reaching it, and they are lightened in the liquid by the weight of the liquid contained in a volume equal to that of the solid body.

"If a body lighter than a liquid (for the same volume) is released in the liquid, the ratio of its weight to the weight of the same volume of liquid will be equal to the ratio of the submerged part to the total."

In his treatise, Archimedes then determines the equilibrium of diverse solids in the form of spheres, hyperboloids of revolution, etc.

The future of the discoveries of the 3rd century BC

As we have shown, Lagide Egypt undergoes a period of troubles and reduced prosperity from the 2nd century, a situation not at all propitious for development. Ptolemy VIII (called Physcon – i.e. the vain) hunts down the Alexandrian intellectuals in 145 BC and later sends his mercenaries to attack this city that had revolted. Far from being lost, the discoveries of the 3rd century BC reappear in the hydraulic projects that we are going to describe subsequently. They comprise a patrimony shared by the Roman engineers and the scientists of the school of Alexandria from the first centuries of our era - the school poised for another fruitful period under Roman domination.

The implementation of new hydraulic technologies in the Hellenistic kingdoms, from the 3rd to 1st century BC

Even though Pergamon and Alexandria dominated intellectual life, the thread of innovation runs throughout the entire Hellenistic world in this period, from Egypt to the Black Sea. The first examples that we will describe are linked to the intensive urbanization that developed in Asia Minor. These are new cities in a mountainous country, and they demand new principles of water supply.

Hellenistic water delivery and the generalization of the siphon

As we have seen in Chapter 4, the Greek aqueducts most often use clay pipes that follow the slope of the terrain, as would a free-surface canal. During the Hellenistic period, in the new cities of Asia Minor and Palestine, the technology of inverse siphons is developed to permit an aqueduct to cross a valley and return to a higher elevation. These siphons are schematically in the form of a "U" with depth of from 15 to 75 m (i.e. from the top to the bottom of the "U".) The portion of the conduit that is under pressure (a "forced main") is most often constructed from massive stone. Numerous pipe sections of this type have been recovered, often of cubical external shape, and fitted with ferrules so that individual elements can be connected to form a conduit (Figure 5.7). Often the upper portions of these elements have holes in them, probably serving as air vents, cleaning holes, or perhaps pressure-surge relief valves.

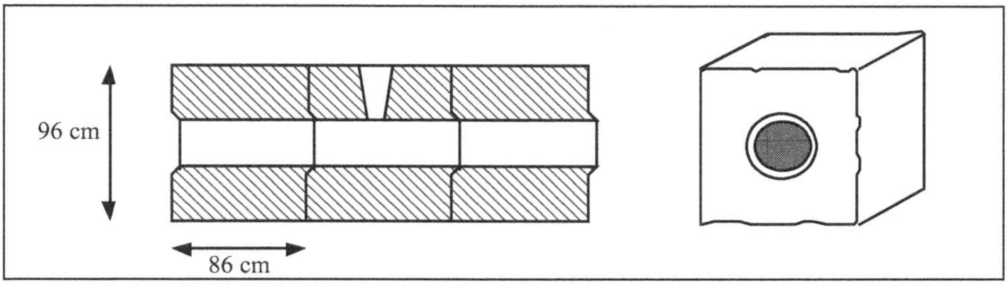

Figure 5.7 Example of a forced main in stone (Laodicée) (after Hodge, 1995)

Inverse-siphon technology made it possible for Hellenistic water-delivery systems to cross valleys without the engineering structures (bridge-aqueducts) that the Romans later used for such crossings. Their development coincides with an improved knowledge of the effects of fluid pressure, as we have seen in the previous section. But it is difficult to say if this knowledge resulted from the technological development, or vice-versa. Table 5.1 summarizes some of the Hellenistic inverse siphons.

Table 5.1. Inverse siphons of Hellenistic technology (Hodge, 1995)

City	Depth	City	Depth
Ephesus	15 m	Apamea Kibotos	28 m
Antioch on the Meandre	15 m	Magnesia	30 m
Blaundos	20 m	Trapezopolis	40 m
Philadelphia	20 m	Prymnessos	40 m
Patara	20 m	Tralleis	75 m
Laodicea of Lycos	25 m	Smyrne	158 m
Akmonia	25 m	Pergamon	200 m (see below)
Antioch on Pisidia	28 m		

The water delivery system of Pergamon: the first large forced main

Lysimachus, who had received Thrace in the partition of Alexander the Great's empire, imprudently left part of his war spoils in the custody of Philetairos, in the Asia Minor citadel of Pergamon. This citadel occupied a rocky spire that overlooked the plain from 300 m above it, and about 30 km from the sea. After the secession of Philetairos, in 282 BC, Pergamon rapidly became the capital of a kingdom, then an intellectual center that sought to rival Alexandria with a great Library (some 200,000 rolls of papyrus) and a School of original thought.

The provision of a supply of water to these citadels perched on hills always posed a problem in Antiquity. Initially, cisterns were built to store rainwater. Later, tunnels (or *sinnors*) were often dug to ensure access, especially during sieges, to underground cisterns fed by springs located on the flanks of the hills; this was the solution adopted at Mycenae, and at Jersusalem.

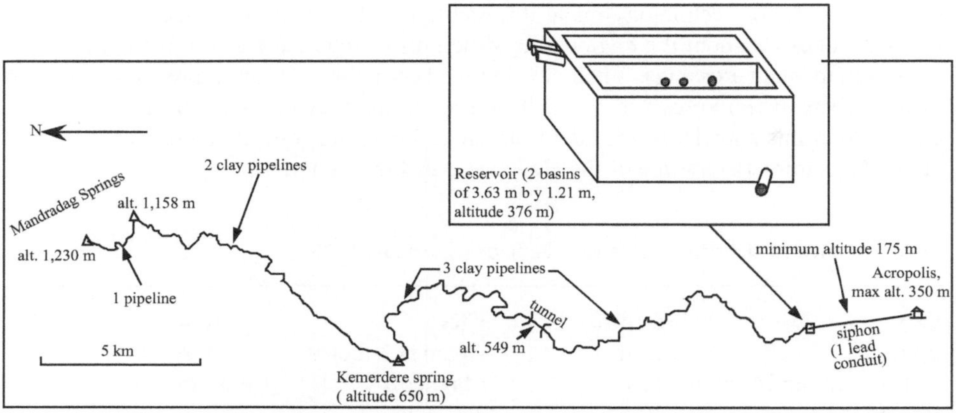

Figure 5.8 The Hellenistic aqueduct of Mandradag for water delivery to Pergamon (after Garbrecht, 1983)

Having become a powerful city, Pergamon needed water, and this need was met with an unprecedented hydraulic installation. The project was founded on the understanding of the hydraulic concept of a siphon and the experience of the first Hellenistic applications, as well as on the mastery of the metallurgy of lead. This project is known to us through site studies carried out by a German team between 1968 and 1972.[16] Its construction was probably carried out during the reign of King Eumene II (197 – 159 BC). This water delivery system comprises two parts (Figure 5.8). The upstream portion brings water from the Mandradag spring (captured at an altitude of 1,230 m) as well as from other springs, some as far away as 25 km as the crow flies, to a reservoir located 3 km from the city across from the citadel, at an altitude of 376 m (i.e. about 26 m higher

16 Gunther Garbrecht (1983) gives an excellent overview. See also Trevor Hodge (1995).

than the citadel). This original project includes parallel buried pipelines (three of them downstream of the Kemerdere spring) assembled from about 200,000 connected clay sections (Figure 5.9). These sections are from 50 to 70 cm long, with interior diameters from 16 to 19 cm, and wall thicknesses of about 4 cm. The watertight joints between the sections are built up from a mixture of sand, mud, and clay, including certain organic matter such as petroleum or greases.[17] The parallel pipelines follow the slope of the land along a total length of more than 40 km. They do not flow under pressure, and thus in principle belong to the family of water delivery systems developed earlier in Greece (see the end of Chapter 4) – but on a larger scale.

Figure 5.9 The three clay pipelines of the Mandradag aqueduct (photo of G. Garbrecht).

It is the second section of the pipeline that, although much shorter, is of revolutionary conception (Figure 5.10). It conveys water from the reservoir that we just described, at an altitude of 376 m, down to the citadel at 350 m, in a straight-line distance of only 3 km. But this section crosses a valley whose lowest elevation is only 175 m, i.e. nearly 200 m below the reservoir. The inverse siphon comprises a pressure conduit made of lead, with an outside diameter that appears to have been 30 cm, the inside diameter

17 According to Gunther Garbrecht, many of these elements were secretly sold to collectors, but the three pipelines are quite visible at certain locations.

appearing to be the order of 20 cm.[18] The conduit is not buried, but rests on above-ground stone supports. No visible traces remain of the forced main itself (the metal, having considerable value, was ultimately recovered for other uses). But the conduit's support blocks, with their 30-cm holes, have been recovered, along with massive anchor blocks on the two high points of the profile, to withstand static and hydrodynamic forces. Traces of lead have been found on the ground along the course of the conduit. The longitudinal profile of the pipeline (Figure 5.10) is in the form of a W, unlike the earlier U-shaped Hellenistic siphons. It seems that the designers of the facility, concerned with the effects of hydrostatic pressure corresponding to 200 m of elevation difference, sought to limit the length of pipe sections subject to the greatest pressure. They appeared to have done so by intentionally routing the conduit over intermediate high points within the depression. At these intermediate high points, there is a risk of the formation of air pockets that can endanger a pipe system in several respects. Air release vents were very likely installed at these locations.

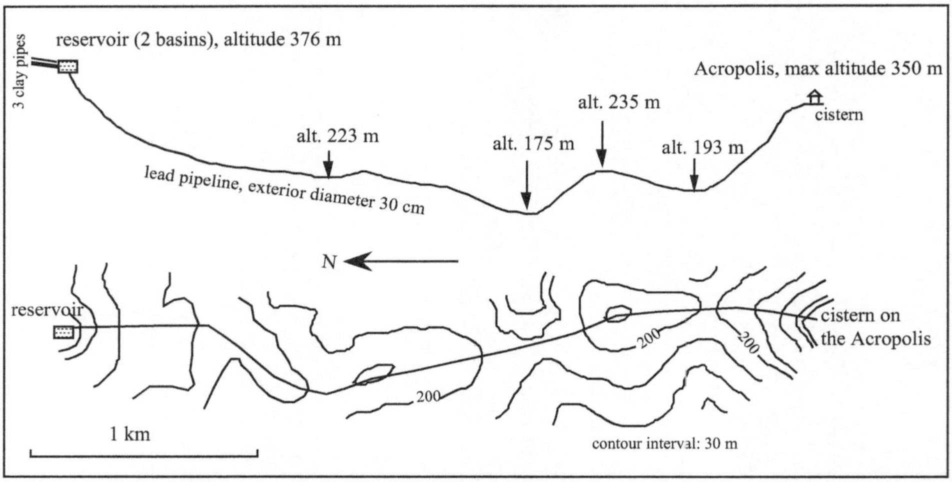

Figure 5.10. Longitudinal profile and routing of the inverted siphon of Pergamon: the first large forced main (Hodge, 1995).

The discharge in this system has been estimated to be 45 l/s (i.e. nearly 3,900 m^3 per day). Later, urbanization spreads to the low areas situated at the foot of the hill. New aqueducts will be built, in particular during the period of Roman domination, to supply these low areas. But none of these aqueducts can rival the audacity of the Mandradag pipeline, the only one to provide water to the summit of the Acropolis. No subsequent Roman aqueduct will ever approach the bold technical audacity of this forced main.

---

18 22 cm according to Gunther Garbrecht (1979), only 17.5 cm according to Trevor Hodge (1995). These dimensions are suppositional, no element of this conduit having been recovered. The exterior diameter of 30 cm is, on the other hand, well established on the basis of the diameter of holes in the blocks that supported the conduit.

Egypt under the Lagide rulers: maritime commerce

In Chapter 3, devoted to Egypt of the pharaohs, we mentioned several projects that represented completion or termination of efforts that the Ptolemites had begun earlier. In their development of commerce with distant partners in Arabia, and even as far as India according to Strabo, the Ptolemites needed to develop the infrastructure to access the Red Sea and to launch a fleet for this purpose. Necho's canal, linking the pelusiac branch of the Nile with the Gulf of Suez, passing through Lake Timsah and the Bitter Lakes, is maintained or brought back into service. As we have seen in Chapter 3, a device (a single gate?) making it possible to accommodate water-level changes in the Gulf of Suez is installed. A new city, Arisone,[19] named for the sister-spouse of Ptolemy II Philadelph, is founded at the outlet of the canal. A new port on the Red Sea is constructed at about the latitude of Aswan, 320 km southeast of the ancient Egyptian port of Gawasis;[20] this port, called Berenice after the name of the mother of Ptolemy II, is connected by a trail to the region of Edfou.

Egypt under the Lagide rulers: development of irrigated agriculture

A constant preoccupation of the Lagide kings, pressured by their politics of prestige and expansion, was to increase agricultural productivity. Each region (or *nome*) is under the responsibility of an economist (the Greek name is *Oikonomos*), charged, among many other tasks, to "control the delivery canals across the fields, from which the peasants draw water conveyed to their cultivated fields; to verify that the feeder canals have the prescribed depth, and that their interior space is sufficient."[21] Every retention basin has its irrigation controller (*catasporeos)* responsible for water distribution.

Under Ptolemy II, the region of Fayoum, already made productive during the era of the pharaohs, is brought under a new development policy. In Chapter 3, we have discussed the legendary Lake Moeris, and the works realized since the time of the pharaohs (Figures 3.6 and 5.11). The capital of the region, the ancient Shedet of the Egyptians that the Greeks called Crocodilopolis,[22] is renamed Arisone, again in honor of the queen. The ancient hydraulic systems are renovated, new projects undertaken, and the lake level lowered. A chief engineer named Cleon is charged with management of the new developments of Fayoum.

---

19 Several cities are given with the name of this queen, who was first married to Lysimachus, the king of Thrace.

20 On a site that had perhaps already been used during the period of the Pharaohs: it is the site called "Head of Nekheb" in Figure 3.1.

21 Citation from Fabienne Burkhalter (1992). See also Bonneau (1993).

22 There were crocodiles in the region, and the ancient Egyptians venerated the sacred crocodiles there. This city is today called Medinet el-Fayoum.

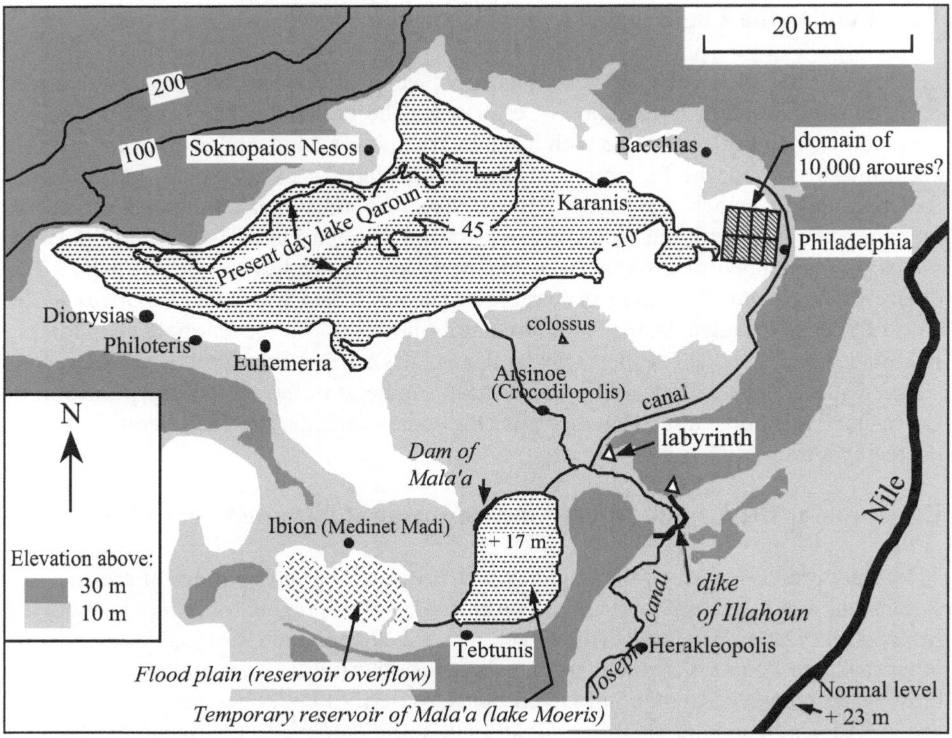

Figure 5.11. Fayoum in the Hellenistic era. We have shown the hydraulic works identified by Garbrecht and Jaritz (1992), and shown the level of Qaroun Lake at the elevation –10 m.[23] The level reached by the Nile in flood is about 30 m.

Thus, when the minister of the Treasury, Apollonius,[24] receives from the king Ptolemy II Philadelph a concession for the development of a vast domain of 2,700 hectares in Fayoum, he relies on the services of Cleon to establish the economic planning and the layout of the hydraulic infrastructure (Figure 5.12). The correspondence of a certain Zenon of Caunos, who is successively secretary for Apollonius then manager of this domain, gives us a vivid illustration of certain aspects of water resource management in the region. In 258 BC Apollonius has the upper zone of the basin, to the north-east, developed as the "domain of ten thousand aroures". A new city called Philadelphia is founded nearby. Certain conflicts of interest were inevitable between Cleon the engineer and Zenon the exploiter, as evidenced in the correspondence of Zenon:

"Zenon to Cleon, greetings! The water in the canal has not risen more than a cubit, so the

23 In effect, Claude Orrieux (1983) situated the "domain of the 10,000 aroures" between the altitudes +20 m and –10 m. Other studies have situated this elevation at –2 m (Garbrecht, 1996).

24 One can consider that Apollonius served the king in an intermediate function, between that of a Finance Minister and Prime Minister, according to our references. He very closely followed the agricultural performance, this being the principal source of fiscal revenue.

land cannot be watered from it. Therefore you would be well advised to open the gates to water the land. Stay well! (October 258 BC)

"Panakestor to Cleon, greetings! We sent you a letter on the 19th, asking you to provide us with a team to do maintenance on the bends of the small canal. Well, it seems that you have left us aside in going toward the Small Lake. Instead of avoiding us as you have done, your duty was to meet with us briefly, and having seen for yourself that the land is not being watered, to ask yourself why. Your job is not only to direct the infrastructure works in the region of the Small Lake, but also here. So, at least come meet us tomorrow at the lock and sketch out for us the path canal bends should take, for we do not have this experience. We will provide you with the labor and other facilities, whatever you command. But if you do not come, we will be obliged to write to Apollonios that his land is the only land not to be irrigated in the region of the Lake. So, we are more than ready to make all needed facilities available for you. Stay well. (October 257 BC).[25]

Is it possible that the "Small Lake" in this letter is the reservoir of Mala'a (the lake Moeris of Strabo) to the southeast of Fayoum? This seems to be the chief engineer's main preoccupation, to the point that he neglects the domain of the minister. It is indeed thought that the works controlling this lake, the dam of Mala'a in particular, were constructed at about this time.[26] The "domain of 10,000 aroures", to the northeast, is irrigated by a derivation from the Joseph canal (or Bahr Youssouf), the derivation works quite probably located near the Labyrinth. This domain includes parcels that are located on higher ground compared to the other cultivated land of the region. The lower lands must be irrigated first, which causes friction with the villagers outside the domain who think that their water has been confiscated:

"Psenemous to Zenon, greetings! The outlying peasants have taken out their (mules and shovels) and opened the irrigation ditches at the ends of the ten thousand aroures. People from Philadelphia attacked them, (chasing away) the mules and breaking (the shovels). I sent Pelôis, son of Pachôs, to (tell you of this). But I presume that you already know of these ugly incidents. In order that this business be cleared up as soon as possible, you would do well to order that (their land) be supplied with water. [...][27]

Thanks to irrigation, numerous new fruits are adapted and cultivated in Fayoum: olives, pears, apricots, figs, etc. In addition, during this period an attempt is made to develop grains yielding two annual harvests: the first planted at the falling flood, in October, and the second irrigated artificially. This is the sense of a written instruction send to Zenon:

"Apollonius to Zenon, greetings! The King has ordered us to cultivate the land a second time. Therefore, as soon as you have harvested the early grain, quickly water the land by hand. In case this is not possible, install as many irrigation machines as you can, but do not leave water on the land more than five days. Then dry the land and as quickly as possible plant the three-month wheat. Write to me personally when you are ready to harvest the

25 After Claude Orrieux, The papyrus of Zenon (1983), Chapter VI.
26 Garbrecht (1996).
27 Orrieux (1983), Chapter VI.

grain. Go well" (December 256 BC).[28]

The "irrigation machines" of this letter are perhaps balance beams (shadufs), or perhaps the very first models of manual waterwheels that appear in the 3rd century BC.

Two centuries after these accounts, at the beginning of the Roman domination around 25 BC, Strabo sojourns in Egypt and travels up the Nile to Aswan, in the company of the prefect, newly named by Augustus. Twice in his accounts he speaks of irrigation by "machines" (always powered by men or animals) and, in particular, the use of

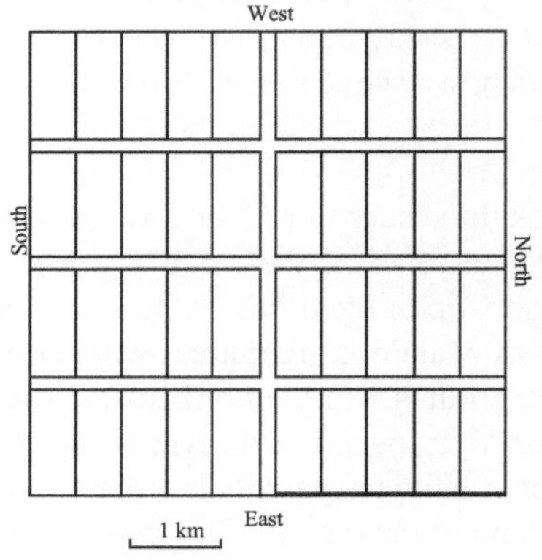

"Sketch by Stothoetis [...] for Appollonius in the year 27, the [...] of the month of Phaophi *(..., i.e. December 259 BC)* under the responsibility of Zenon (?) and the control of Diodore.

"The perimeter of the ten thousand aroures is 400 schoenes *(this clearly refers to hundredths of schoenes)* along four dikes. In the interior, from south to north, there would be three dikes, separated one from another by 25 schoenes, and in addition, from east to west, nine transversal levees, separated by 10 schoenes. There would be, therefore, within the ten thousand aroures, forty basins of 250 aroures each *(69 ha)*, measuring 25 by 10, as shown on the plan, which represents sixteen dikes each of length 100 schoenes, i.e. 1600 schoenes in all, for which it would be necessary to excavate the fill dirt.

"The width of a ditch is 4 cubits, the depth 2; we estimate that one will be able to extract enough dirt to raise the dikes to the desired height, which will be done. [...] As for the four supply canals that must be added to the four existing ones, they will each be 100 schoenes long. [...] If this work is completed during the winter, we estimate that it will be costly. [...]"

Figure 5.12 Plans for development of the "domain of 10,000 aurores" and reproduction of a descriptive sketch (Orrieux, 1983; Burkhalter, 1992)

Archimedes Screws and wheels; first in the region of Memphis, then in the region of Aswan.

"There is a ridge extending from the encampment *(the cantonment of a Roman legion)* even as far as the Nile, on which the water is conducted up from the river by wheels and screws; and one hundred and fifty prisoners are employed in the work; and from here one can clearly see the pyramids....[29]

"The Nile has very many islands scattered along its course, of which some are wholly cov-

28 Ibid., Chapter V.

29 Strabo, The Geography of Strabo, William Heinemann Ltd., London, 1932, XVII, 1, 30, transl. Horace Leonaard Jones.

ered at its risings and others only partly; but the exceedingly high parts of the latter are irrigated by means of screws." [30]

The appearance of the water mill

The water mill, the first energy source to replace muscle power, appears in the Hellenistic cultural sphere at the end of the 2nd or beginning of the 1st century BC. The region of origin of this important invention, somewhere in Asia, is not well known. The first traces are claimed to be from the kingdom of Pontus, at Cabeira, by Strabo (who is a native of that region), in the proximity of the new palace of Mithridatus VII Eupator, king of Pontus from 111 to 63 BC. He fought against the Roman expansion in the region, but was finally defeated by Pompey in 63 BC.

"... at their junction *(i.e. of the two rivers Lycos and Iris)* is situated a city which the first man who subjugated it called Eupatoria after his own name, but Pompey found it only half-finished and added to it territory and settlers, and called it Magnopolis. Now this city is situated in the middle of the plain, but Cabeira is situated close to the very foothills of the Paryadres mountains about one hundred and fifty stadia farther south than Magnopolis, the same distance that Amaseia[31] is farther west than Magnopolis. It was at Cabeira that the palace of Mithridates was built, and also the water-mill; and here were the zoological gardens, and, near by, the hunting grounds, and the mines." [32]

Another piece of written evidence, essentially from the same time as the text of Strabo, is attributed to Antipatros of Thessaly. It speaks poetically of the new pleasures to be enjoyed by the miller, liberated by hydraulics from the need to turn the mill with his own muscles. It also suggests that this is a recent invention:

"Women who toil at the querns, cease now your grinding;
Sleep late though the crowing of cocks announces the dawn.
Your task is now for the nymphs, by command of Demeter,
And leaping down on the top of the wheel, they turn it,
Axle and whirling spokes together revolving and causing
The heavy and hollow Nisyrian stones to grind above.
So shall we taste the joys of the golden age
And feast on Demeter's gift without ransom of labour."[33]

Some see in this sentence the proof that the mill to which the text refers (the one at Mithridatus? Another mill?) has a vertical axis and horizontal wheel (the Nymphs turn the axle in bounding to the top of the wheel). But this would appear to be thin proof at best. It is only under the Roman Empire, and in particular during the 1st and 2nd cen-

30 Ibid., XVII, 1, 52.

31 Amaseia is the native city of Strabo. The ancestors of Strabo participated in the dynastic unraveling of the kings of Pontus, and in the wars between the Mithridates and the Romans.

32 Strabo, The Geography of Strabo, William Heinemann Ltd., London, 1932, XII, 3, 30, transl. Horace Leonard Jones.

33 Greek Anthology, IX, 418 (Loeb ed. Vol. 3 p 233).

turies AD, that the use of the water mill spreads into the Roman provinces of Asia and the West. We will return to this in more detail in Chapter 6. Note that in China we find hydraulic energy applied to complex industrial uses as early as the 1st century AD, (Chapter 8).

The Nabatians of Petra, hydraulicians of the desert

Antigonus, the old one-eyed general of Alexandria, sought to solidify its domain in the Near East, between the domains of Ptolemy in Egypt and that of Seleucos in Mesopotamia. He coveted the wealth of the semi-nomad "barbarians" who frequented the routes of caravans carrying spices, myrrh, and incense from Maryab in Arabia Felix, to the south of the Dead Sea. In 312 BC, he sends his friend Athenes on an expedition toward the "formidable citadel", the "rock" (*petra* in Greek) where these "barbarians" store their riches. Athenes does not come back alive. Antigonus then sends his son Demetrius to lay siege to Petra, again without success. Later, the Seleucids, new masters of Syria, attempt the same exploit and also fail. These "barbarians", objects of such envy for the successors of Alexander, call themselves the *Nabatu*. They likely came from central Arabia in the 5th century BC, and settled peacefully in the south of Palestine, left unpopulated by the deportation of the Assyrians and the Babylonians.

The Nabatian civilization becomes sort of a synthesis of the civilization of the ancient East and the new Hellenistic influences, since some degree of normal exchange continues during the wars with the Seleucids. Its golden age is at the very beginning of the Christian era before the Romans, new masters of Egypt, establish a direct maritime route toward the land of incense through Alexandria and the port of Myos Hormos on the Red Sea. In 106 AD, on the orders of Trajan, the kingdom of the Nabatians is peacefully integrated with the Roman Empire, henceforth becoming the province of Arabia. The prosperity of the Nabatians is founded not only on the spice trade, but also on the technical prowess that enables them to farm the desert. This echoes the situation of the kingdom of Sheba, as we discussed in Chapter 3. Here, the main problem is to create arable land in the mineral wasteland of rocks and stones of the Negev desert. Weirs that partially block watercourses retain not only the annual floodwaters, but also and perhaps more importantly, they retain the silt conveyed by the floodwaters. This silt, accumulated flood after flood, creates lands that eventually can be cultivated. This technique is surely inspired by earlier practices in Arabia Felix. But the reader may also recall the very ancient developments of Jawa and Khirbet el-Umbashi, in the Syrian-Jordanian desert (Figures 2.6 and 2.8). The technique will later be used by the Romans for the development of North Africa.

Under Nabatian control, the Negev becomes a land of immense orchards, farms, and villages, with roads, numerous water mills, and even cities like *Oboda* (today Advat) and *Mampsis* (Kurnub).

If Petra, the capital, is able to resist all attempts at conquest, it is because it is located on a site that is a natural fortress. The city is constructed in the valleys of the wadis, and served, thanks to the Nabatian techniques, by small dams. It is deeply entrenched in a tall sandstone massif, contoured by the combined actions of water and wind:

"The capital of the Nabatians is Petra; this is what it is called, for it is situated on a site that is regular and level, but that is fortified all around by a rock (πετρα), the exterior parts of the site are steep cliffs, whereas there is water in abundance, both for domestic use as well as for watering the gardens."[34]

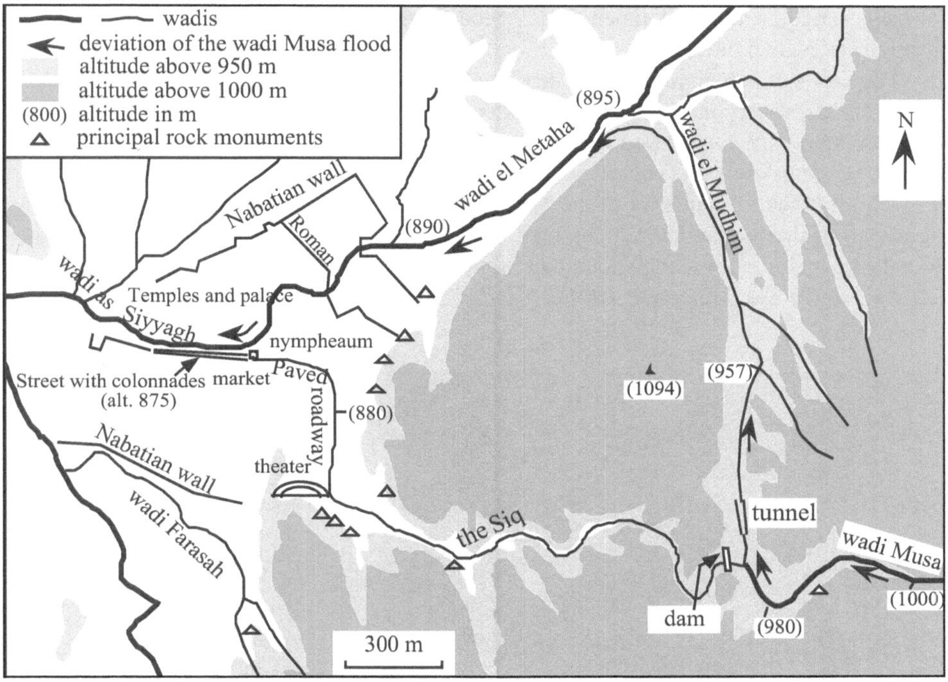

Figure 5.13 The site of Petra, and the hydraulic works to redirect the floodwaters of the Musa wadi. The steep slope of the wadis can be appreciated from the altitudes shown on this map (after Deletie, 1995). See a photo of the tunnel fig 5.13b.

The site is accessible only through narrow gorges. The main route for access and communication, called *Siq*, is the gorge through which the Musa wadi, whose source is several kilometers to the east, penetrates the massif. This gorge is 1,500 m long, but only several meters wide (sometimes less than 3 m), and about a hundred meters deep. The wadi is usually dry, but the winter flood can be very sudden. To keep the *Siq* safely dry, once it had been transformed into a paved road, the Nabatians constructed a dam on the Musa wadi at its entry into the narrowest portion of the gorge. They also construct a tunnel, 9 m high and 6 m wide and 88 m long, to redirect the flood waters toward the el-Mudhim wadi, then toward the el Metaha wadi, the flood waters rejoining their natural course at the city center (Figure 5.13).

This project was very likely built at the beginning of the 1st century BC, at the same time as the planning for the urbanization of the city. The dam itself is 14 m high and 43 m long. The violence of the torrential rain, combined with the slope of the gullies and

34 Strabon, Géographie, XVI, 4, 21 (Translation of F. Lasserre), Les Belles Lettres, 1981, adapted

wadis (the bed of the Musa wadi, itself, has in the *Siq* a steep slope, around 40 m per kilometer), explains the rapidity of the floods. A recent hydrological study estimated that the maximum discharge of the Musa wadi flood must be about 200 m^3/sec, reached in only one hour.[35] This dam was reconstructed in 1964, following the death of 24 tourists carried away by a sudden flood. Since then, the floodwaters have again drained through the Nabatian tunnel.

The city's water supply is provided by aqueducts as well as by extended systems of dams, canals, reservoirs and cisterns collecting stormwater runoff. The first aqueduct was a channel which captured the spring water of the Musa wadi valley, and was built in the beginning of the 1st century BC, then destroyed by a flash flood of the Musa wadi. It is rebuilt in the third quarter of the 1st century BC as a terracotta conduit following the right wall of the Siq, and complemented by the end of the 1st century AD by five other aqueducts, including a second aqueduct in the Siq (see photo fig 5.13c), carved along the left wall of the Siq (Bellwald, 2006).

Figure 5.13b. The tunnel constructed by the Nabateans to redirect the flood waters of the Musa wadi towards the el-Mudhim wadi. The tunnel is, at its entrance, 9 m high and 6 m wide (photo by the author).

35 Deletie and collaborators, 1995.

Figure 5.13 c. The Siq, which is the main access route to Petra, and which was the course of the Musa wadi before the dam and tunnel were constructed. On the right, the channel containing the terracotta aqueduct (dated by 25 BC), and on the left the channel of the other aqueduct constructed in the middle of the 1st century AD (photo by the author).

The science of fluids at Alexandria under the Roman domination

Heron of Alexandria and the "pneumatic" machines

The contributions of Heron of Alexandria belong for the most part to the continuum of work of Ctesibios and Philon of Byzantium. It was believed for quite some time that he lived in the 1st century BC. Now, Heron, in his work *Dioptra*, describes how to estimate the distance between Rome and Alexandria through observation of a lunar eclipse – an eclipse that took place in 62 AD.[36] The work of Heron therefore must be dated from the second half of the 1st century AD. The importance of this detail will appear in Chapter 6 in the context of understanding Roman treatises on aqueducts. Like Ctesibios and Philon, Heron is the author of a treatise on *Pneumatics*. Acknowledging his debt to earlier authors, but claiming some of his own originality, he describes a number of machines. These include the fire pump or pump of Ctesibios[37] (with a single modifica-

36 After sources cited by Gilbert Argoux (1994).
37 This type of pump is used as a fire pump, as well as a bilge pump, up until the beginning of the 20th century (Figure 5.6).

tion compared to what is shown in Figure 5.5: the intermediate reservoir C is deleted), and automatic devices that "solicit astonishment and admiration", and therefore are essentially toys (today we would call them "gadgets").

The principle of these automatic devices is based on the effects of *pressure* in fluid. Some of them use the siphon and connected chambers, as mechanisms for the automatic filling of a vase, mixing of two liquids, etc. Other devices are powered by the effects of gaseous expansion: the most remarkable invention is surely "Heron's steam ball" (Figure 5.14), or *éolipile*, a device in which water brought to a boil emits steam, the pressure of which turns a ball around an axis. This is the principle of a *steam engine*, nothing less! But the technology (or perhaps society?) was not up to the task of industrial exploitation of this invention.

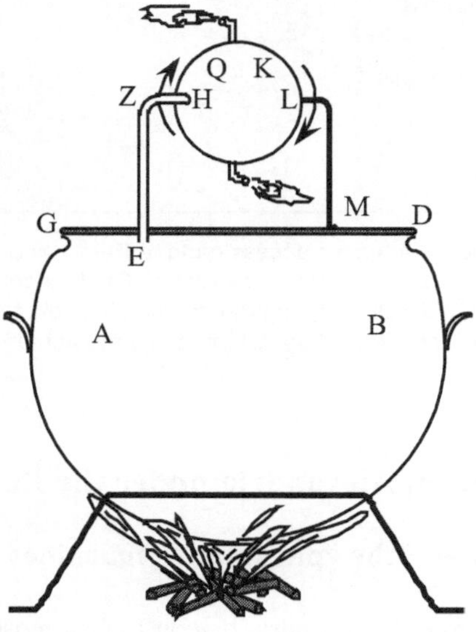

Figure 5.14 "Heron's steam ball", the principle of a steam engine. "Let the cauldron AB be placed over a flame, with water; its opening will be closed by the cover GD, through which passes the bent tube EZH, its end penetrating into the small hollow sphere QK; diametrically opposite the extremity H, one mounts the pivot LM, which rests on the cover GD. One adds to the sphere two small bent tubes, welded to the sphere, diametrically opposed to each other and bent in opposite directions; these elbows must be at right angles, and the tubes perpendicular to the line HL. The following occurs when the cauldron is heated: the steam passes into the tube EZH to go into the sphere, and it leaves by the small bent tubes in the wall, and causes the sphere to turn, like figures that dance."[38]

38 Heron, Pneumatics, II, 11.

Heron of Alexandria and the first expression of the volumetric discharge of a canal

Before Heron, no correct notion of the *discharge* of a canal, pipe, or river had been correctly formulated. Indeed, the notion of *velocity* was essentially unknown in Greek mechanics. The quantity of water delivered by an aqueduct or canal was quantified uniquely by a measure of the flow area. It was Heron who formulated for the first time the notion that the volumetric discharge, i.e. the volume of water delivered in a unit of time, is the product of the flow area and the velocity. One finds the following in his work *Dioptra*:

> "It is to be noted that in order to know how much water the spring supplies it does not suffice to find the area of the cross section of the flow... It is necessary also to find the speed of the flow, as the swifter the flow is, the more water the spring supplies."[39]

The importance of the current velocity in calculating the discharge is thus established, but Heron did not have any means of measuring this velocity. So he also proposed another means of calculating the quantity of water delivered in a day:

> "One should therefore dig a reservoir below the stream and note with the help of a sundial how much water flows into the reservoir in a given time, and thus calculate how much will flow in a day... The amount of water will be clear from the measure of the time."

Coming too late to be useful to the Romans, who were the great constructors of this period, the concept is destined to be gradually forgotten over time. It is only in the West during the Renaissance that it will once again be formulated.

Galien of Pergamon and the beginnings of biomechanics

There is another branch of fluid mechanics that sees some early development in this period: this is the knowledge of blood circulation. Whereas it was believed that the arteries contained only air prior to this period, Galien of Pergamon (129 – 200? AD)[40] is the first to describe arterial circulation and to study seriously the circulation of blood in the heart. For this he relies on an intense practice of dissection.[41] His only error is in believing that the blood passes directly from the right ventricle to the left ventricle.

The first discovery of the resistance to motion through the air

We mentioned in Chapter 4 the dominant theory of Aristotelian Greek science on the movement of objects in air. This theory held that air actually entrains the movement of a body (a thrown spear or an arrow) rather than slowing down this movement. In this

39 Dioptra, cit. after Gunther Garbrecht (1987).

40 Galien is born at Pergamon, studies at Smyrna, Corinth, and Alexandria, and then lives in Rome as a renowned doctor (Lloyd, 1973, Chapt. 9).

41 In Egypt, due to the practice of embalming, human dissection is not taboo as it is in the Greek world. It is thus generally permitted at Alexandria, leading to numerous anatomical discoveries.

theory, the air displaced by the front of the projectile comes back to the rear and pushes the object in its flight. Jean Philopon of Alexandria (in the 6th or beginning of the 7th century BC), in his *Critique of the Physics of Aristotle*, strongly rejects this theory:

> "How could it be that the air, pushed by the arrow, does not move in the direction of the impulse that has been given to it, but instead does an about-face, as if ordered to do so, and backtracks? Moreover, how could it be that this air, in this about-face, does not disperse into space, but instead returns to strike precisely the notched end of the arrow, continuing to push it and stick to it? Such a conception totally lacks plausibility, and smacks of fiction."[42]

Later in the same work, Jean Philopon suggests that it is indeed the thrower who "provides the motive force for the rock" (which is what will later be called the *kinetic energy* or *momentum*). He also says that "if one imparts an unnatural movement, or a forced motion, upon an arrow or a stone, the same degree of motion will be attained more easily in a vacuum." Continuing his discourse through the description of experiments with falling bodies, Jean Philopon observes that the time of fall depends very little on the weight. He shows finally that the air does indeed exert a *resistance* to the advancement of the body in motion.

Who burned the Library of Alexandria?

The legend that attributes the arson of the Library of Alexandria to Julius Caesar is questioned today. We have seen clearly, to the contrary, that the intellectual movement of Alexandria knew a second fruitful period under the Roman Empire. In this period, it is a *procurator supra museum et ad Alexandrina Biblioteca*, named by the Roman prefect, who administers the Museum and the Library.[43] It is generally accepted that the period of greatest creativity at the institution was in the 2nd century of our era.

However, we should remember that it is only a short time before the capture of the city by the Arabs, in 642, that one finds Jean Philopon's most virulent critique of the *Physics* of Aristotle. Thus to the extent that periods of intellectual decadence are reflected in the absence of a critical view of the treatises of the "ancient greats", Alexandria remained a great intellectual center to the very end.

But in a world that insists on rushing toward a precipice, important discoveries and ideas are destined for oblivion. The legendary burning of the Library is a symbol of this headlong rush: did the books merely burn, or was it the intellectual drought of troubled times that prevented the survival of the theories of Heron, Jean Philopon, and so many others?[44] There certainly were episodes of book destruction during the period of the last centuries of Roman domination, either as acts of war, or as the ravages of religious fanaticism. In 290 AD, during the re-taking of Alexandria after it had been conquered by Zenobia, queen of Palymyra, an entire section of the city is set aflame. In 391, the

---

42 Commentary on the Physics of Aristotle, 639, 30, adapted from the translation of Cohen and Drabkin, cit. after Lloyd (1973).

43 Sartre (1991), p. 418.

44 Archimedes, Philon of Byzantium and several others will be translated into Arabic. Certain of these works owe their survival to this translation.

bishop Theophilos intentionally destroys the library of Serapeion. And in 640, after the taking of Alexandria by the Arabs, the remaining 54,000 books are burned as fuel for the public baths, on the order of the caliph Omar. According to the tradition reported by Arab historians, Jean Philopon tried in vain to persuade the conquerors to spare the books.[45]

45 The reader can consult the work of Luciano Canfora (1986).

6. Hydraulics in the Roman empire: Driving Force of development and symbol of civilization

The Etruscan hydraulic heritage and the beginnings of Rome

Civilization does not truly begin to develop in the western Mediterranean until the beginning of the 8th century BC. This begins according to the legend when the Phoenicians, led by Elissa, princess of Tyr, found Carthage on the Tunisian coast. Then the Greeks establish colonies in Sicily and in the south of the Italian peninsula in the middle of the 8th century BC. These colonies become a cultural ensemble called *Greater Greece.* But the capital event for Italy during this same period is the arrival of yet another people who settle to the north of the Tiber. According to Herodotus, these new arrivals came from Anatolia. In response to an extended period of famine, the king of the Lydians has his son Tyrrhenios lead half of his people out of their homeland. They take to the sea, traveling westward to Tuscany and Umbria where they become a new people: The *Etruscans* – or the Tyrrhenians as they are called by the Greeks. In the 6th century BC they spread onto the plain of the river Po, and to the south as far as Campania. Their beautiful urban civilization profoundly marks the Italian countryside; the cities have their own water supply, the streets are straight and aligned at right angles, with gutters and extended underground sewage networks, following the Aegean and Oriental traditions. The Etruscan economic miracle rests on three pillars: the mastery of commerce, the exploitation of iron mines, and the development of land for agriculture.

The principal obstacle to development of Etruria is the accumulation of water in the numerous marshy depressions of the valleys. During the entire period from the 7th to the 4th century BC, the Etruscans drain large regions of Italy by means of dense networks of underground drainage galleries, called *cuniculi.*[1] These galleries, from 300 m to several kilometers in length, are about 1.5 m high and 0.5 m wide. They normally follow the courses of the valleys they drain, aligned slightly off the valley centerline (and usually to the right) at a depth of some 30 meters. Vertical shafts are regularly spaced at 30 to 35 m, extending from the galleries up to the surface. These shafts facilitate the initial digging of the drainage galleries, and then provide aeration. These *cuniculi* follow the valley down to its mouth, or sometimes pass under a ridge to connect to a neighboring valley.

This preoccupation with land drainage is accompanied by a need to control the supply to, and level of, the many lakes of Etruria. Underground drainage works are built in the lakes of Burano, Nemi (near Rome), and Albano. Lake Albano has a 1.5 m wide tunnel that varies in height from 2 to 3 m. It is 1,200 m long, and passes underneath the present city of Castel Gandolfo. There are also artificial reservoirs, excavated and then sealed through paving with a mixture of clay and chalk. These reservoirs are used in the

---

1 See Goblot (1979), pp. 188-192.

wet season to store water, and in the dry season to supply water for crops through terracotta pipelines.[2]

In their conception and construction of underground conduits, the Etruscans show a marvelous mastery and skill in hydraulic works. They are also good miners. Their know-how in both areas could well have been developed locally, but they surely brought much of their expertise in hydraulics from the Orient. It is interesting to note certain similarities between the Etruscan *cuniculi* and the *qanat,* broadly spread within the Persian Empire for water supply (see Chapter 2, Figure 2.19). Recall too that the Mycenaeans used natural grottos to drain marshy depressions (see Chapter 4, in particular Figure 4.10).

The Etruscans are not only miners and peasants, but also sailors: they give their name to the Tyrrhenian Sea. In the 6th century BC, they share maritime domination of the western Mediterranean with the Carthagenians. Their united forces defeat the Phoenicians of *Massilia* (Marseilles) at the naval battle of Aleria (in Corsica), in about 535 BC. Maritime commerce flourishes, and numerous ports are developed. But in 474 BC, the Syracuse fleet defeats the Etruscans, pushing their maritime commerce back toward the Adriatic. The large port of Spina on the Adriatic is built on a lagoon that is connected to the sea, some 3 km distant, by a 30-m wide canal. It is likely that this port resembled modern Venice, with its canals connecting to the sea and to the nearby estuary of the Po.[3]

Simple huts occupy the site of the seven hills of Rome From the 8th century BC, around the marshy depression where the *forum* will later be built. This site, at the southern boundary of Etruria, is easily fortified, and the Tiber River, navigable by small boats, adds to the attractiveness of the site. It is here that one finds the first ford of the Tiber, and soon the first bridge, that of Sublicius. In 575 BC, a city is founded here by Tarquin (Tarquin the Elder), a rich Etruscan.

Quite logically, the first hydraulic project in the city is a drainage canal (Figure 6.1). Tarquin the Elder constructed the *cloaca maxima* to drain the depression of the *forum* and reclaim its land. At the end of the 6th century BC, Tarquin the Superb, third and last of the Etruscan kings of Rome, had the *cloaca Maxima* covered over since its physical presence had become an obstacle to further development of the city. With this, the canal began to also serve as a sewer. It had been generously dimensioned by its initial builders, and was not further modified until the 1st century BC under Augustus. Pliny the Elder left us a few words describing this work:

> "(...) there are seven rivers, made, by artificial channels, to flow beneath the city. Rushing onward, like so many impetuous torrents, they are compelled to carry off and sweep away all the sewerage; and swollen as they are by the vast accession of the pluvial waters, they reverberate against the sides and bottom of their channels. Occasionally, too, the Tiber, overflowing, is thrown backward in its course, and discharges itself by these outlets: obstinate is the contest that ensues within between the meeting tides, but so firm and solid is the masonry, that it is enabled to offer an effectual resistance."[4]

---

2 Keller (1976), p. 52-54, 274.

3 Heurgon (1989), p. 170.

4 Pliny the Elder, The Natural History, Book XXXVI, 24, Translation ed. John Bostock, M.D., F.R.S., H.T. Riley, Esq., B.A. http://www.perseus.tufts.edu/cgi-bin/ptext?lookup=Plin.+Nat.+toc

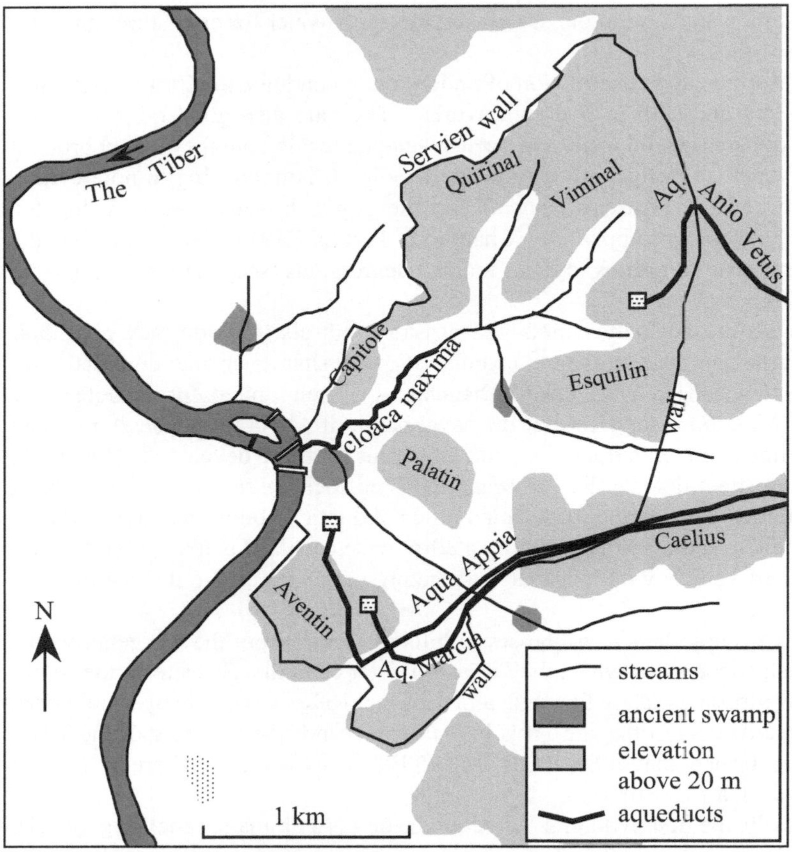

Figure 6.1 The city of Rome under the Republic, the cloaca maxima, and the arrival into the city of the first aqueducts. The Aqua Applia passes underneath the Aventin. The Marcia, carried on arches in the countryside, is at a higher elevation, and arrives at Aventin through the siphon of Caelius.

Rome became a republic in about 509 BC, Tarquin having been run out. After the incursion of the Gaels into Italy, about 390 BC, Rome becomes the main power of central Italy. The *Aqua Appia*, first of the Roman aqueducts, is built in 312 BC. Its dimensions are similar to those of the Etruscan *cuniculi* (0.69 m by 1.69 m), and almost all of it is buried:

> "For 441 years after the founding of Rome, the Romans made do with water that they drew from the Tiber, from wells, or springs. The memory of the springs is still a fond one today. (….) Under the consulship of M. Valerius Maximus and of Pl. Decius Mus, thirty years after the start of the Samnite war, the Aqua Appia was brought into the city by the censor Appius Claudius Crassus, later known as Caecus (The Blind). He is the one who also established the Appian Way from the Capene Gate to the city of Capua. (….). The Appia is fed from a loca-

tion in a domain of Lucullus, on the Praenestina road, between the seventh and eighth milles. (...). From the saline spring, a named place near the Porta Trigemina, the conduit has a length of 11,190 paces, of which 11,130 are an underground canal, and, above ground, on supporting walls and arcades, 60 paces near the Capene Gate."[5]

The Alexandrian heritage

Carthage is defeated in 202 BC, at the end of the Punic wars. This leaves Rome without a rival in the western Mediterranean, so she immediately begins her expansion toward Greece and the Orient. This evolution is inexorable, despite some temporary setbacks due to resistance such as that of the king of the Pontus, Mithridate Eupator (in whose land the remains of one of the first water mills has been found, as noted in the preceding chapter). The annexation of Egypt by Augustus in 31 BC effectively ends the Roman expansion toward Asia. After the occupation of the coast of North Africa at the end of the 1st century AD, the Mediterranean becomes the *mare nostrum*, a sea that is entirely bordered by Roman lands.

One can clearly see the appearance of the Alexandrian heritage in Roman techniques during the Augustin period. The monumental work *On Architecture* of Marcus Vitruvius Pollio, who lived in the 1st century BC under Julius Caesar and Augustus, paints a vast tableau of techniques for the information of the new emperor. This broad-brush panorama integrates the skills of the Alexandrian School from the 3rd century BC. It describes the siphon, whose use, when implemented with Roman know-how, had already made it possible for water from the *Aqua Marcia* aqueduct to reach Capitol and Palatain (in 144 BC). It also describes use of the Ctesibios pump, lifting water wheels, etc. During the four centuries of prosperity of the Empire – until its economic decline of the 3rd century AD and the fall of the western Roman Empire in 410 AD - indelible symbols of the power of Rome were left in the development of water supply, water use in the cities, agricultural productivity in the provinces, and the development of maritime commerce. Many of the countless hydraulic structures and installations that were constructed remain with us today as symbols of that power.

The great aqueducts of Roman cities

Water is at the very top of the scale of values of Roman civilization. Water "not only services and satisfies the needs of the public, but also satisfies their pleasures."[6] Numerous public fountains flow constantly in the city of Rome. Some individual users are granted a special concession for drawing water. Under the Republic this service is paid for, and it later becomes a free service granted by the Emperor. But the thermal baths, becoming widespread from the period of Augustus, are the most important water users.

5 Frontinus, Aqueducts of the city of Rome, IV-V, adapted from the translation of P. Grimal. We present this author further on.
6 Frontinus, Aqueducts of the city of Rome, XXIII, adapted from the translation of P. Grimal.

Figure 6.2 A Roman public fountain at Ostia (photo by the author)

When water is in short supply, basic needs (e.g. public fountains, and flushing of sewers to maintain hygiene) must take priority over pleasure use. The architect Vitruvius describes a design giving priority to the public fountains (see Figure 6.4):

> When they *(the channels)* are brought home to the walls of the city a reservoir (castellum) is built, with a triple cistern attached to it to receive the water. In the reservoir are three cisterns of equal sizes, and so connected that when the water overflows on either side, it is discharged into the middle cistern, in which are placed pipes for the supply of the public fountains; in the second those for the supply of the baths, thus affording a yearly revenue to the people; in the third, those for the supply of private houses. This is to be so managed that the water for public use may never be deficient."[7]

There remain only a few traces of the Roman distribution networks, one of the few being at Pompeii (Figure 6.3) whose *castellum* corresponds rather well to the scheme described by Vitruvius.[8]

---

7 Vitruvius, On Architecture, VI, 1 and 2, translated by Joseph Gwilt, London: Priestley and Weale, 1826. http://penelope.uchicago.edu/Thayer/E/Roman/Texts/Vitruvius/8*.html (adapted)

8 For water in Pompeii, see the two publications of Hans Eschebach (1983). Note that one does not find the distribution scheme described by Vitruvius in the castellum of Nîmes, constructed in the middle of the first century AD under the emperor Claudius (Figure 6.17).

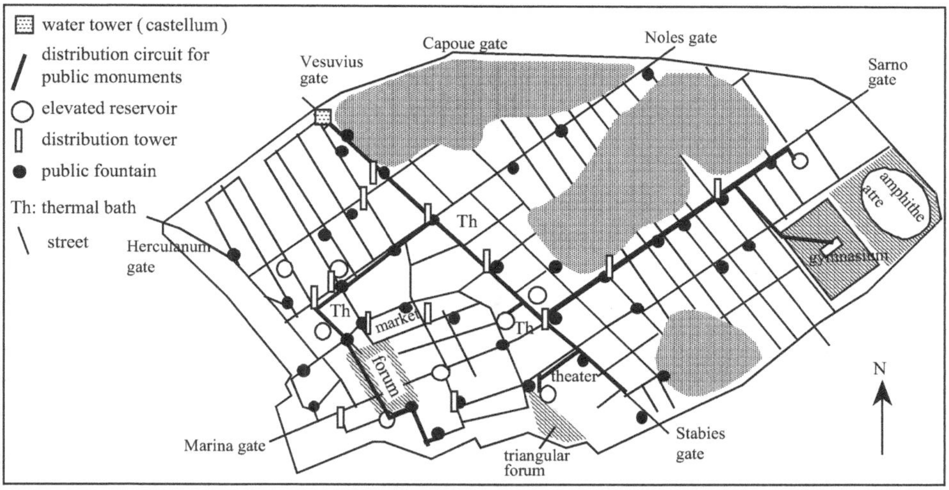

Figure 6.3 Water in a Roman city, Pompeii. The Figure shows the map of public fountains, the layout of water supply to public monuments and the intermediate works in the various circuits: elevated reservoirs and distribution towers. The distribution towers are fitted with a small elevated reservoir, whose level makes it possible to adjust the pressure in the downstream installations. The castellum of the Vesuvius gate (Figure 6.4) is situated at the highest point of the city, receiving its water from the aqueduct of Serino (Escheback, 1979-1983). In August 79 AD Pompeii is destroyed by an eruption of Vesuvius.

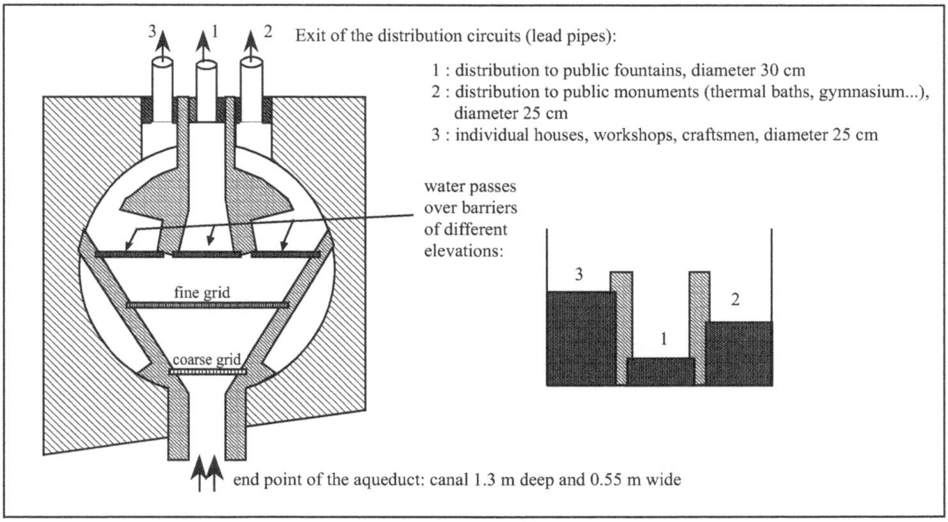

Figure 6.4 Castellum schematic at the Vesuvius gate of Pompeii, terminal point of the aqueduct, feeding the three distribution circuits. Water to these three circuits is supplied by overflow weirs at different levels, thus providing for a distribution hierarchy as recommended by Vitruvius.

Figure 6.5 A drainage inlet at Ostia (photo by the author).

Figure 6.6 The Anio Novus, one of the greatest Roman aqueducts, supported by arches in the Anio valley, between Tivoli and Castel madama. The canal, still lined with opus signinum, is well conserved. This aqueduct was built in 52 AD near the end of the reign of emperor Claudius (who also was responsible for the Aqua Claudia aqueduct), and is some 87 km long. It carries water captured from the Anio river, in the Sabine mountains near Subiaco. (photo by the author).

As we can see through these examples, Roman cities benefit from an abundance of water, a bounty unequalled until the 20th century AD. Creation of this abundance requires that water be brought to the cities from springs or other supply points, where in general a storage reservoir is built, serving also as a settling or clarifying basin. The water is conveyed by aqueducts to the city's *castellum*, or water tower, from which emanate the local distribution circuits. The aqueducts are often lasting monuments in which the Romans take great pride. A highly placed Roman official, whom we introduce later on, even compares these aqueducts to the Pyramids and to Greek temples:

> "With such an array of indispensable structures carrying so many waters, compare, if you will, the idle Pyramids or the useless, though famous, works of the Greeks!"[9]

Evolution of aqueduct techniques

The Cretan, Greek, and later the Hellenistic aqueducts primarily use terra-cotta pipes. The Romans, following the Etruscan heritage, build their aqueducts as masonry canals, usually rectangular in section, and covered over by a vault or stone slabs. The aqueducts are fitted with openings at regular intervals (from tens to hundreds of meters apart) to facilitate inspection and maintenance. We have seen earlier that the *Aqua Appia*, the first Roman aqueduct built in 312 BC, is nearly entirely underground. This practice is surely inherited from the Etruscan techniques, but also has a military dimension, since Italy is still far from being pacified at this time – this is only eleven years after the death of Alexander.

The Romans stay with this concept throughout the long period of aqueduct construction in the Empire, i.e. from the 4th century BC to the 3rd century AD. A canal *(specus)* is, depending on the local topography, sometimes laid on the surface or on a supporting wall, or sometimes buried underground, occasionally passing through true tunnels, and, when necessary, supported on arched structures. The longitudinal profile of an aqueduct is driven by two constraints: first to maintain sufficient slope to convey the water, as regular as possible to avoid local break points; and second to deliver water at a high enough elevation to supply the city's water tower (*castellum*) and thus enable distribution of water to the highest areas of the city. Thus the arch structures that one often sees near the cities are needed to keep the canal at a sufficiently high elevation. And it is of course these arch structures, still visible today in numerous locations of the Roman Empire, that one commonly associates with the Roman aqueducts. However one should not forget that these structures represent only a small fraction of the total length of an aqueduct. The slopes of the canals are quite variable, as we will see in examples presented further on, but most often are of the order of one meter per kilometer. Occasionally, where there is a need to drop to a lower elevation, there are local chutes comprising short sections of steep slope, or even true cascades.

---

9 Frontinus, "The Aqueducts of Rome", 16, Translation of Charles E. Bennett in the Loeb edition, 1925 http://penelope.uchicago.edu/Thayer/E/Roman/Texts/Frontinus/De_Aquis/text*.html

Table 6.1 Characteristics of several known Roman aqueducts (after Hodge, 1995; Leveau, 1979; and other sources).

Location	City	Length (km)	Average slope (m/km)	Estimated discharge (m^3/day)
Italy	Rome	11 aqueducts; see Table 6.2		1,127,220
	Bologna	20	1.00	35,000
France	Lyon (*Lugdunum*)	4 aqueducts; see Table 6.3		
	Metz (*Divodurum*)	22	0.56	22,000
	Nîmes (*Nemausus*)[10]	50	0.25	40,000
	Arles (*Arelate*)	48	0.625	8,000
	Strasbourg (*Argentorate*)	20	3.13	2,160
Germany	Treves (*Augusta Treverorum*)	13	0.6	25,450
	Cologne (*Colonia Agrippina*)	95.4	3.89	27,000
	Mayence (*Mogontiacum*)	28.6		
Switzerland	Geneva	11	0.55	8,640
	Neuchatel (*Aventicum*)	8	4.4	2,880
Spain	Segovia (*Segovia*)	15	16.4	1,728
	Tarragona (*Tarraco*)	35		
	Toledo (*Toletum*)	50		
England	Dorchester (*Durnovaria*)	15		5,200
Africa	Cherchell (*Caesarea*)[11]	40	2	38,000
	Carthage (*Cartago*)[12]	90.4 (Zaghouan) 118 (Djoukar)	2.9	17,280
Syria	Apamea-on-Orontes[13]	150	1.3	
Turkey	Pergamon (*Pergamum*)[14]	50	0.3	20,000

10 Fabre, Fiches, Leveau, Paillet (1992).
11 Leveau and Paillet (1983).
12 Rakob (1979); Clamagiraud, Rais, Chahed, Guefrej, Smaoui (1990).
13 Balty (1987).
14 This is of course not the Hellenistic aqueduct of Mandradag (Chapter 5), but rather the Roman aqueduct of the Kaikos valley. It comes from the east; even though its elevation is high enough to supply only the lower quarters of the city, it transports a much larger quantity of water than the Hellenistic aqueduct.

Let's once again listen to Vitruvius:

"Water is conducted in three ways, either in streams by means of channels built to convey it, in leaden pipes or in earthen *(terra cotta)* tubes (*that is, according to the Greek process cited earlier*), according to the following rules. If in channels, the structure must be as solid as possible, and the bed of the channel must have a fall of not less than half a foot to a length of one hundred.[15] These channels are arched over at top, that the sun may strike on the water as little as possible.

"(....) If hills intervene between the city walls and the spring head, tunnels under ground must be made preserving the fall above assigned; if the ground cut through be sandstone or stone, the channel may be cut therein, but if the soil be earth or gravel, side walls must be built, and an arch turned over, and through this the water may be conducted."[16]

A masonry canal is not inherently watertight. Therefore it is necessary to plaster the useful (or "wetted") walls of the canal, both to minimize leakage and to protect the masonry walls themselves from infiltration damage. The Roman plaster (*opus signinum*) is of very high quality. It is a mortar of crushed tile solidified with thick lime, also containing crushed brick and pottery shards, and sometimes other additives.[17]

The most imposing and important structures are those necessary to carry the aqueducts across valleys. These are the famous bridge-aqueducts such as the Pont du Gard near Nîmes (47.8 m high), or the bridges of Segovia and Tarragona in Spain or, similarly, the bridge of Chabet Ilelouine on the Cherchell aqueduct in Algeria, all three some thirty meters high. There are also the inverted siphons echoing the technology of the Hellenistic world. These expensive siphons are reserved for valleys more than 50 m deep, a depth beyond which the cost of a bridge becomes prohibitive. Further on we describe the siphons of Lyon, the largest ones in the Roman world, but it is first useful to provide a glimpse of this Roman technology.

Roman inverted siphons begin with a head tank, receiving water from the aqueduct's canal, then distributing the water into one or more parallel lead pipes leading out of the basin. These pipes descend to the bottom of the valley and cross it on a bridge-siphon, and then come back up on the other side to an exit, or escape, basin. This exit basin then delivers water into the continuation of the aqueduct (Figure 6.7). Vitruvius emphasizes the need to provide a rather long and rectilinear length of siphon at the low point of the valley (the "venter"):

"(...) when it arrives at the bottom, let it be carried level by means of a low substruction as great a distance as possible; this is the part called the venter, by the Greeks *coelia*. When it arrives at the opposite acclivity, the water therein being but slightly swelled on account of the length of the venter, it may be directed upwards.

"If the venter were not made use of in valleys, nor the level substruction, but instead of that the aqueduct were brought to an elbow, the water would burst and destroy the joints of the

15 This indication is somewhat uncertain (5 cm/km?) It does not correspond to actual slopes in the aqueducts.

16 Vitruvius, On Architecture, VIII, VI, 1 and 3, translated by Joseph Gwilt, London: Priestley and Weale, 1826. http://penelope.uchicago.edu/Thayer/E/Roman/Texts/Vitruvius/8*.html

17 See for example Leveau (1979), Fabre, Fiche, Leveau, Paillet (1992).

pipes. Over the venter long stand pipes *(collivaria)* should be placed, by means of which, the violence of the air may escape. Thus, those who have to conduct water through leaden pipes, may by these rules, excellently regulate its descent, its circuit, the venter, and the compression of the air."[18]

One can appreciate the value of a rectilinear "venter" (usually implemented as a bridge-siphon) in reducing the hydrodynamic force that would be concentrated on a small-radius elbow at the valley floor. But the reasons given by Vitruvius (i.e. to avoid the excessive "swelling" of the water) somewhat miss the mark. There has been considerable questioning of what is meant by the *colliviaria*. Some have interpreted them as purges to eliminate air. According to Henning Fahlbusch,[19] they are towers supporting an elevated intermediate reservoir, acting like a modern surge tank to purge air from the siphon's "venter". Such towers are useful when, in an inverted siphon, there are elbows or high points favorable to the formation of air pockets. The Tourillon de Craponne, on one of the aqueducts of Lyon that we describe further on, could serve as an example. Possible remains of others can be found at Aspendos, in the south of Anatolia.

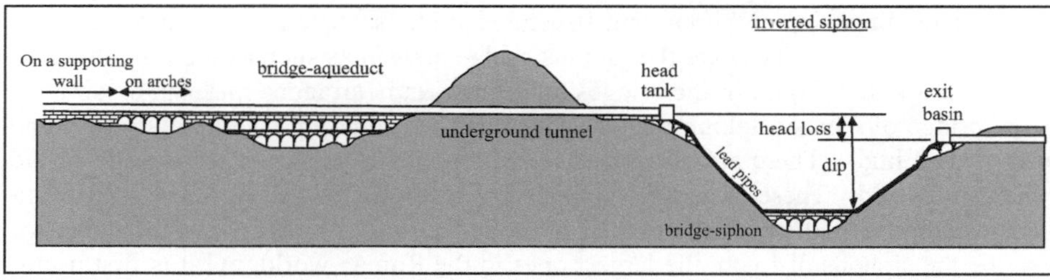

Figure 6.7 The types of works found on Roman aqueducts: bridge, arches, tunnel, inverted siphon

Table 6.1 presents a synthesis of data for a number of known aqueducts in the Roman Empire. It shows that the aqueducts could reach quite significant lengths, often up to 50 or 100 km. But the length of an aqueduct is not, by itself, a good indicator of its grandeur. The construction of bridges, arcades, tunnels, etc. in effect reduces the length of an aqueduct, limiting the numerous detours that would be necessary if the structure had to follow the local topography along the ground surface. Certain aqueducts are actually shortened during renovation, thanks to the construction of such projects. This is notably the case for the *Aqua Marcia*, one of the aqueducts of Rome:

"But now, whenever a conduit has succumbed to old age, it is the practice to carry it in certain parts on substructures or on arches, in order to save length, abandoning the subterranean loops in the valleys."[20]

---

18 Vitruvius, On Architecture, VIII, VI, 5 and 6, translated by Joseph Gwilt, London: Priestley and Weale, 1826. http://penelope.uchicago.edu/Thayer/E/Roman/Texts/Vitruvius/8*.html
19 See Fahlbusch, 1979, 1987.
20 Frontinus, "The Aqueducts of Rome", 18, Translation of Charles E. Bennett in the Loeb edition, 1925 http://penelope.uchicago.edu/Thayer/E/Roman/Texts/Frontinus/De_Aquis/text*.html

The aqueducts of the city of Rome at the end of the 1st century AD

The accession of Nero to supreme power marked the beginning of a somber period for Rome.[21] Civil war followed his assassination in 68 AD, then came the too-short reigns of Vespasianus and Titus. After a promising start, Domitian becomes morbidly suspicious, and sheds the blood of those near to him and his collaborators only to end up being assassinated himself in 96 AD. Then Nerva, an old senator, is chosen as the new emperor in a climate of restoration of the old Roman virtues. We have noted earlier that the Romans placed high symbolic value on water; among the early preoccupations of the new emperor is the restoration of water distribution in the city. For this he calls on a former collaborator and experienced administrator: Sextus Julius Frontinus, 61 years old, former governor of Britain, i.e. Great Britain, and former preconsul of Asia. In 97 Nerva names Frontinus as *curator aquarum*, translated as "high commissioner for water" of the city of Rome. Frontinus is a methodical man who takes his responsibilities seriously. At the beginning of his service, Frontinus prepares a written description of the aqueducts of the city for his own information and then continues with written descriptions of his observations, measurements, and actions during his tenure. This treatise has come down to us, and we have already cited several extracts. We can learn much about both the characteristics and the history of these aqueducts from Frontinus' writings (Table 6.2).

The principal aqueducts at the time of Frontinus came from the region of the Sabine mountains; the four largest ones, *Anio Vetus, Marcia, Claudia* and *Anio Novus*, came from the valley of the Anio, upstream of Tivoli (the Roman *Tibur)* (see Figure 6.8). Where this river flows out onto the plain of Rome, at Tivoli, the aqueducts leave the valley of the Anio and trace a large arc in the direction of the Alban mountains, thus maintaining their elevation as they near the city of Rome. On the Frascati heights, they are rejoined by the *Tepula* and *Julia* aqueducts, and complete their trajectory to the city on arches. At their entrance into Rome, the *Anio Novus* flows above the *Claudia*, on the same arches (Figure 6.9). Similarly, the canal of the *Julia* is above those of the *Tepula* and the *Marcia*, all three superposed on the same arches. These five aqueducts flow on a slope of about 1.3 m/km in their final reaches.[22]

What surely motivated Nerva to put things in order was that the amount of legally distributed water was far less what it was expected to be, as dictated by the written imperial records. The new *curator aquaraum* therefore took it upon himself to make discharge measurements in the aqueducts, both at their origins and at their points of arrival in the city. To his surprise, he found that now the measured discharges were well above those specified in the official registers. There could only be one conclusion: there had been fraud, greater fraud than had originally been suspected!

Here we must pause to note the significance of discharge for the Romans. Not having understood or recognized the physical concept of water velocity, the Romans found it sufficient to measure the flow cross-section at spots where the water velocity was

---

21 Some authors mention that after the fire of Rome, Nero, to his credit, took a certain number of positive measures making it possible to use the water in the aqueducts for fire fighting.
22 Fahlbusch, 1987).

Table 6.2 The nine aqueducts of Rome in the days of Frontinus (about 100 AD)[23]
Two other aqueducts, the Trajana and the Alexandrina, are constructed later. The indicated discharges are estimated from the equivalence: 1 quinariae = 40.6 m^3 per day. The total discharge (all but Alsietina) is 977,00 m^3/day.

Name	**Construction Date**	**Length (km) and % underground**	**Avg Slope (m/km)**	**Discharge (i.e. Area) (quinariae)**	**Estimated Discharge (m^3/day)**	**Origin and quality of water**
Appia	312 BC	16.5 (95%)	0.6	1,825	73,000	Spring in the Anio valley. Excellent
Anio Vetus	272 BC	64 (99%)	3.6	4,398	175,900	Anio River. Somewhat muddy
Marcia	144 BC	91 (88%)	2.7	4,690	187,600	Captured from springs Excellent
Tepula	125 BC	18 (46%)	5	445	17,800	Captured from groundwater. Warm water
Julia	33 BC	22 (54%)	12.4	1,206	48,200	Springs. Excellent
Virgo	19 BC	21 (92%)	0.2	2,504	100,200	Springs. Excellent
Alsietina	2 AD	33 (99%)	6	392	15,700	Lake Alsietinus nonpotable (used for water games)
Claudia	47 AD	69 (78%)	3.8	4,607	184,300	Springs. Excellent
Anio Novus	52 AD	87 (84%)	3.8	4,738	189,500	Anio River. Muddy

judged to be neither excessive nor too weak. The Roman unit of discharge, the *quinaria,* is actually just a measure of area (4.2 cm^2), whereas everyone knows today that volumetric discharge is the product of the velocity and the cross-sectional area of the canal. What Frontinus really measured is the depth of water in well-defined locations in the canal, from which he deduces the area and therefore calculates his estimate of discharge

23 From Frontinus and the commentaries of P. Grimal.

in *quinariae*. The discharges shown in table 6.2 are based on a fixed relation between *quinariae* and real discharges. We will discuss this further at the end of this chapter, regarding the hydraulic knowledge of the Romans. As imprecise as it was, this system nevertheless enabled Frontinus to detect the important deficits and differences described above:

> "I do not doubt that many will be surprised that according to our gaugings, the quantity of water was found to be much greater than that given in the imperial records. The reason for this is to be found in the blunders of those who carelessly computed each of these waters at the outset. Moreover, I am prevented from believing that it was from fear of droughts in the summer that they deviated so far from the truth, for the reason that I myself made my gaugings in the month of July, and found the above-recorded supply of each one remaining constant throughout the entire remainder of the summer. But whatever the reason may be, it has any rate been discovered that 10,000 *quinariae* were intercepted, while the amounts granted by the sovereign are limited to the quantities set down in the records.
>
> "Another variance consists in this, that one measure is used at the intake, another, considerably smaller, at the settling-reservoir, and the smallest at the point of distribution. The cause of this is the dishonesty of the water-men, whom we have detected diverting water from the public conduits for private use. But a large number of landed proprietors also, past whose fields the aqueducts run, tap the conduits; whence it comes that the public water-courses are actually brought to a standstill by private citizens, just to water their gardens."[24]

This observation is rather severe. Numerous wildcat taps provide a clandestine water supply to those living adjacent to the aqueducts, and support a parallel, unofficial water market. Nearly half of the water delivered to Rome by the aqueducts escapes the official accounting. But what also upsets Frontinus is that water from the different aqueducts is senselessly mixed together without taking into account the quality of the different sources. The last aqueduct built under Claudius, (predecessor of Nero), the *Anio Novus*, serves to complete the aqueduct system, and its additional supply was useful in obscuring the pilfering:

> "The two Anios *(aqueducts)* are less limpid, for they are drawn from a river, and are often muddy even in good weather, because the Anio, although flowing from a lake whose waters are very pure, is nevertheless made turbid by carrying away portions of its loose crumbling banks, before it enters the conduits (....)
>
> "One of the Anio *(aqueducts)*, namely Old Anio *(Anio Vetus)*, running at a lower level than most of the others, keeps this pollution to itself. But New Anio *(Anio Novus)* contaminated all the others, because, coming from a higher altitude and flowing very abundantly, it helps to make up the shortage of the others; but by the unskillfulness of the water-men, who diverted into the other conduits oftener than there was any need of an augmented supply, it spoiled also the waters of those aqueducts that had a plentiful supply, especially Claudia, which, after flowing in its own conduit for many milles, finally at Rome, as a result of its mixture with Anio, lost till recently its own qualities. And so far was New Anio from being an advan-

---

24 Frontinus, "The Aqueducts of Rome", 74-75, Translation of Charles E. Bennett in the Loeb edition, 1925 http://penelope.uchicago.edu/Thayer/E/Roman/Texts/Frontinus/De_Aquis/text*.html

tage to the waters it supplemented that many of these were then called upon improperly through the heedlessness of those who allotted the waters. We have found even Marcia, so charming in its brilliancy and coldness, serving baths, fullers, and even purposes too vile to mention."[25]

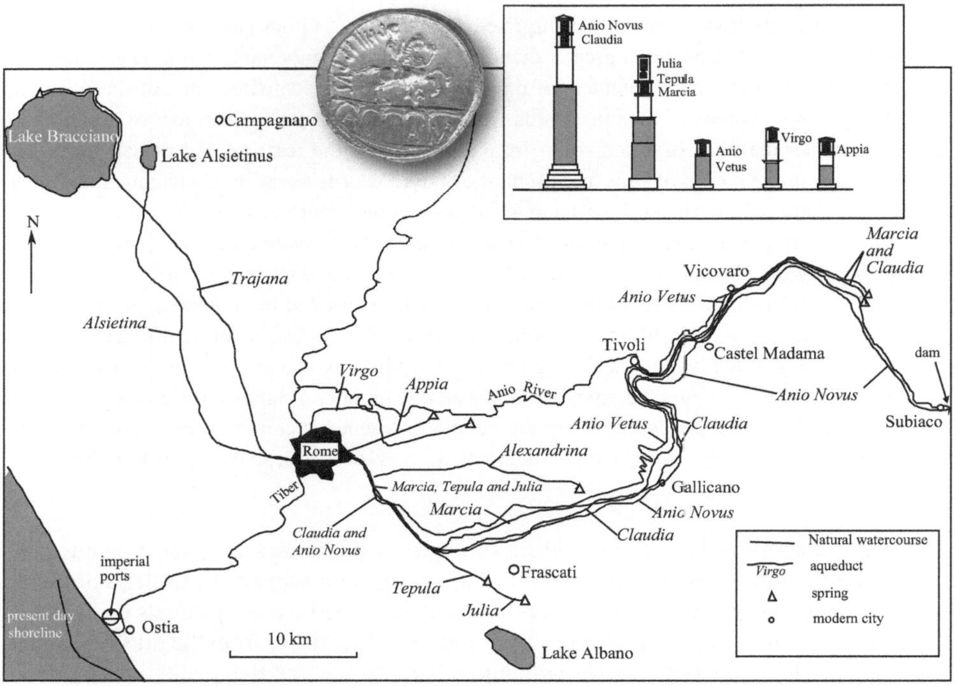

Figure 6.8 Layout of the aqueducts of Rome. This map also shows the Trajana and the Alexandrina, built after the magistracy of Frontinus. See Figure 6.36 for the imperial ports. Adjacent: elevations of the aqueducts at their points of arrival in Rome. The Aqua Marcia is shown on a Roman coin (author's collection).

Frontinus devotes the three years of his mission to reducing these clandestine diversions, and improving the distribution system, both quantitatively and qualitatively. He restores respect for the quality of consumable water, and moves the intake of the *Anio Novus* further upstream to capture clearer water. He also improves the reliability of supply to the public fountains, supplying each with two outlets so as to assure continuous supply when work is being done on one of the aqueducts.

25 Ibid., 90-91.

Figure 6.9 The Agua Claudia and the Aqua Anio Novus on the same arches, at the Porta Maggiore of Rome. The aqueduct was integrated into the new city wall under the emperor Aurelius, in the 4th century AD (photo by the author).

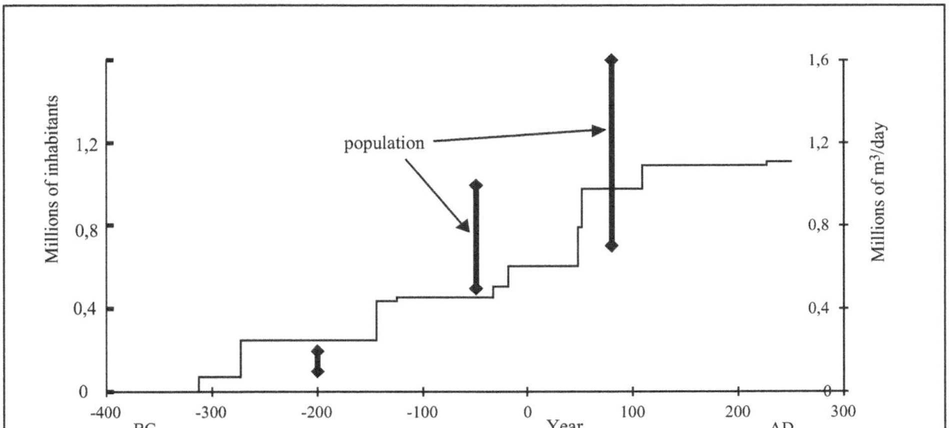

Figure 6.10 Evolution of discharge delivered into Rome by the aqueducts, compared with the evolution of the city's population.[26] The vertical error bars reflect uncertainty in present-day estimates of the city's population.

---

26 The discharges are those of table 6.2, after Frontinus. The demographic data are taken from Christol and Nauny (1997).

Whatever may have been these problems in managing the system, they did not reflect any fragility in the supply of water to the city. Figure 6.10 shows how the increase in discharge delivered to Rome accompanies the demographic growth. With nearly a cubic meter of water per person per day, a Roman inhabitant had nearly a hundred times more water than his or her counterpart in Paris of the 19th century!

Later, two additional aqueducts are constructed, but they are less well known to us since they came after the writings of Frontinus. They are the *Trajana*, constructed around 110 AD under the grand Emperor Trajan, successor of Nerva; and the *Alexandrina*, built under Alexander Severus in 226 AD. From the 2nd century AD, Rome falls into a kind of recession, and its population probably stabilizes. This second century is the century of the provinces, insofar as expansion is concerned. The construction of grand aqueducts begins again in these provinces, in particular in Gaul, Spain, and Africa.

The Roman aqueducts in Gaul

The growth and prosperity of the period of peace between 97 and 180 AD directly benefits the Roman provinces. All Gallo-Roman cities of importance benefit from one or several aqueducts during this period, under the reigns of Nerva, Trajan, Hadrian, Antonius, and Marcus Aurelius, then until 235 AD under Severus. Aix-en-Provence and Lyon have four aqueducts; later on we will examine those of Lyon in detail. At Vienne, where the sources are very near,[27] we know of eleven aqueducts; at Poitiers (*Lemonum*) and Périgueux (*Vesone*), three. Some of these aqueducts are quite simple in design, with a very regular slope, as at Périgueux and Rodez. Others have quite a variable slope, most often steeper upstream than downstream, and with quite a variety of hydraulic works: tunnels, bridges, drops, siphons. The average slope itself can be quite variable from one aqueduct to another; it is usually between 0.2 and 1.6 m/km. One notable singularity is the Craponne aqueduct at Lyon, whose average slope is 16 m/km.

The dimensions and discharges of the aqueducts are also quite variable. The aqueducts of largest cross-sectional area are at Vienne and Nîmes. At Lyon and Nîmes, the lengths rival those of the aqueducts of Rome itself (50 to 75 km). What is striking from the outset is, as shown in table 6.3, the diversity of configurations. Yet the dates of construction, when known, are most often in the 1st and 2nd centuries AD, the period of prosperity, Roman peace, and the concomitant economic climate favorable for development of the provinces. Samples of the broad panorama of Roman techniques of this period, adapted to local conditions, can be found in Gaul. We too rarely know the history of these aqueducts, histories that are sometimes quite dynamic[28] as conditions change. The needs for water can increase over time, or the discharge available from the sources can decrease. Calcareous water can clog the conduits with deposits.

At Sens (*Agedincum*), the Fauconderie spring (on the left bank of the Vanne) is initially the sole water source for the aqueduct. But when the spring dries up, four other springs are tapped successively until the flow is restored to, or slightly exceeds, the ini-

27 Pelletier (1983).
28 The reader can consult the article of Marcel Bailhache (1979).

tial supply. The aqueduct is increased in length from 6.1 km to 14.2 km, so as to reuse both the canal between the former Fauconderie spring and Sens, and the bridge on which the aqueduct crosses the valley of the Vanne.

At Saintes (*Mediolanum*), after some fifty years the capacity of the aqueduct is reduced due to the increase in roughness caused by calcium deposits. Therefore the aqueduct is completely rebuilt, being replaced by one of larger cross-sectional area but flatter slope, capturing water from new springs and delivering it to the city at a higher elevation. This reconstruction, like that of Sens, was based on re-use of large art works – the two bridges called the Arcs (27 arches, 298 m long) and the Hautmont (62 arches, 400 m long) as well as the tunnel of Neufs Puits. On these bridges, the canal is simply raised. But along underground segments, major construction efforts are devoted to modification of the cross section. However, the water from new sources is even more calcareous than the original. According to the estimates of Marcel Bailhache, the discharge, which had been increased from 4,000 to 22,600 m^3/day by the reconstruction, subsequently decreases to a value of only about 8,000 m^3/day.

At Toulouse (*Tolosa*), the capturing of supplementary sources increases the discharge of the Ardenne aqueduct. But the resulting water depth (0.66 m) ends up exceeding the depth of the 40-cm canal lining, thus causing serious degradation of the canal walls.

Water quality is a strong determinant of the longevity of the aqueducts. If the water is calcareous, lime encrustation can reduce the flow area and increase the roughness to the point of reducing the discharge capacity by a factor of two in a few decades. Without maintenance, the canal can become completely blocked. On the other hand, if the water is not calcareous, the canals can have a very long lifetime. The aqueduct of Sens, for example, will be used up until the middle ages.

The most important works are those of Lyon and Nîmes.

Table 6.3 A collection of synthesized data on the principal known Gallo-Roman aqueducts, listed in decreasing order of length.[29]

Note: The useful depth indicates the height of impermeable lining opus signinum. *The discharges in parentheses are those that we have been able to calculate when there is a constant slope on a sufficiently long reach, assuming a roughness height of 3 mm for the canal walls.*

Name	**Length** (km)	**Avg Slope** (m/km)	**Est. Max. Flow** (m3/day)	**Total Depth** (m)	**Useful Depth** (m)	**Width** (m)	**Remarks**
Lyon: Gier Aq.	86	1.5	15,000	1.7	1.33	0.6	4 siphons; 11 tunnels
Lyon: Brevenne Aq.	70	5.3	10,000	1.4 1.7	0.94 0.54	0.55 0.75	1 siphon; 8 drops

29 Sources: Grenier (1960), Bailhache (1979), Burdy (1979, 1996), Jeancolas (1983), Pelletier (1983), Fabre et al (1992), Andrieu (1997), Andrieu and Cazal (1997), Ardhuin (1997), Jaccotey (1997), Lefebvre (1997), Provist and Leprêtre (1997), Rigal (1997).

continued from page 145

Name	Length (km)	Avg Slope (m/km)	Est. Max. Flow (m^3/day)	Total Depth (m)	Useful Depth (m)	Width (m)	Remarks
Arles: Eygallière Aq.	55		8,000	1.3		0.65	
Nîmes	49.7	0.25	(40,000)	1.85	1.2	1.2	Passes over Pont du Gard
Reims	44	0.5	22,000 (34,000)	1.4	0.9	0.7	3 tunnels (800 m; 1,850 m; and 900 m); regular slope
Fréjus	40	12?		1.07	0.67	0.7	Several bridges
Béziers: Gabian Aq.	37		2,500 to 5,000	1.05	0.8	0.42	1,300 m tunnel
Cahors	33	1.4	3,800	1.6	0.5	0.2 to 0.8	Trapezoidal section (largest below the top)
Rodez	30	1.6	38,000 (32,000)	1.4	0.7	0.55	Terminal siphon; regular slope
Lyon: Mont-d'Or Aq.	26	3.2	6,000	0.9	0.6	0.45	Two siphons
Carhaix	27	0.3	4,000	?	0.8	0.6	800 m tunnel
Lyon: Craponne Aq.	27	16.8	8,000	?	0.6	0.5	1 double siphon; vortex drop shafts
Poitiers: de Fleury Aq.	25	0.123	6,700	1.21	0.75 0.46	0.75 0.9	Likely dates from 2nd century
Aix: Traconnade Aq.	24	8		1.1		0.69	
Metz	22	1	22,000	1.6	0.92	1.1	Bridge over the Moselle, 1300 m long and 30 m high
Narbonne: Cabezac Aq.	22	1.4	8,500	1.5	1.2	0.59	Only upstream 11 km known
Paris: Arcueil Aq.	15.7	0.56	2,400	0.5	0.45	0.37	Water depth does not exceed 0.28 m (deposits)
Sens	6.1 14.2	1	39,000	1.56	0.9	0.56	Dates from the 2nd century; later lengthened
Vaison-la-Romaine	12			1		0.6	Siphon under the Ouvèze
Besançon: Arcier Aq.	10.3	2	(69,000)	1.5	0.8	0.8	
Toulouse: Lardenne	9.5	1	19,000	1.15	0.4	0.65	
Saintes: 1st canal reconstructed canal	5 7.5	0.85 0.87	4,000 22,600	0.5 0.7	0.5	0.2 0.45	Two siphons (2 km)
Lisieux	7				0.7	0.5	
Sant-Bertrand-de-Comminges	2.7	1	13,600 (18,000)	1.4	0.5	0.7	5-meter cascade
Périgueux: Grand-Font Aq.	2	0.66	6,200 (4,200)	0.66	0.66	0.37	Water depth = 0.33 m (deposits); regular slope

The aqueducts of Lyon

Lucius Munatius Plancus founds *Lugdunum* (Lyons) in 43 BC on the Fourvière hill, 130 m above the waters of the Saône and Rhône rivers. Thirty years later, Augustus makes Lyon the capital of Gaul. It is often the case that such cities are initially established on high ground for strategic reasons, and consequently not well supplied with water. Then the cities rise in importance to become regional capitals, necessitating significant efforts to provide adequate water supply. This was the case of Pergamon as we described in Chapter 5.

At *Lugdunum*, it was necessary to cross a valley some hundred meters deep to reach the Fourvière hill, only its western facade being accessible. As at Pergamon (but several centuries later), large inverted siphons, the largest in all the Roman world, are used to bring water across the deep valley.

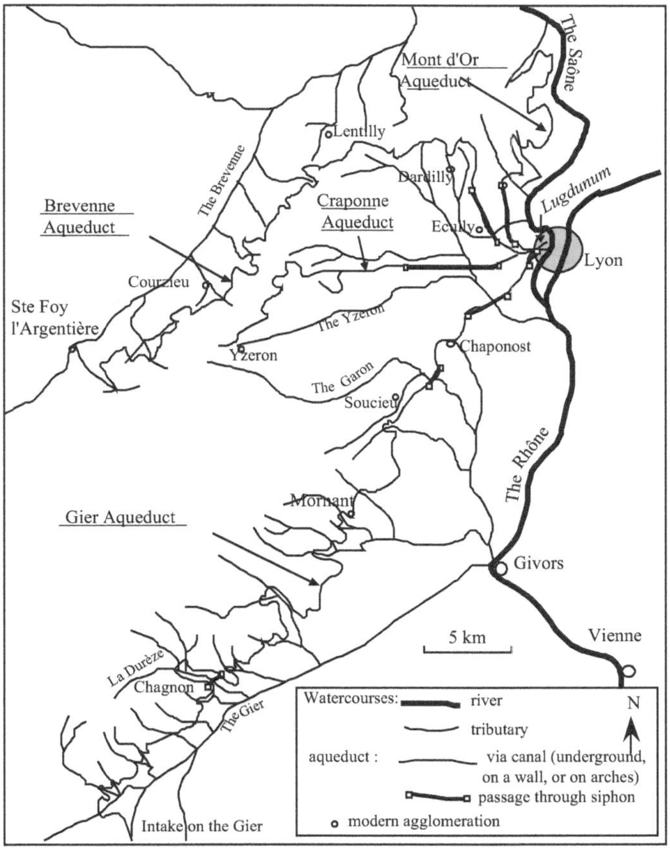

Figure 6.11 The four Roman aqueducts of Lyons.

Figure 6.12 Longitudinal profiles of the four Roman aqueducts of Lyon (Burdy, 1979, 1986); the dark squares represent known points.

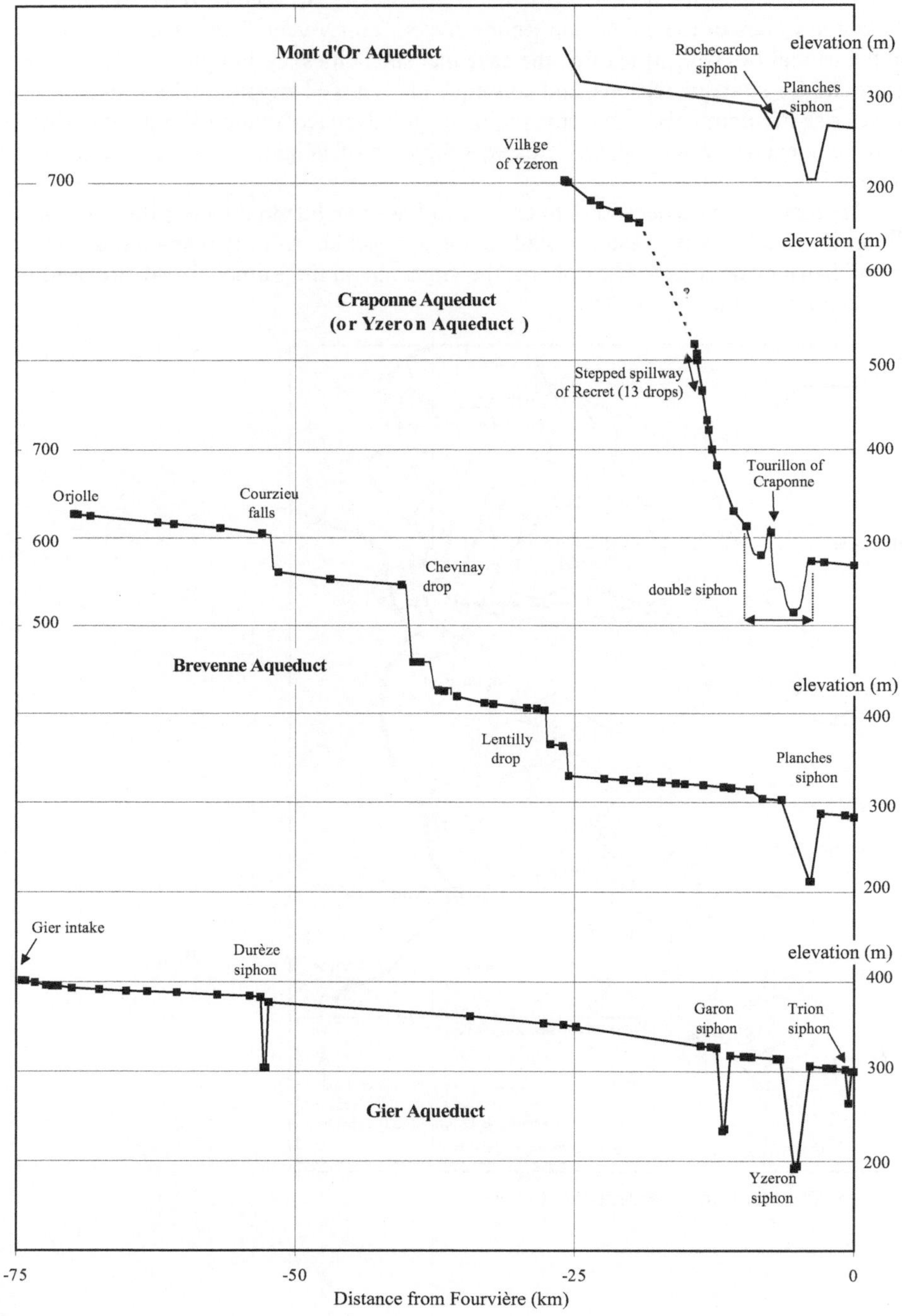

Four large aqueducts supply water to the Roman *Lugdunum*. They have been the subject of numerous archaeological investigations[30]. Their remains are still visible, but unfortunately are degrading over time; however some additional remains are surfacing during modern construction projects. The construction chronology of these aqueducts is not known with certainty, indeed it is based on hypotheses. One of these hypotheses is that the Mont-d'Or aqueduct, issuing from the mountain of the same name, was the first to be constructed, about 10 BC by Agrippa, the son-in-law and close collaborator of Augustus. (Agrippa was a great aqueduct builder - Rome owes the *Julia* and the *Virgo* to him). The Mont-d'Or aqueduct is 28 km long, with two siphons. It is the smallest of the aqueducts in terms of dimensions and discharge, and only a few traces of it remain today.

It is very likely that the Craponne aqueduct (sometimes called the aqueduct of the Yzeron) was built next from catchments developed upstream of the village of Yzeron at 700 m altitude, perhaps under Augustus.[31] This canal is noteworthy for its slope of nearly 17 m per kilometer on the average, i.e. five times steeper than that of the Mont-d'Or aqueduct. Vortex drop shafts constitute what amounts to "hydraulic stairways". Though the Craponne is of similar length and dimensions to the Mont-d'Or, it is distinc-

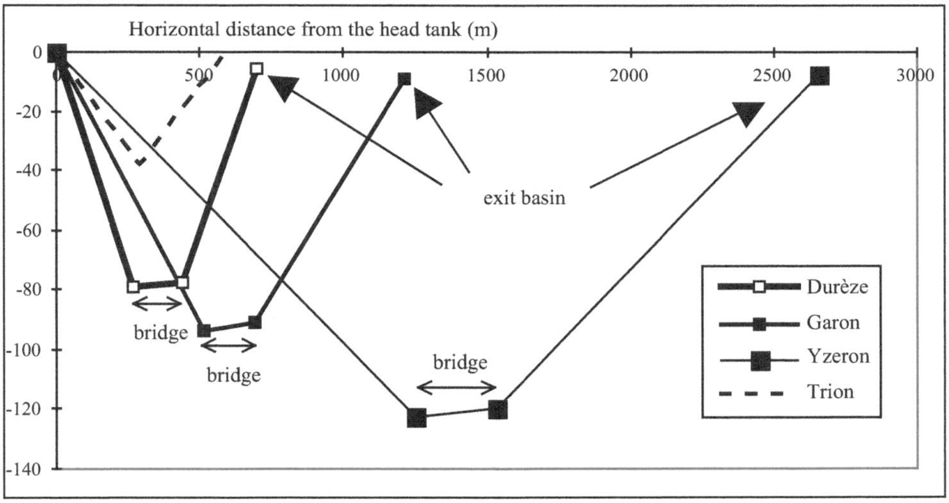

Figure 6.12b. Comparison of profiles along the four siphons of the Gier aqueduct. The Durèze siphon has the most pronounced slope, which can be explained by its later construction (hypothetical) by the Chagnon loop. Note the slightly rising profile (1% ca.) of the siphon bridges, which allows evacuation of air pockets downstream -discovered by the work of Burdy (1996).

30 One reference study, unfortunately disappeared, is that of Germain de Montauzan (1909). Here we have based our discussions primarily on the study of Louis Jeancolas (1977, published in 1983), on the synthesis of Jean Burdy (1979), and on the monograph of Jean Burdy on the Gier aqueduct.

31 This is the "traditional" dating estimated by Germain de Montauzan; Jeancolas, based on the remains of a tomb that is said to have preceded the aqueduct, estimated that the aqueduct could date from the second part of the 2nd century AD. In this case, the aqueduct would be the most recent of the four.

tive in having a double siphon.

To understand this double siphon, it is useful to reconsider, yet again, the great siphon of the Pergamon aqueduct, whose longitudinal profile shows a high point (Figure 5.10) with a risk of air accumulation. The Craponne includes a free-surface reservoir that stands 15 m above the ground on an intermediate plateau (the Tourillon de Craponne). This reservoir enables the aqueduct to cross two valleys over a total length of 5.5 km, avoiding the problems of a high point. It serves as both an exit basin, or outlet box, for the upstream portion of the siphon and a head tank for the longer downstream portion, which dips through nearly 70 m of elevation.

The Brevenne aqueduct is a very large installation 66 km in length, whose construction could date from the Emperor Claudius in the middle of the 1st century AD. It issues from the Lyon mountains and is buried for the first forty kilometers. Then it follows the valley of the Brevenne and, after some twelve kilometers, has an increased cross section to accommodate intermediate catchments. This aqueduct also has particular distinctive features. It crosses significant elevation changes in short distances at four locations. At Courzieu, it drops nearly 44 m in less than 200 m; then at Chevinay, there is a drop of 87 m in 300 m, and again three other drops of from 30 m to 40 m each. Downstream of the drop of Courzieu, there is a small basin some 45 cm deep and 80 cm long, likely intended to provide energy dissipation as well as to trap transported sand and gravel. Moreover, there is a contraction of the canal at the location of the last drop, at Lentilly; the normal width of 75 cm decreases to 44 cm. It can be shown that in the reaches of steep slope, the flow is supercritical.[32] As for the other aqueducts, the Brevenne terminates at a siphon. It probably merges with the Craponne aqueduct at its entry into *Lugdunum.*[33]

The fourth aqueduct, Gier, issues from an intake on the river of the same name and is the longest at 74.5 km. The Gier is the aqueduct that was most carefully constructed, and its art features, bridges, and arcades reflect the majesty of the all-powerful Empire at its peak. Ample remains of the Gier are still visible today, in particular its bridges and parts that were supported on arcades. Traditionally the Gier is attributed to the time of Hadrian (beginning of the 2nd century AD), but recent indices (in particular a fountain bearing the name of Claudius) lead us to date the works from the reign of this emperor. The Gier also has two curious features. In parallel with the first siphon, that crosses the Durèze over a length of 900 m, there is a large derivation into a canal of small slope (only 0.5 m/km). This derivation canal goes around the village of Chagnon (the "loop of Chagnon") for a distance of 11.5 km (making the total length of the aqueduct some 86 km). The purpose of this derivation is not obvious. Along the path of this loop, there is an inscription on a stone noting the rules for riparian use of the aqueducts:

> "By order of the Emperor Cesar Trajan Hadrian Augustus, no one has the right to work, harvest, or plant in this space that is intended to provide protection for the aqueduct."[34]

---

32 the water velocity is greater than that of the wavespeed. For more detail, see Chanson (2000).

33 This is the hypothesis of Louis Jeancolas, from his observation of a piling of the Craponne aqueduct that is particularly reinforced, suggesting that it could have supported the structure at the junction of two aqueducts.

34 "Chagnon Stone", discovered in 1887, visible under the playground of the Chagnon school. A second inscription, apparently identical, was discovered in 1996 along the main path of the aqueduct, but further downstream (Burdy, 1996).

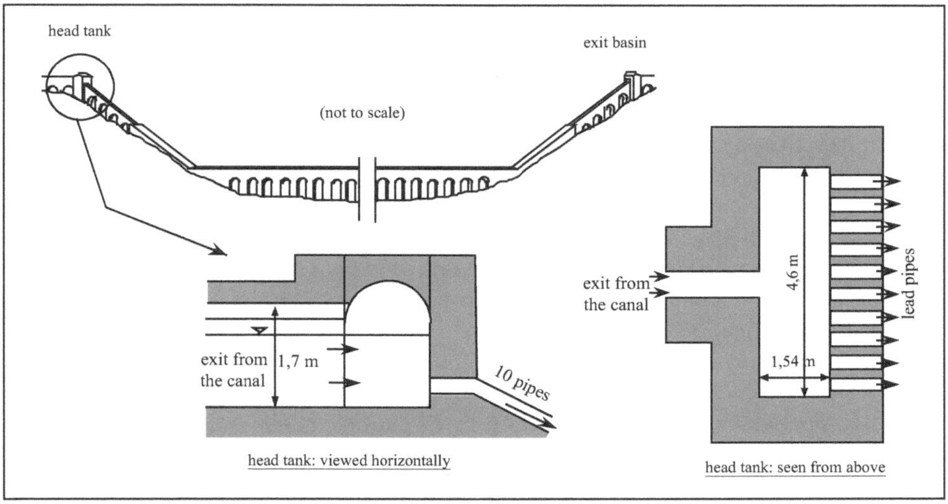

Figure 6.13 The head tank of the Garon siphon at Soucieu (Gier aqueduct); longitudinal and horizontal sections at the level of the openings, and seen from downstream, with the four openings that remain intact (photo by the author). This reservoir is raised 4 m above ground level. Its walls are 80 cm thick.

There has been much discussion as to the relative age of the loop and the siphon. If indeed the aqueduct dates from the reign of Claudius, the inscription shows that the derivation canal, in service under Hadrian, postdates the siphon. This may be because the siphon had maintenance or operational problems caused by the steep slopes in the valley of the Durèze (see Figure 6.12b).

Upstream of the first siphon there is also evidence of an abandoned earlier alignment along 40 km, dug deeply into the rock parallel to the aqueduct, but 8 to 10 m higher. Could this reflect a surveying error?

Further downstream, the Gier aqueduct passes under the village of Mornant in a one-kilometer curved tunnel more than 20 m underground. It has four siphons in all (table 6.4). The third of these siphons crosses the Yzeron valley. 2,660 m long and dropping 122 m, it is the largest siphon of all those in the Lyon complex.

There are eight siphons in all in the four aqueducts of Lyon. Following the specifications in the manual of Vitruvius, each siphon has an exit basin, a head tank, and at the bottom of the valley, a bridge-siphon. The Tourillon de Craponne comprises what Vitruvius might call a *colliviaria*. Whereas at Pergamon the Hellenistic siphon has a single pipe, here we see the use of veritable batteries of parallel lead pipes. There are no less than nine pipes of average exterior diameter 23 cm (likely 18 cm interior diameter) side by side for the Durèze siphon of the Gier aqueduct, and ten for the Garon siphon (Figure 6.13). There are no remains to indicate the number of parallel conduits for the Yzeron siphon; but taking into account the large size of its reservoir, there were likely 11 or 12 similar pipes (Figure 6.40). The bridge-siphons were particularly wide at around seven meters, to support numerous parallel pipes.

Table 6.4 Approximate characteristics of the eight siphons of the Lyon aqueducts (the siphons of the Gier aqueduct are the best known; see Burdy, 1996).

Aqueduct	**Mont-d'Or**		**Craponne (double siphon)**		**Brevenne**	**Gier**			
Valley crossed	Rochecardon	Planches	Corvelet	Charbonnières	Planches	Durèze	Garon	Yzeron	Trion
Length	500 m?	3,500 m	2,200 m	3,600 m	3,500 m	700 m	1,210 m	2,660 m	575 m
Drop	25 m?	66 m?	33 m?	91 m	91 m	78.5 m	93 m	122.3 m	38 m?
Head loss[35]	4 m?	11 m	7 m?[36]	33 m	15 m	5.8 m	8.8 m	7.9 m	2.3 m
Number of pipes?	?	?	6 or 7		?	8 or 9	10	11	9

The spring water conveyed by the siphons of Lyon is generally not calcareous, and therefore there is not much encrustation. An exception is the Mont-d'Or aqueduct, which does show evidence of some deposits. The aqueducts arrive into the city of *Lugdunum* at different elevations. It is thought that the Mont-d'Or aqueduct supplied the thermal baths whose remains have been found in the Minimes area. No traces remain of the water distribution system in the city itself, but it is likely that the conduits passing

35 Difference in water level between the head tank and exit basin.

36 These data have an unexplained anomaly. An elementary hydraulic calculation shows that the ratio of the head losses must be equal to the ratio of the lengths of the two parts of the double siphon, if the pipes are identical and of the same number.

under the Saône through siphons supplied the peninsula between the two rivers.

The Nîmes Aqueduct and the Pont du Gard[37]

The Roman *Nemausus* (Nîmes) is much older than Lyon, founded in the 6th century BC, even before the conquest of Gaul by the Romans. Nîmes was the capital of the Arecomic Volques, a Gallic people often allied with the Romans. Nîmes is naturally well supplied with water, from wells and especially from the Fontaine spring, abundant and perennial. Under Augustus the spring was the subject of major construction, including a masonry canal and basin at its outlet. The need for an aqueduct only came later under the pressure of urban development (20,000 inhabitants in the Gallo-Roman *Nemausus*), with the objective of supplying water to the highest areas of the city above the level of the Fontaine (at an elevation of 51.1 m). Under Claudius, in the middle of the 1st century AD,[38] the Eure fountain spring is tapped near Uzès at an altitude of 72 m.[39]

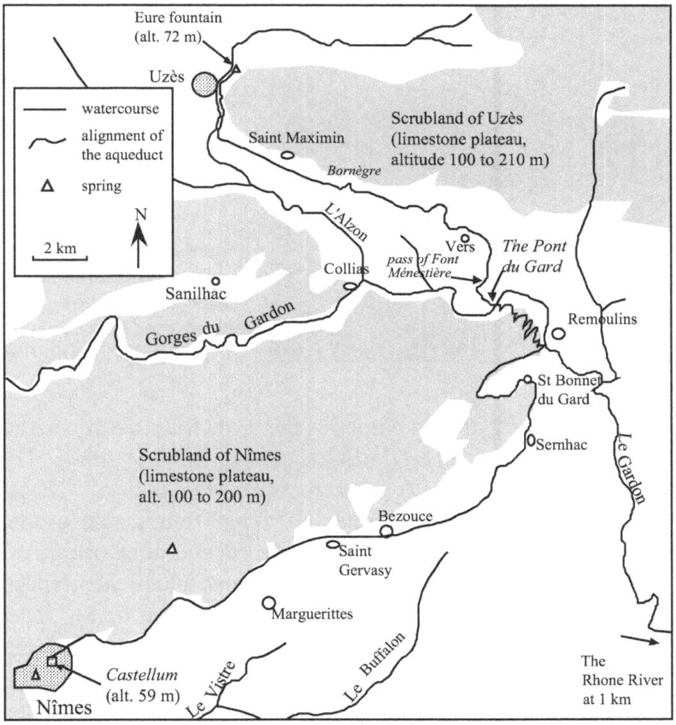

Figure 6.14 The Roman aqueduct of Nîmes (after Fabre, Fiches, Leveau, Paillet, 1992).

37 Our principal sources here are the article of Victor Lassalle (1979), conservator of the museum of Nîmes, and the work of Guilhem Fabre, Jean-Luc Fiches, Philippe Leveau, and Jean-Louis Paillet (1992) who coordinated, from 1984 to 1990, a program of research on the aqueduct of Nîmes and the Pont du Gard.

38 This new date is that established by Fabre, Fiches, Leveau and Paillet (1992). Earlier, the aqueduct had been attributed to Agrippa, the son in law of Augustus, also presumed father of the first aqueduct of Lyon (around 19 BC). The Brevenne aqueduct is also attributed to the era of Claudius.

Figure 6.15 The Pont du Gard, supporting the aqueduct of Nîmes across the Gardon (photo by the author).

Although it is only some 20 km from the source to Nîmes as the crow flies, the aqueduct had to wind around the vast scrubland plateau of Nîmes, whose elevation is above 100 m (see Figure 6.14). There are numerous obstacles to be crossed on this plateau. These obstacles include ravines that are dry in summer but subject to violent floods, such as the Bornègre ravine with its three-arch bridge; passes, with the two-level arched bridge of Font-Ménestière; and notably the sunken valley of the Gardon, across which the highest bridge-aqueduct of the Roman Empire is built, the Pont du Gard (Figure 6.15). The bridge is about 360 m long, and carries the canal of the aqueduct at an altitude of 65 m, some 48.4 m above the bed of the river. Just upstream of the bridge the canal has a basin provided with a gate and a discharge canal enabling diversion of the discharge of the aqueduct (or its excess) into the Gardon, if necessary. Downstream of the bridge, the canal traverses the rough terrain along the edge of the limestone plateau by means of multiple switchbacks, crossing ravines on small bridges. The canal passes through three tunnels of about 400 m in length, further downstream near Sernhac, then again near

39 The ancient writings refer to an elevation of 76 m for the tapping of the Eure fountain. We have adopted here the more recent estimate of Fabre et al. (1992), namely 72 m.

Nîmes. The terminal point of the aqueduct is the water tower in the city (*castellum*), at an elevation of 58.95 m.

All of these bridges are designed to handle the strong floods typical of the Mediterranean climatic regime. The bases of the bridge piers are protected by shaped prows on the upstream face. The bridges all leave a very large opening for flood passage; the widths of the openings of the arches of the Pont du Gard are 24.5 m and 19 m.

The *castellum* of Nîmes (Figure 6.17) is one of the rare Roman water towers still conserved in more or less its original state. The aqueduct dumps water into the tower's

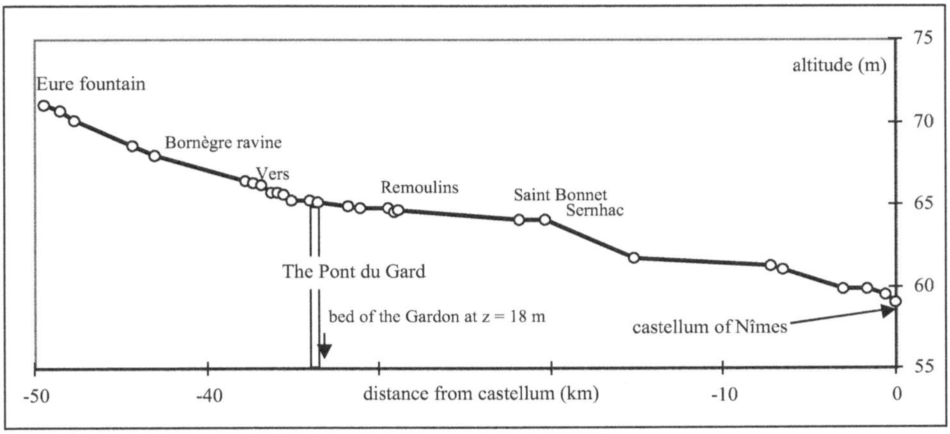

Figure 6.16 Longitudinal profile of the Nîmes aqueduct, from the data of Fiches (1991).

Figure 6.17 The castellum of Nîmes: this distribution basin is the terminal point of the Nîmes aqueduct. It is visible in the city, on rue de la Lampèze (photo by the author).

circular basin of 5.5 m inside diameter and depth 1.4 m. Issuing from the basin are ten circular distribution pipelines of 0.4 m diameter. Valves enable the isolation of one circuit or another, and drains in the bottom make it possible to empty the basin.

The aqueduct has several slope changes;[40] upstream of the Pont du Gard it is only 38 cm/km on the average, which is relatively small compared to other aqueducts. But downstream of the bridge, the slope is only 8 cm/km along a particularly sinuous segment of more than 10 kilometers length. The theoretical maximum discharge for this slope can be estimated at around 40,000 m^3/day. Even though this aqueduct may not attract the same admiration as the Pont du Gard, it is all the more impressive for its incredibly small slope, and the precision of surveying and construction that this implies.

These slope changes have hydraulic consequences that were not well understood or mastered by the Roman engineers. In a canal of constant width, the depth of water is larger when the slope is small; yet the Nîmes aqueduct was initially constructed for an essentially constant depth. Early on, it became necessary to raise the canal walls in several locations (of flatter slope) to avoid overflow.

Another problem is the fact that the water from the Eure fountain is very calcareous. Over the years, the useful cross-section of the aqueduct's canal becomes considerably reduced by deposits, effectively reducing the discharge to a value that was probably only about 20,000 m^3/day. The canal had to be scoured out on several occasions.

Other aqueducts in the Roman Empire

The panorama of Gallo-Roman aqueducts that we have just described represents the diversity of situations and solutions adopted throughout the Roman Empire. Table 6.1 gives an incomplete list of the numerous Roman aqueducts that have been discovered and studied. It would be impossible to describe all of them. To the best of our knowledge, the longest one is at Apamea-on-Orontes in Syria, built in 116 or 117 AD as part of the broad reconstruction after the earthquake of 115 AD.[41] But another aqueduct also merits our attention. It is the aqueduct of Carthage, one of the marvels of Roman architecture in Africa, and among the longest of all the Roman aqueducts in the western world (Figure 6.18).[42]

Carthage is destroyed by the Romans at the end of the Punic wars, then rebuilt in 29 BC by Augustus for the purpose of granting property to legionnaires at the end of their service, and then later becomes a capital of the province of Africa. With completion of the great thermal baths of Antonio, it becomes necessary in 162 AD to build an aqueduct to supply numerous public fountains, and also to build cisterns to store rainwater. But the usable water sources are some distance away, in Djebel Zaghouan. The closest source is 56 km away as the crow flies, with obstacles in the way including a lagoon to the east of Tunis, and a saline lake somewhat further. At the wide valley of the wadi Milliane there is a crossing of 4.5 km on arches that are some 20 m high in places, and

---

40 It is probably to reduce the height of the Pont du Gard that the aqueduct's slope is steeper upstream than downstream of it.

41 It is about 150 km long; see Figure 6.32 below (Balty, 1987).

42 Rakob (1979); Clamagirand, Rais, Chahed, Guefrej, Smaoui (1990).

a two-level bridge 126 m long and 34 m high.[43] In this plain of Carthage there are 17 km of crossings on high arches in all. Later on, very likely under the Severus emperors, the aqueduct is lengthened to 132 km, to capture the even more distant springs of Aïn Djoukar.

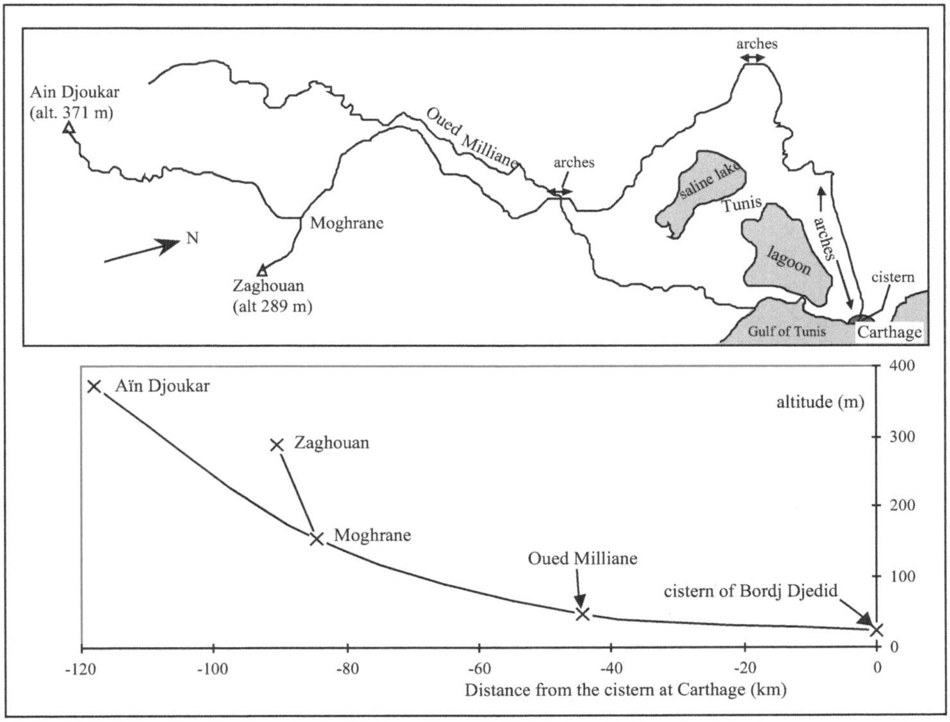

Figure 6.18 The aqueduct of Carthage: map of the alignment and schematic longitudinal profile (after Rakob, 1979). Only the points indicated (x) are exact, the remainder of the profile is deduced from simple interpolation. The maximum capacity is estimated at 25,000 m^3/day.

Like the Pont du Gard in Europe, the aqueduct of Carthage and its appurtenances attract the admiration of African observers:

> "This distance, from the source *(Aïn Djoukar)* to the cisterns *(the 24 cisterns of Carthage)* was covered thanks to an infinite number of aqueducts, in which water flowed at a constant level. These aqueducts were composed of stone arches. When these were on the ground, they were low and when they were in the depressions or valleys, they were extremely high. This was one of the most remarkable things to be seen on the surface of the earth." [44]

The upstream portion of the aqueduct is still used today by the Tunisian water service.

43 The bridge has disappeared, but the alignment of the arches of wadi Milliane is still well preserved
44 Al-Idrissi (12th century), III, 2 Translation of Jaubert.

Paddle wheels and water mills in the Roman world, the beginnings of industrial use of water

In the provincial countryside some ten kilometers from Arles in the direction of Saint-Rémy-de-Provence, a hiker can come upon the remains of Roman aqueducts on arches. These remains are even indicated by a sign. Looking at the remains closely, the tourist can easily see that there are in fact two parallel aqueducts, side by side. If the hiker follows the path alongside these aqueducts, he or she comes upon a deep notch cut in a rocky outcrop. The canal of the left aqueduct passes through this notch (Figure 6.19).

After passing through this notch, our hiker then comes out at the top of an escarpment, 20 meters high along a length of 60 meters, beyond and below which is a broad plain with no visible trace of the aqueduct. On the slope one can recognize the ruins of walls and structures in the form of stairways (Figure 6.22). These ruins were identified in 1935 by Fernand Benoît as those of a Roman hydraulic flour mill, the mill of Barbegal, the very first of this kind known to us.

Figure 6.19 Barbegal: the remains of the canal of the left-hand aqueduct, and the notch in the rocky outcrop that marks its end (the right-hand aqueduct, though it cannot be seen in this photo, turns sharply before the rocky outcrop and continues its route toward Arles) (photo by the author).

In Chapter 5 we mentioned the appearance of the water mill in Asia. This was only weakly suggested by an allusion of Strabo to the mill in the palace of Mithridate, king of Pontus, conquered by the Romans in the middle of the 1st century BC. The appearance of this device marked a major step in the history of technology, but passed almost unnoticed. With the Romans, the first written evidence of a mill is found in book X of Vitruvius' work. Let us examine how Vitruvius describes the new technology to get a clearer idea of how it appeared.

Water lifts, paddlewheels, and water mills in the world of Vitruvius

Vitruvius provides the oldest known description of a water lift powered by hydraulic force, or *noria*, and of a water mill. This description comes immediately after that of manual water lifts (drum wheel, bucket wheel: see Figure 6.20):

> "Wheels on rivers are constructed upon the same principles as those just described *(manual lift wheels)*. Round their circumference are fixed paddles *(pinnae)*, which, when acted upon by the force of the current, drive the wheel round, receive the water in the buckets, and carry it to the top with the aid of treading; thus by the mere impulse of the stream supplying what is required.
>
> "Water mills *(hydraletae)* are turned on the same principle, and are in all respects similar, except that at one end of the axis they are provided with a drum-wheel, toothed and framed fast to the said axis; this being placed vertically on the edge turns round with the wheel. Corresponding with the drum-wheel a larger horizontal toothed wheel is placed, working on an axis whose upper head is in the form of a dovetail, and is inserted into the mill-stone. Thus the teeth of the drum-wheel which is made fast to the axis acting on the teeth of the horizontal wheel, produce the revolution of the mill-stones, and in the engine a suspended hopper supplying them with grain, in the same revolution the flour is produced."[45]

Figure 6.21 is an attempt to reconstitute the devices described by Vitruvius. We

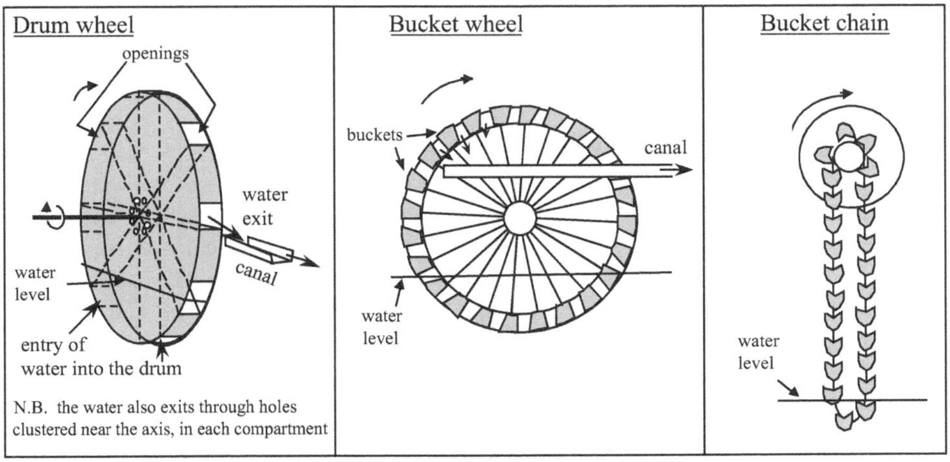

Figure 6.20 The manual water lifts described by Vitruvius, the possible origins of the noria and water mill. The lift height increases from left to right.

---

45 Vitruvius, On Architecture, X, 5, 1 and 2, translated by Joseph Gwilt, London: Priestley and Weale, 1826. http://penelope.uchicago.edu/Thayer/E/Roman/Texts/Vitruvius/8*.html

recall that he was a contemporary of Augustus (about 25 BC), and therefore came after the reign of Mithridate, though by only a few decades. He doesn't enlighten us on the details of the innovation much better than did the Greek sources that we cited in Chapter 5. These details and their origins therefore remain somewhat obscure. However, two clues enable us to form at least a rough idea of how the devices operated.

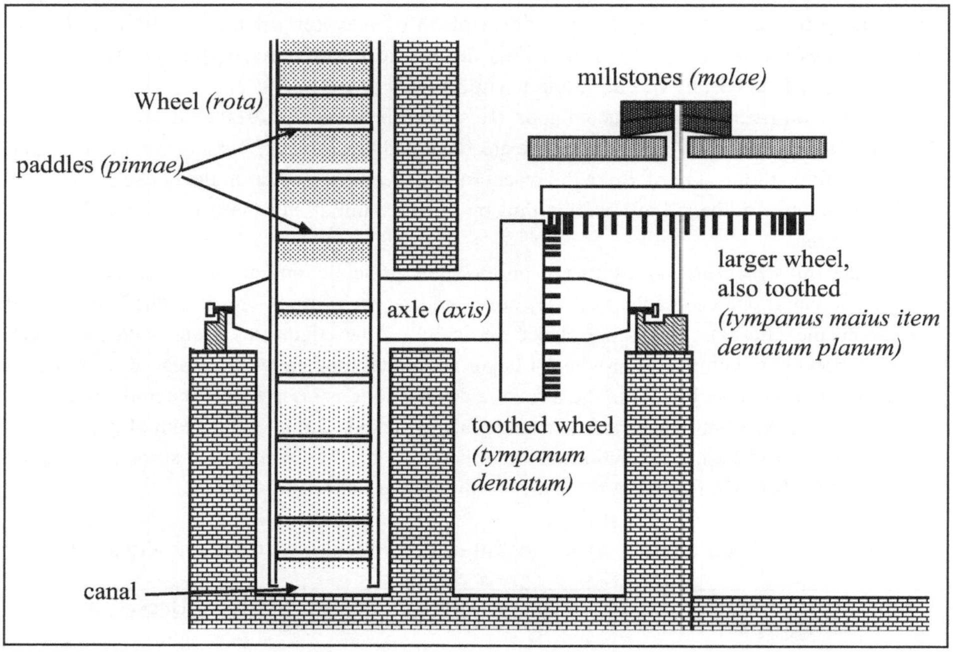

Figure 6.21 Reconstitution of the water mill described by Vitruvius.

The first clue is Vitruvius' outline, the order in which he describes the different machines. He puts his description of the water mill in between those of the water lifts – as in Figure 6.20. He first describes the drumwheel, then the bucket wheel, and finally the bucket chain. All three machines are powered by human muscular force, and raise water to increasing heights in the order they are listed. It is after these descriptions that Vitruvius describes the paddle wheel powered by the force of the current and equipped with buckets to lift the water, and then finally the mill that turns a grinding stone using hydraulic force.

Vitruvius' work then comes back to lifting machines, presenting two inventions of the 3rd century BC that we have already seen: the Archimedes screw or *limaçon*, and the pump of Ctesibios.[46] Certain authors[47] believe it is possible that the order in which these machines are described, an order whose logic is not at all obvious, must correspond

46 Although Vitruvius does not cite Archimedes in this context, he does very explicitly refer to Ctesibios, as shown by the extract that we cited in Chapter 5.
47 See the notes of Louis Callebat in his edition of Vitruvius' book X.

to a technical genealogy of these wheels. The manual bucket wheel could have been perfected through the addition of paddles, so that it could be powered by the force of the current. In this case the paddle wheel could have been used to lift water (this is the *noria*, an invention destined to have a glorious future in the Orient), before it was realized that the rotation of the wheel could, through appropriate gearing, be used to turn a millstone. Other authors[48] propose that the vertical-axis mill (having a horizontal millstone) was invented first, since it does not require reduction gearing to turn the stone. Moreover, this mill appears in China at about the same time, as we see further on in Chapter 8.

A second clue as to the first use of the paddle wheel can be found a bit further on in the same book X of Vitruvius. There, he describes procedures "passed down to us from our ancestors" (which ones?) to measure travel distances over land or water; these would be "hodometers". After describing the adaptation of the idea to chariots, Vitruvius then describes an analogous procedure for measuring the distance traversed through the water by boats. It is the paddle wheel that comprises the essential part of this instrument:

> "In navigation, with very little change in the machinery *(i.e. the hodometer for wheeled chariots)*, the same thing may be done. An axis is fixed across the vessel, whose ends project beyond the sides, to which are attached wheels four feet in diameter, with paddles *(pinnas)* to them touching the water. [....]
>
> "Thus, when the vessel is on its way, whether impelled by oars or by the wind, the paddles of the wheels, driving back the water which comes against them with violence, cause the wheels to revolve, whereby the axle is also turned round, and consequently with it the drum-wheel, whose tooth, in every revolution, acts on the tooth in the second wheel, and produces moderate revolutions thereof."[49]

We know that on modern ships a common instrument for measuring the speed is a small paddle wheel. However the procedure described by Vitruvius appears curious to say the least, and it seems doubtful that its use was widespread.

Development of the water mill in the Roman Empire

About fifty years after Vitruvius, under the reign of Vespasian and Titus, Pliny the Elder[50] wrote in his book, *The Natural History*:

> "All the grains are not easily broken. [....] Throughout the greater part of Italy, however, they employ a pestle that is only rough at the end, and wheels turned by water, by means of which the corn is gradually ground."[51]

---

48 See, for example, Jacques Bonnin (1984). See Viollet (2005) for more detail.

49 Vitruvius, On Architecture, X, 9, 5 and 7, translated by Joseph Gwilt, London: Priestley and Weale, 1826. http://penelope.uchicago.edu/Thayer/E/Roman/Texts/Vitruvius/8*.html

50 Pliny was born in the summer of 23 or 24 AD, and was, around 70 AD, in the Orient with Titus, to whom he dedicated his book, The Natural History, in 77 AD. He died in 79 AD during the eruption of Vesuvius, attempting to awaken the inhabitants of Pompei.

51 Pliny the Elder, The Natural History, Book XVIII, 23, Translation ed. John Bostock, M.D., F.R.S., H.T. Riley, Esq., B.A. http://www.perseus.tufts.edu/cgi-bin/ptext?lookup=Plin.+Nat.+toc

The Roman development of the water mill has been consistently underestimated; but new archaeological findings are becoming more and more numerous.[52] One can find remains of simple mills on small rivers near Hadrian's Wall at the Scottish border.[53] In eastern Tunisia, at Chemtou (the Roman *Simitthus*) the remains of an installation comprising three horizontal wheels set side by side have been found near a dam constructed by Trajan on the river Medjerda.[54] And in North Africa, there are still other horizontal-wheel mills. Vertical-wheel mills are found in Gaul from the 1st or 2nd centuries; in the villas of Var in the 2nd century; on the Janiculum Hill at Rome from the 3rd century; and at the agora of Athens in the 5th century. There is a depiction of a *noria* on a mosaic dating from 469 AD at Apamea-on-Orontes; this date comes after the fall of the Roman Empire in the west, but the representation suggests that the use of the *noria* had largely spread to the east during the Roman period. And then, there is the flour mill of Barbegal.

Let's now describe this installation,[55] the one we used to begin our discussion of

Figure 6.22 Roman flour mill at Barbegal: view from the summit of the rock outcrop where the aqueduct arrives; to the left and the right, in the form of a V, one can see the two walls that carried the two divergent lateral canals. These canals delivered water to the two "mill" canals that, in turn, cascaded down the slope and supplied the two banks of water wheels (photo by the author).

52 See the census of remains established by J.-P. Brun, in Brun and Borreani (1989), p. 308-309, or the census of Wilson (2002), Wikander (2000), or Viollet (2005).

53 Scare (1995), p. 128; Wikander (2000).

54 Hodge (1995).

55 The first description comes from the discoverer of the site, Fernand Benoît, in 1935. We have used here the results of the more recent study of Sellin (1979) for the flour mill itself, and of Leveau (1995) for its water supply.

Figure 6.23 The remains of the Roman hydraulic flour mill, comprising two series of eight paddle wheels. View from below, remains of the right-hand canal. The arrows indicate the flow path (photo by the author).

Roman mills. We have seen that the aqueduct that is dedicated to the supply of the flour mill lies parallel, in its final stretch, to the aqueduct that supplies the city of Arles. Both aqueducts come out of a junction basin which is supplied, in its turn, by two canals delivering water from different remote sources.

At the mill, another distribution basin[56] conveys water into two parallel headrace canals that descend from the hill with a discharge that modern studies have estimated at about 0.15 m^3/sec. Along each of these canals, eight vertical wheels, each about 2 m in diameter, are aligned from the top to the bottom of the slope. Each wheel has a sill or weir immediately downstream of it, controlling flow into a small drop that feeds water onto the wheel below it. Adjacent to each wheel, near its center, is a chamber enclosing the reduction gears, with grinding wheels likely set on a platform above each of these chambers.

One of the astonishingly modern aspects of this installation is its engineered, non-natural water supply, using one of the two branches that earlier had come together to supply Arles. The mill was built at a very convenient location, benefiting from a steep slope and yet providing ready access from the plain below, obviating the need for the installation to accommodate the vagaries of a natural river. The discharge is regular, with no risk of erosion. This installation probably dates from the beginning of the 2nd century AD.[57]

56 The first studies of the site supposed that this distribution system was a triangular reservoir. Sellin showed that it was more likely two canals, laid out in the form of a "V", supported on walls that are still visible today, conveying water into the headraces. We have adopted this hypothesis in Figure 6.24.
57 Leveau (1995).

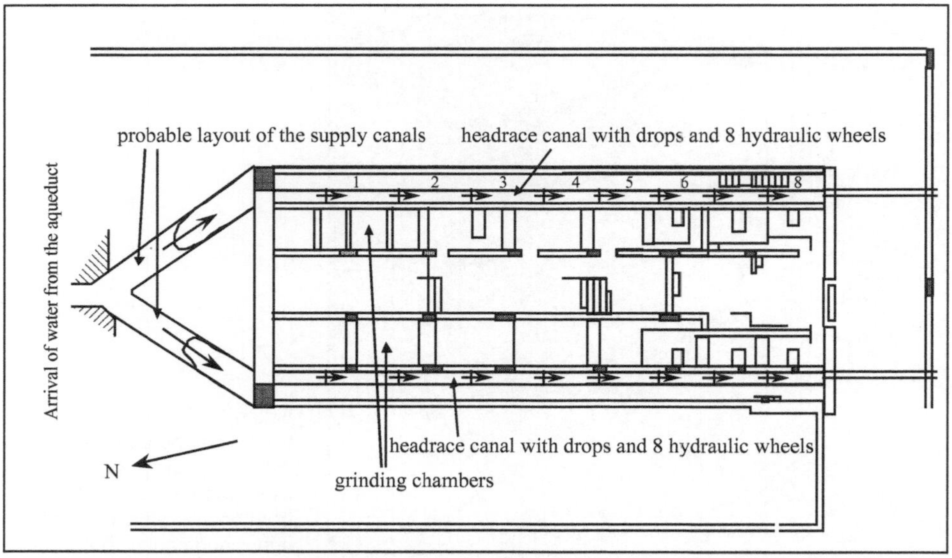

Figure 6.24 Reconstitution of the plan of the flour mill of Barbegal (after Sellin, 1979). The light lines indicate assumed structures.

Economic development of the provinces: dams for irrigation and industry

The abundance of water and water infrastructure in the city of Rome not only provides for essential needs, but also serves as a cultural background for the urban life and pleasures of Rome's citizens. Water plays an equally important role in the economic development of the Roman provinces. We have already described the development of the water wheel as it appeared in all of the Empire. But water management and exploitation went much further than this. Water had to be removed from the deepest galleries of mines, and massive quantities of water were needed to obtain lead, silver, and gold from their ores; the remnants of such ore-washing installations have been found in many of the provinces. In addition, the textile industry relied upon a steady supply of water. There is also of course agriculture, for which irrigation is required to support plentiful yields in southern and eastern regions. It is in these provinces, where water is less abundant and often scarce, that there is the most plentiful archaeological evidence of water acquisition and use. The most spectacular evidence is that of the Roman dams constructed in Spain, North Africa, and the Orient. Before describing these projects, as well as projects other than dams, let us first introduce Roman dam-construction technology.

Romans and dam technology

In about 60 AD the Emperor Nero built his villa at Subiaco, on the river Anio upstream of Tivoli (Figure 6.8). He formed lakes for his personal pleasure by damming the river. The largest of the structures he built for this purpose is across a natural gorge and at 40 m, is the highest dam of all the Roman Empire. Rectilinear, 80 m long at its crest and 13.5 m thick, it is what one would call today a gravity dam, relying on friction at its base to resist the force of water in the reservoir behind it. This dam remained standing until 1304, when the monks of a neighboring monastery dismantled it to recover the stones for other use.[58] Forty years after Nero, Frontinus built the new intake for the *Anio Novus* aqueduct on one of these lakes to improve the quality of its water, as we have seen earlier.

Nero's dam at Subiaco is not only the highest, but also one of the first dams constructed by the Romans, and the only one in Italy. Since dams were not essential to the development of Italy, dam technology is not a traditional Roman discipline (Vitruvius says nothing of dams in his treatise *On Architecture).* We have seen in the first part of this book that numerous dams were built in the Orient from the IVth millennium BC. It is undoubtedly through their domination of the Orient (Syria-Palestine, Egypt) and through their military expeditions (Yemen) that the Romans acquired this technology and then further developed it, especially from the 1st century AD.

Like those of the Orient, Roman dams are almost always of the gravity type, the one at Subiaco being a good example. These dams most often are built of simple rectilinear walls of masonry or concrete, supported by earth fill or buttresses, as we will see further on. However the Romans also invented the arch dam, a structure whose shape enables it to transmit the pressure force of impounded water to the lateral valley walls, just as the arch of a bridge transfers the load to its supporting piers.

In the valley of the Baume, several kilometers south of Saint-Rémy-de-Provence in France, are the remains of a structure that is now buried under a dam built in 1891. A provincial scholar named Esprit Calvet fortunately discovered these remains in 1765 before they were buried. The principal vestiges of the dam are the keyways in the valley walls where the dam abutments were anchored. The shape and alignment of the keyways, which extend somewhat above the level of the modern dam, reflect and reveal the curvature of the structure (Figure 6.26).

A recent study led to reconstitution of the dimensions and function of this Roman dam.[59] It is likely during the time of Augustus that the Romans built the dam in the narrow gorge of the Peyrou to supply an aqueduct leading to the nearby Roman city of Glanum. This structure, the first known arch dam, is nearly 15 m high, 23.8 m wide at its crest, and has a radius of curvature of about 30 m. The dam is keyed into the nearly vertical rock walls of the valley at its two ends. It apparently is built of either two faces of quarried rock blocks with an impermeable fill, or perhaps a solid mass of mortared blocks with a watertight joint.

---

58 Schnitter (1994), p. 59.
59 Augusta-Boularot and Paillet (1997).

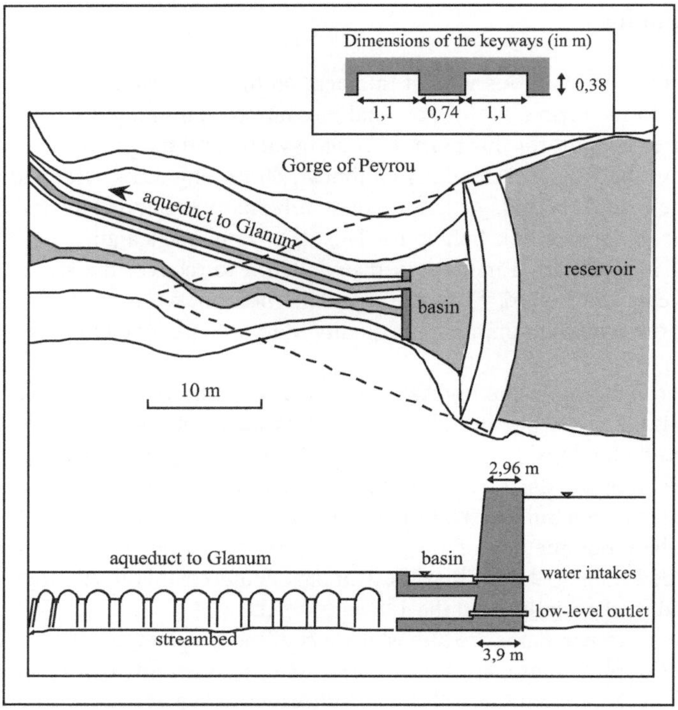

Figure 6.25 The dam in the Baume valley, near Sant-Rémy-de-Provence: the oldest known arch dam (from the reconstitution of Agusta-Boularot and Paillet, 1997). The water transported by the aqueduct is destined to supply a massive fountain at Glanum, perhaps a watering station for migrating herds. The aqueduct is 500 m long, with a slope of 3.2 m/km. The canal is 0.59 m wide, and 0.89 m deep.

Figure 6.26 In the Baume valley, site of the Gorge of Peyrou, seen from the reservoir. The arrow shows the location of the keyway of the Roman dam, at the right extremity of the present-day dam (photo by the author).

To the south of Evora in Portugal there is another small arch dam: Monte Novo. It is less than 6 m high, but is probably of Roman heritage.[60] The only other known example of an arch dam revealing Roman techniques is the one built at Dara (Anatolia) in the Byzantine period (Chapter 7).

The dams built for Nero on the Anio were expressly for the personal pleasure of an emperor. On the other hand, the dams that we describe now respond to economic needs; the provinces had to produce enough food to supply the Empire. In Spain, in North Africa, and as always in the Orient, irrigation was essential to the development of agriculture.

Dams of the Iberian peninsula

Spain is one of the oldest Roman provinces – and it is the native province of Trajan and Hadrian. At the end of the Punic wars in 202 BC, Spain is taken by the Carthaginians, who found Cartagena (*Carthago Nova*). The south becomes rapidly romanized, but the pacification of the northwest is not fully achieved until 19 AD. Toward the end of the 1st century AD the conjunction of the emperor's protection, along with generalized economic development in the western provinces and the relatively recent Roman competence in dam technology lead them to build increasing numbers of dams in the area. Today the remains of some 80 structures are known, either dams or weirs, between 1 and 19 m high and of equally variable lengths up to 700 m.[61] Most of these structures serve

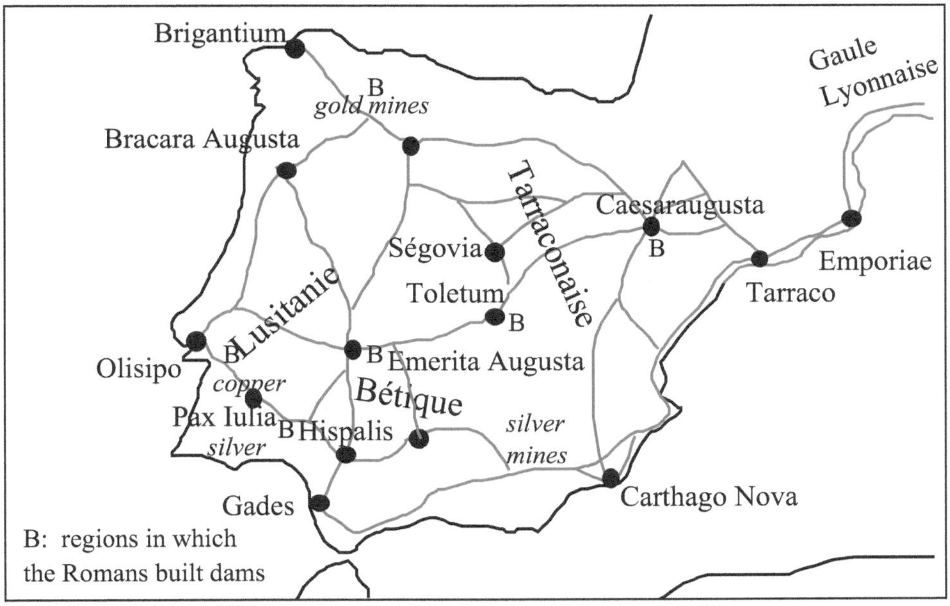

Figure 6.27 Spain and Portugal in the Roman era. The fine lines represent Roman roads.

61 Ibid., p. 60.

Figure 6.28 The dams of Prosperina (above) and Cornalvo (below), near Merida (Emerita Augusta). These are among the oldest dams still in use, seen from upstream (photos by CEHOPU (CEDEX), Miguel Otero).

the needs of irrigation as well as those of urban and industrial development.

The structures are grouped in rather well defined regions (Figure 6.27). There are nine dams around the northern metropolis of Saragossa (*Caesaraugusta*), and 15 around Toledo (*Toletum*), a city situated on the Roman road that links the north of Spain to Lusitania. But the largest number of Roman dams are in the south of the province of Lusitania: twelve around the provincial capital Merida (*Emerita Augusta*), both for industrial and urban water needs as well as irrigation; and another twenty or so, especially for irrigation, around the cities of Evora (*Ebora*), Beja (*Pax Iulia*), and Lisbon (*Olisipo*), a region of cereal production.

The three largest dams in Spain represent different techniques whose comparison is

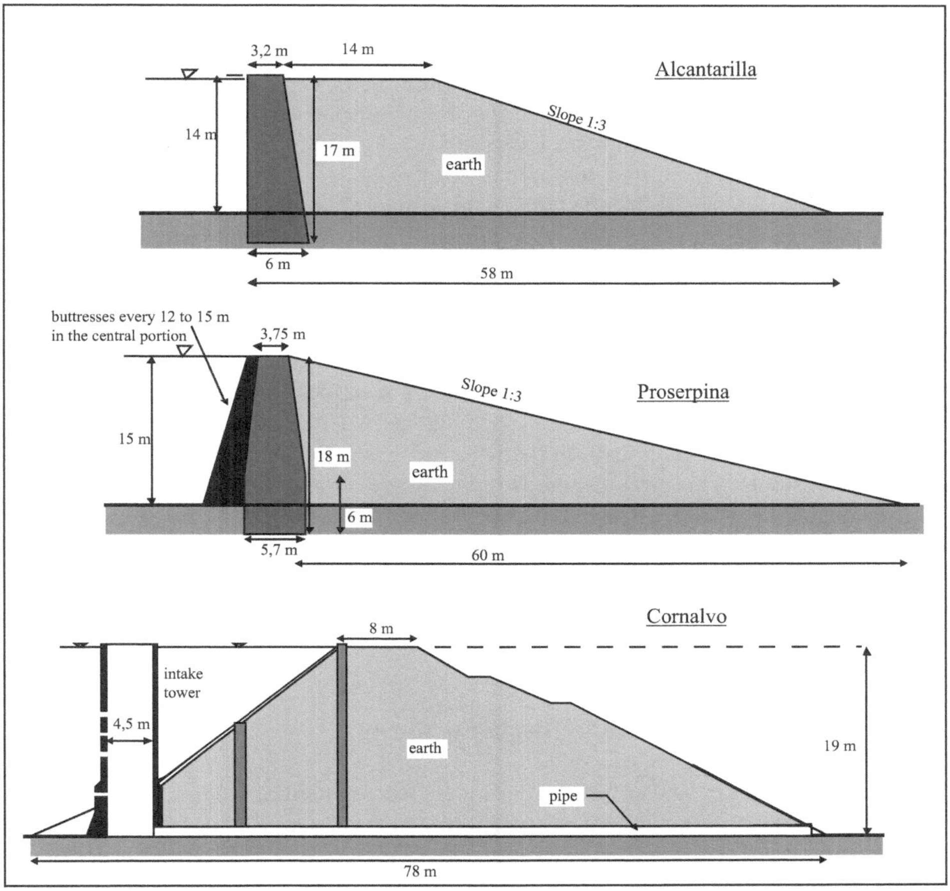

Figure 6.29 Structures of the three highest Roman dams in Spain, in the order of their dates of construction (after Fernandez Ordonez, 1984, and Schnitter, 1994).

interesting (Figure 6.29).[62] The oldest of the dams is thought to be that of Alcantarilla, situated at the head of a 50-km long aqueduct that supplies the city of Toledo and crosses the valley of the Tagus through an inverted siphon. Closely associated with the city of Toledo, the dam's history may reach back to the 2nd century BC. The structure is 14 m high and 557 m long, and comprises a fairly simple wall supported on its downstream face by buttresses and an earthen embankment to resist the water pressure. The dam failed at an unknown date, for reasons that are not difficult to imagine since debris from the wall seems to have been pushed toward the upstream. It was perhaps after a rapid emptying of the reservoir, or perhaps at a moment when the reservoir was dry and/or rainfall cut channels into the earthen embankment, that the pressure of this embankment

62 Smith (1992); Schnitter (1994); Fernandez Ordonez (1984).

against the dam wall caused it to fail toward the upstream, there being no support on that side.

The two other structures, near Merida (Figure 6.28), clearly are designed to avoid this kind of accident. The dam of Prosperina, 15 m high and 426 m long, is located some 6 km north of Merida. It comprises a masonry wall supported on its downstream side by an earthen embankment like that of the Alcantarilla dam. But the Prosperina dam is also supported on its upstream side by nine thick masonry buttresses. The water intakes are accessible thanks to two shafts in the earthen embankment, right up against the dam wall itself.

The most recent of the three dams is that of Cornalvo, situated 13 km northeast of Merida. The dam is composed of an earthen dike 220 m long, and its maximum height is 20.8 m (19 m to the right of the intake). The upstream face of the dike is compartmentalized by a series of braces and protected by a revetment on its upper portion. The water intake tower, fitted with intake openings at two levels, is displaced some ten meters upstream of the dam wall, in the reservoir itself, to accommodate the upstream embankment slope. This arrangement is typically seen in modern dams.

It is difficult to assign a date to the construction of these dams. It has been proposed that Prosperina dates to the period of the reign of Trajan, around 100 or 110 AD, or 75 years after the founding of Merida. Cornalvo is thought to date from the period of Hadrian, around 120 or 130 AD. The Prosperina and Cornalvo dams are still in service today, nearly in their original state, thanks to their particularly wise design and good maintenance.

Mines and gold mining on the Iberian peninsula

Spain and Portugal are lands of mines during the Roman period: gold in the northwest, copper and silver in the southwest, silver in the southeast. Quite a panoply of hydraulic machines are used to evacuate water from deep galleries in the peninsula's Roman mines: Archimedes screws, Ctesibios pumps, water wheels. Water is also used to wash sediments so that heavy metals, like gold and silver, can be settled out and recovered. This is the classic technique of gold miners working on rivers, a technique that has come down to us from Antiquity. After having visited silver extraction installations in the region of Cartagena (*Carthago Nova*), Strabo writes:

> "[...] as for the silver ore collected, [...] it is broken up, and sifted through sieves over water; that what remains is to be again broken, and the water having been strained off, it is to be sifted and broken a third time. The dregs which remain after the fifth time are to be melted, and the lead being poured off, the silver is obtained pure."[63]

The exploitation of Spain's richness in gold is of high importance to the Roman Empire since gold is one of the bases of its currency. In the northwest of the peninsula, there are huge surficial deposits of gold-bearing sediments from ancient river deposits,

---

63 Strabo, Geography, book III, 2, 10, Translation of H.C. Hamilton, Esq., W. Falconer, M.A. http://www.perseus.tufts.edu/cgibin/ptext?doc=Perseus%3Atext%3A1999.01.0239&query=head%3D%2319.

moraines, etc. Enormous quantities of water are needed to wash these sediments and remove the slag, and likely also to wash down entire hillsides of gold-bearing alluvial sediments. The necessary water is brought to the mining sites through a network of aqueducts and canals, sometimes supplied by the capture of rivers, and sometimes by dams. The water is typically stored in large reservoirs next to the excavation sites. Pliny the Ancient, who visited some of these sites (in the second half of the 1st century AD), was visibly impressed:

> "Another labour, too, quite equal to this, and one which entails even greater expense, is that of bringing rivers from the more elevated mountain heights, a distance in many instances of one hundred milles perhaps, for the purpose of washing these debris. [....] The fall, for instance, must be steep, that the water may be precipitated, so to say, rather than flow; and it is in this manner that it is brought from the most elevated points. Then, too, valleys and crevasses have to be united by the aid of aqueducts, [...] The water, too, is considered in an unfit state for washing, if the current of the river carries any mud along with it. The kind of earth that yields this mud is known as "urium;" and hence it is that in tracing out these channels, they carry the water over beds of silex or pebbles, and carefully avoid this urium. When they have reached the head of the fall, at the very brow of the mountain, reservoirs are hollowed out, a couple of hundred feet (*c. 60* m) in length and breadth, and some ten feet (*3* m) in depth. In these reservoirs there are generally five sluices left, about three feet (*1* m) square; so that, the moment the reservoir is filled, the floodgates are struck away, and the torrent bursts forth with such a degree of violence as to roll onwards any fragments of rock which may obstruct its passage."[64]

Remains of this system have been discovered in a region of the present province of Leon called *La Valduerna.* Claude Domergue describes a reservoir 170 m long, with a variable width from 33 m to 70 m and a maximum depth of 2.8 m, separated from upstream to downstream into two compartments of 3,500 m^3 and 9,825 m^3 in volume. Another reservoir of more than 10,000 m^3 has been found in the same region, above a circular deposit some 2 km in diameter. Additional reservoirs that are more than 250 m long[65] have been found on the cliffs overlooking the mining sites.

Irrigation works in North Africa

Lucius Septimus Severus was born at *Leptis Magna*, capital of Cyrenaica (today's Libya), at the end of the 2nd century AD. After the death of the Emperor Commodius in 192 AD, Severus emerges victorious from the civil wars, and becomes emperor in his turn, founding the dynasty of the Severians. North Africa had already been a prosperous region for some time; we have mentioned earlier the great aqueducts of Carthage and Cherchell *(Lol Caesarea)*, testaments to the prosperity of these cities. But the new emperor adorns North Africa even further, battling the desert nomads, building roads and fortifications, and creating the new province of Numedia to the west of present-day

---

64 Pliny the Elder, The Natural History, Book XXXIII, 21, Translation ed. John Bostock, M.D., F.R.S., H.T. Riley, Esq., B.A. http://www.perseus.tufts.edu/cgi-bin/ptext?lookup=Plin.+Nat.+toc
65 Domergue (1986).

Tunisia. Severus had a natural attachment to his native country, but was also motivated by the desire to support new colonists and to create new lands for production of cereals to help feed the Empire.

The Romans built some 130 dams for irrigation in North Africa, counting only those structures with known remains.[66] The dams were likely constructed in the 2nd century

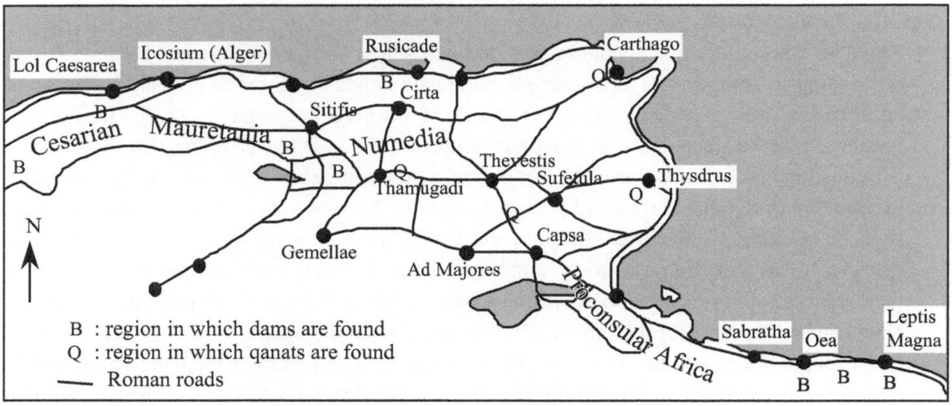

Figure 6.30 Northeast Roman Africa: provinces of Numedia and its neighbors

AD (i.e. somewhat later than the dams in Spain). Some of the dams were intended to trap alluvial sediments and thus help create arable land, following the ancient technique of the Nabatians. One of these dams, near *Leptis Magna*, was built to divert flood waters of the wadi Labda, a watercourse at whose mouth the city's port was developed (Figure 6.38). As far as we know, the highest of the dams was some ten meters; the longest is about 260 m. These are gravity earth dams, sometimes with buttresses, that can be submerged by violent floods of the wadis without suffering damage. Some 70 dams are thought to have been identified in the region of *Leptis Magna* and Tripoli (*Oea*); thirty in the region of Constantine (*Cirta*), Sétif (*Sitifis*) and Rusicade; and fifteen around Cherchell (*Lol Caesarea*). The largest dam is that of Kasserine in Tunisia, on the wadi Derb, in a grain region not far from Sbeitla (*Sufetula*). This dam includes a wall 7 meters thick at its base and 4.9 m at its crest, 10 m high, and about 130m long. It has since been replaced by a modern dam on the same site.

There are also *qanats* in the Roman provinces of proconsular Africa and Numedia.[67] In North Africa, they are later called *kettaras*, or *foggaras*. We have seen in Chapter 2 how this technique was spread into numerous sectors of the Acheminides Empire by the Persians. We encountered it again in the oases of Egypt (Chapter 3). As we will see further on, it is obvious that the Romans maintained and developed *qanats* in the Orient. It is therefore no surprise that *qanats* are found in the eastern part of the African provinces: at El-Djem (*Thysdrus*), between Tebessa (*Thevestis*) and Gafsa

66 Schnitter, 1994, p. 70.
67 Goblot (1979), pp. 117-125.

(Capsa), near Carthage and, more to the west, in the region of Timgad *(Thamugadi).* The latter are attributed to the period of Commodius, i.e. the end of the 2nd century AD. Latin inscriptions sometimes attest to the Roman origin of the *qanats*, as in the region between Tebessa and Gafsa, or as at Timgad:

> "Facility for the collection and delivery of underground water."[68]

The Roman origin of the qanats is assumed in many other situations. The above observations make it tempting to accept the hypothesis of Henri Goblot, according to whom it was indeed the Romans who imported the *qanat* into Africa. Much later, the Arabs diffuse the technique even further, toward Spain and Morocco.

Development of the Roman Orient

During the last years of the Republic, Greece, Anatolia, Syria-Palestine, Egypt, and Cyrenaica had become Roman. This process began with the bequeathing of Pergamon to the Roman Republic, and was marked by Augustus' taking possession of Egypt in 31 BC, and continuing through the tumult of the many wars during the last years of the Republic. There is not much to say of Greece, for she was knocked flat by wars and never got back on her feet economically. The reader may recall the development works at lake Copais undertaken in the Mycenaean era (Figure 4.10), and somewhat restored by Alexander the Great. Several sources mention that during the Roman period the lake's dikes are no longer maintained, that the adjacent cities are subject to flooding, and that the best land must be abandoned.[69]

Other lands of the Orient fare differently. Trajan reaches the Persian Gulf in 116 AD after having conquered the Parthians, but in the end Rome is able to hold onto only the extreme northwest of Mesopotamia. Thereafter, the Orient provides Rome with emperors: Septimus Severus marries the daughter of the sun-god priest at Homs (*Emesa*), and his son Caracalla is therefore half Syrian. His successors, Elagabal and Severus Alexander, from the maternal side of Caracalla's family, are entirely Syrian. In the middle of the 3rd century AD, when the Empire is threatened along all of its borders, the Sassanide Persians reach Antioch and, to add to the humiliation, take the Emperor Valerius and all of his army as prisoners, in 260 AD. At Palmyra, queen Zenobia seizes the opportunity to launch offensive military expeditions, taking Alexandria in 270 AD. But in 272 AD the new Emperor Aurelius retakes control of all of the Roman Orient, and it remains peaceful for an extended period. Diocletius begins to create an autonomous Roman Orient, in dividing political power among four emperors and establishes his own capital at Nicomedia, near Byzantium. In 330 AD the Emperor Constantine takes personal control of all of the Empire, but especially favors the Orient and transforms the Greek city Byzantium into Constantinople, the new capital of the Orient. When Rome falls in 410 AD, the Roman Orient remains standing.

---

68 Inscription found at Timgad: opus aquae paludensis conquiriendae concludenaeque, citation after Goblot (1979).
69 Sartre (1991), Chapter 5.

Egypt is arguably the least Roman of the lands of the Empire, and it retains its identity under Roman domination. Its governance continues under the Ptolemaic tradition, though the peasants are under increasingly oppressive fiscal pressure. Over and above direct taxation, they must meet a number of other obligations. Just as in the Ptolemite period (and also probably in the Pharaonic era), they are obliged to work on maintenance of the dikes.[70] During the time of the Pharaohs, Egypt had only to feed herself. But under the Ptolemites, she must in addition produce the surpluses necessary for the grand political ambitions of her leaders. Then under the Romans, Egypt must also feed the great cities of the Empire (and the city of Rome first and foremost), a heavy burden shared with only a few other provinces such as Numid.

In response to these needs, the Egyptians further develop the Fayoum irrigation system – dropping the lake level down to its present level to reclaim additional land, and cleaning out the canals. Moreover, they maintain the *qanats* that had been constructed by the Persians in the Kharga oasis and even build new ones. The small oases of Dakhla, Farafra, and Baharya to the north of Kharga appear to have flourished during the Roman period, and this success is likely attributable to the *qanats* built by the Romans.[71] According to Henri Goblot, it is indeed in Egypt that the Romans learned how to build these very special devices.

The city of Alexandria continues to be a great intellectual center under the Romans, as we have seen in Chapter 5. Alexandria is a cosmopolitan city with a population of some half million,[72] and it is in a state of constant turbulence. It is said that Carcalla, the son of Septimus Severus, had the sword taken to all the young people of the city in the spring of 215 AD after they had publicly criticized him.

Some fifteen Roman dams are found in the Orient - in Anatolia, Syria-Palestine and in the northwest of Mesopotamia. Other hydraulic works are to be built by the Byzantines after the fall the Occidental Roman Empire. The structure thought to be the oldest has a very specific purpose. It is 16 m high and 60 m long (but only 5 m wide, and therefore of precarious stability), and was built in 80 AD to divert floodwaters and protect the port of Antioch on the Orontes.[73] Antioch is the ancient capital of the Seleucids and an opulent capital of the Roman province of Syria.

Earlier in this Chapter we mentioned some of the aqueducts built in the Orient, like the one at Apamea on the Orontes (Figure 6.32), perhaps the longest of all those in the Roman Empire. The water supply of Jerusalem (*Aelia Capitolina* from Hadrian) is one of the most famous ones of the Orient, given its symbolic importance. The pools of Solomon, vast reservoirs 13 km south of Jerusalem that are created by dams, store 200,000 cubic meters of water. An aqueduct begins at a reservoir situated near Hebron, connects this reservoir to the pools of Solomon, and then continues to Jerusalem, passing through the center of Bethlehem, very close to the Cave of the Nativity. These installations have often been incorrectly attributed to King Solomon. In fact, they should be attributed to the Roman governor Pontius Pilate, at the very beginning of the

70 Ibid., chap. 10; Bonneau (1993).
71 Goblot (1979), p. 114.
72 C. Scarre (1995), p. 77.
73 Garbrecht (1991). See also Schnitter (1994), p. 74.

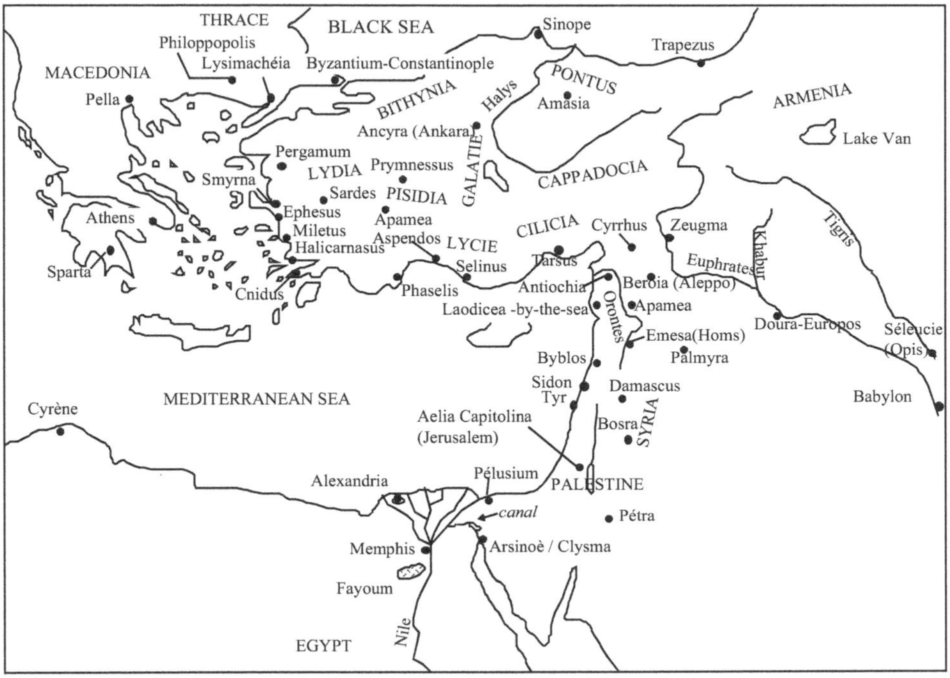

Figure 6.31 The Roman Orient

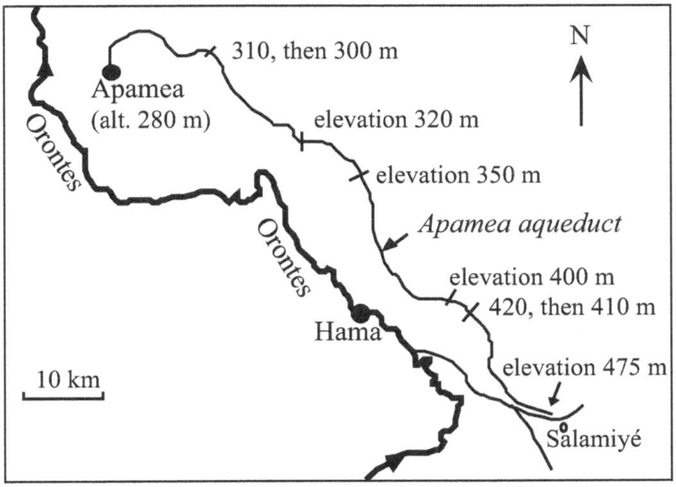

Figure 6.32 Layout of the Roman aqueduct of Apamea, built at the beginning of the 2nd century AD, very likely the longest one of all the Roman world (Balty, 1987).

Christian era.

Other hydraulic projects were built for the storage of irrigation water. Two of them, both located in Syria, merit our attention.

On the Orontes there is a very large lake formed by a dam built around 1935 over the remains of a famous ancient dam, 12 km southwest of Homs (the Roman *Emesa*).[74] In Antiquity, this dam served the same function of maintaining the lake of Homs (at a level three meters lower than the modern dam). According to the study conducted in 1921 by Louis Brossé, the dam was 850 m long[75] and at most seven meters high. The dam is very stable thanks to its broad width, 15 to 20 m in its central portions. Four overflow weirs, two each on the right and left banks, carry water to canals that lead to the city of Homs and agricultural zones.

A curious feature of this installation is that when the dam was first built, the normal course of the Orontes downstream of the dam is not supplied by water from the spillway, but rather by copious percolation through the face of the dam itself, a face obviously not at all watertight. Modern observers attribute this dam to the Roman period, based on its architecture of a fill of rough-stone concrete between walls of quarry stone. More precisely, the dam appears to date from the year 284 AD, first year of the reign of Diocletius. The dam reflects the prosperity of the city of *Emesa* at this time.

But these same observers used to believe, in the 1980s, that the dam was much older, having been constructed by the Pharaoh Sethi I, in the 13th century BC, when Egypt dominated this region. The very ancient city of Qadesh was the site of a famous battle between Egyptians and Hittites. One must also consider the account of Strabo, who situated one of the sources of the Orontes near the "Egyptian wall, toward the territory of Apamea."[76] This account clearly refers to the phenomenon of the Orontes flowing directly from infiltration through the wall of the dam. The most reasonable hypothesis,

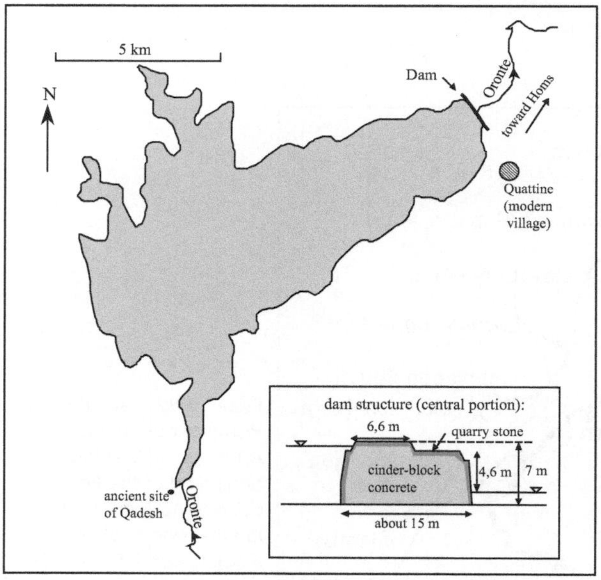

Figure 6.33 The lake of Homs (in 1932), a reservoir of about 90 million m^3, and the structure of the ancient dam, after the measurements of L. Brossé in 1923 (Calvet and Geyer, 1992).

74 Our sources for this dam of Homs are especially Calvet and Geyer (1992), Chapter 3, as well as Schnitter (1994), p. 76.

75 This is the length indicated by Calvet and Geyer (1992), who thought the length of 2,000 m indicated by other authors to be excessive.

76 Strabon, Geography, XVI, 2, 19 (Translation of F. Lasserre), Les Belles Lettres, 1981, adapted.

following Yves Calvet and Bernard Geyer, is that the site of the dam is a natural rocky barrier which, from very early times of Antiquity, was the site of structures built to raise the level of the natural lake. It is therefore probable that the Romans reconstructed or renovated an ancient dam at the time of Diocletius, following local rather than Egyptian practices, much as the modern dam is constructed over the Roman structure. The long life of the dam, and the fact that the impoundment is not filled with sediment, could be explained by the regular flow of the Orontes.

The Orontes is obviously a river that invites the development of hydraulic works, and it continues to do so as we discuss later in Chapter 7. But there is another region whose development necessitates important hydraulic projects: the region of Palmyra, a land of oases in the middle of the harsh desert of Syria, stopover of caravans on the silk road. Many small oases of Palmyra owe their prosperity to *qanats*. These could have been constructed by the Persians during the time of the Achaemindes Empire, but it is certain that the network of *qanats* is further developed and maintained by the Romans.[77] Even more spectacular in this region is the dam of Harbaqa, another project whose purpose is to create a reservoir of water for irrigation. The dam is constructed on a seasonal watercourse, the wadi al-Barda, 70 km southeast of Palmyra.[78] With its height of 21 m and its length of 365 m, this is the largest dam of the Roman world (except possibly for the dam of Nero at Subiaco, which is higher). It constitutes a rectilinear wall whose thickness at the base is 18 m, further reinforced with buttresses. It is undoubtedly thanks

Figure 6.34 In a land of minerals, the dam of Harbaqa across the valley of a Wadi. The impoundment is today completely filled in, but the dam enables the fill to retain sufficient moisture for farming. View from downstream (photo by the author).

77 Goblot (1979), p. 130.

78 Here again, our sources are essentially Calvet and Geyer (1992), Chapter 7, as well as Schnitter (1994).

to this solid construction that the dam remains standing today, minor repairs having been effected during the 1960's (Figure 6.34). The dam was apparently built at the end of the 1st century or the beginning of the 2nd century AD as part of the infrastructural work ordered by Hadrian, who visited Palmyra in 130 AD. Much later, under the Umeyyades, new hydraulic works were built downstream of the dam of Harbaqa (Figure 7.10).

In closing this overview of hydraulic technology used in the Roman Orient, we must mention the technical consequences of a serious Roman defeat. When the Emperor Valerius is taken prisoner by the Sassanide Persians in 260 AD, the civil-engineering competence of the captured army is used to build several multi-purpose structures comprising a dam-spillway and a bridge near the capital city Suse, on the Karun River and its tributaries. This was a kind of forced technology transfer.

Mediterranean ports in the Roman period

Under the Republic, the Roman provinces were resources to be exploited for the sole benefit of Italy. Later on, the emperors came to understand that the cohesion of the Empire depended on the prosperity of the provinces. As the 2nd and 3rd centuries of our era unfolded, the economic situation of the provinces flourished; a number of them officially declared their wish that "the Roman domination should last eternally."[79] Maritime navigation is the most effective link among these provinces, and also the most economical. Transport ships plow the Mediterranean Sea, the *mare nostrum* of the Romans, at least during the warm season – from September to May navigation is officially forbidden on the *mare closum*. In reality, the navigation period is between March and November. The largest merchant ships are those that transport wheat, routinely displacing 200 of our modern tons and measuring 25 m in length – some attain 60 m.

Merchandise circulates widely – from the granaries of Numedia (present-day

Figure 6.35 Harbor scene on a Roman sarcophagus (photo by the author)

79 Sartre (1991), p. 55.

Tunisia), Sicily, and Egypt[80] toward the large Italian cities to be fed; from the regions of oil production, such as Tripolitania, the south of Spain, the north of Syria; from Greece whose wine is highly valued, and also from the regions of Bordeaux (*Burdigala*), Tarragona (*Tarraco*), and the south of Gaul, already renowned for their wine. Some of the goods come from far away, like silk from China, spices from southeast Asia, and incense from Arabia. Under the Empire, merchants based at Alexandria or Antioch forge the first direct relations with China of the Han Dynasty.

All of this merchandise attracts pirates, the restive adventurers of the *pax romana*, often spawned by the upheavals that accompany the progressive annexation of the Orient. Piracy in the Mediterranean is reduced somewhat after the naval victory of Pompeii, and it is eradicated at the beginning of the Christian era, an achievement of Augustus and his son-in-law Agrippa. The Roman military fleet had its beginnings in the Punic wars (3rd century BC). Subsequently, the Empire maintains a military fleet whose squadrons are based at Miseno (near Pouzzuoli, in the Gulf of Naples); at Ravenna, on the Adriatic in the northwest of Italy; and at Alexandria.

Ostia and the imperial ports of Rome

Rome's food supply depends on the chain of maritime transport of wheat from Numidia and Egypt. For a long time Rome could offer only minimal port facilities to cumbersome and large loaded boats needing to enter the Tiber at Ostia. Mooring there was especially dangerous when the *Auster*, an ill-reputed west wind today called the *Libeccio,* blew. Moreover, access by large boats was made problematic by a shoal. Here is how Strabo describes the situation, in about 25 AD:

> "This city has no port, owing to the accumulation of the alluvial deposit brought down by the Tiber, which is swelled by numerous rivers; vessels therefore bring to anchor further out, but not without danger; however, gain overcomes every thing, for there is an abundance of lighters in readiness to freight and unfreight the larger ships, before they approach the mouth of the river, and thus enable them to perform their voyage speedily. Being lightened of a part of their cargo, they enter the river and sail up to Rome, a distance of about 190 stadia."[81]

Boats coming from Alexandria or Antioch at this time, after fifteen to twenty days at sea (and sometimes twice that long), prefer to unload in the bay of Naples, at Pouzzuoli (*Puteoli*). Their merchandise then continues on to Rome either by land or sometimes in smaller boats capable of sailing up the Tiber. Navigation between Pouzzuoli and Ostia is dangerous, since the coast is low and lacks shelter, and therefore Rome is often threatened with shortages. Plans to create a true port at Ostia are proposed under Caesar and Augustus; these plans are often subject to long dissertations. But the first attempt to implement such plans must await Claudius, whom we have already seen as a great hydraulic entrepreneur. A 70-hectare basin (700 m by 1,000 m), excavated

---

80 Rome has a monopoly on the importation of Egyptian wheat, prior to the creation of Constantinople.
81 Strabo, Geography, book V, 3, 5, Translation of H.C. Hamilton, Esq., W. Falconer, M.A. http://www.perseus.tufts.edu/cgi-bin/ptext?doc=Perseus%3Atext%3A1999.01.0239&layout=&loc=5.3.1

behind the shelter of a naturally growing barrier island, becomes a first basin of the port. Two large jetties are constructed to protect it; the most offshore one is built of large limestone blocks tied together by iron grapples cemented with lead;[82] it is 330 m long and 23 m wide. Beyond this breakwater a ship was sunk and filled with sand to serve as a subfoundation for the construction of a four-story lighthouse. This was the great 100-m long ship that had been used by Caligula to bring from Egypt the obelisk that now is in Rome's Place de Saint-Pierre:

> "Claudius created the port of Ostia, constructing two jetties in circular arcs to the right and left, and in quite deep water, a breakwater to block the entrance; to seat this breakwater more solidly, one began by sinking the ship that had brought the great obelisk from Egypt; upon it, one constructed a great number of piles supporting a very high tour, destined, like that of Alexandria, to illuminate with its fire, during the night, the shipping route."[83]

Work on the port of Ostia proceeds from 42 to 54 AD. The project is finished after the death of Claudius and inaugurated by Nero, who associates his name with the achievement, without completely claiming it as his own, in naming it *portus Augusti*: the port of the Emperor. Two canals are excavated from the Tiber, obviously to connect the port to the river, but perhaps also to protect Rome by facilitating the drainage of floodwaters of the Tiber to the sea.[84] Somewhat later, the base of the lighthouse is joined to the west breakwater; other boats are sunk to facilitate the initial construction.

But the works of Claudius at Ostia remain inadequate to meet the needs of the city of Rome. Wheat from Alexandria continues to be offloaded at Pouzzouli, not Ostia. Moreover, the port at Ostia has a tendency to fill with sand from the alluvia of the Tiber brought through the canal(s) that link the river to the port, and from the littoral currents that sweep the alluvia of the mouth of the Tiber toward the north. Therefore Trajan excavates a second basin of 32 hectares from 100 to 112 AD. The basin is in the shape of a hexagon with sides of 358 m, a depth of 5 m, and linked to the first basin (Figure 6.36). A new lighthouse is built at the entrance of the canal linking the two ports. The canal coming from the Tiber is retraced: it constitutes a second mouth of the river and is called today the *Fiumicino*. The new port is connected to this canal, but little of the Tiber's flow and sediment circulate through the port, which therefore minimizes siltation. The overall project eventually will support and receive all maritime traffic with goods destined for Rome. The work is completed, still under Trajan, by creation of an artificial port at Civitavecchia (*Centumcellae*) to the north, and by improvements at Terracino, about halfway between Pouzzouli and Ostia, to provide shelter along this coast.[85]

The vast complex of ports of Claudius and Trajan, called *Portus*, is further improved and maintained under Septimus Severus, and then under Constantine. The complex remains active until the 5th century AD, but then suffers from the decrease in Roman population following the fall of the Empire, and progressively decays. In the 5th centu-

82 Le Gall (1981).
83 Suétone, Claude, XX, 3 (cited after Reddé, 1983).
84 Santa Maria Scrinari (1983).
85 Reddé (1983).

ry the sea level is thought to have risen sufficiently to submerge the barrier island that protects the port of Claudius from the west. At this time work is done to provide protection for the canal that links the two ports.[86] The blocks of Claudius' large jetty to the northwest become partially disconnected, since the structure, lacking the protection of rock armor, was vulnerable to wave attack. There is intense alluvial deposition in the outlets of the Tiber, from the 15th century, and the port complex ends up being landlocked. The remains of the port of Claudius are now beneath the international terminal of the Rome airport. Today's traveler therefore arrives at the same spot as did the ships from Carthage, Tarraco, and Massilia in the first century.[87]

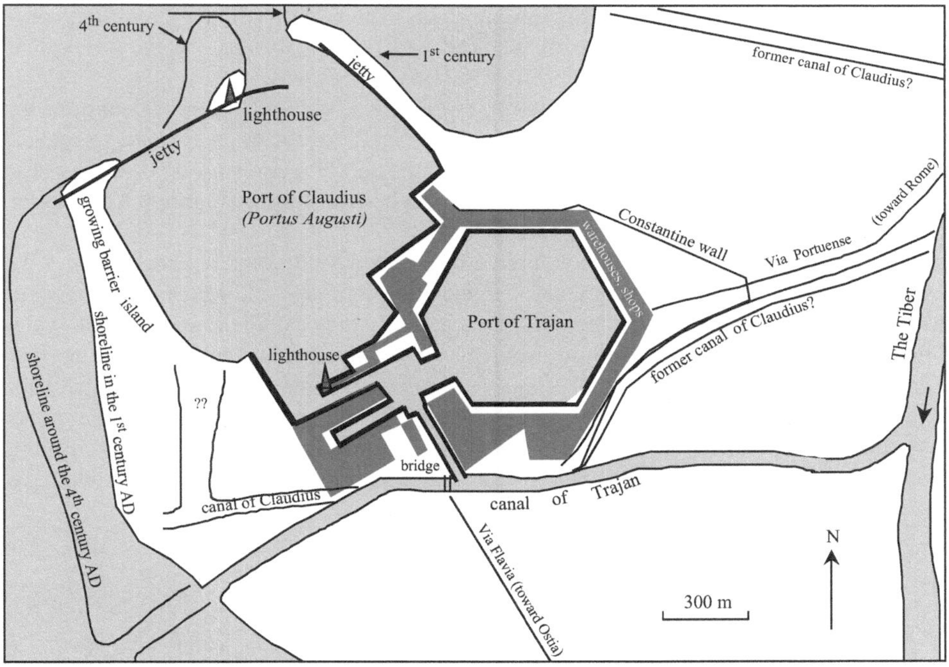

Figure 6.36 The complex of imperial ports of Rome, after Trajan's work (Le Gall, 1981; Reddé, 1983).

86 Le Gall (1981).

87 It was the construction of the Rome airport that led to the discovery of the remnants of the port of Claudius (Santa Maria Scrinari, 1983). Traces of the concrete used to fill Caligula's ship were found in the shell of the hull. The west breakwater is visible. The hexagonal basin of the port of Trajan, which was used as an irrigation storage basin in the 19th century, has been restored.

Other Roman ports and navigation works

We should not forget the role of the port of Alexandria during the Roman period, as it closely follows the Ostia complex in importance. The port was built in the Hellenistic period, and we have already described it at the beginning of Chapter 5 (Figure 5.2). A squadron of Roman warships is based at the port, a point of departure for Egyptian wheat bound for Rome (via Pouzzuoli before completion of the port of Trajan). But Alexandria is also an intermediate stop for products coming from Sudan, Arabia, and the extreme Orient. The Romans continue to use and maintain the canal of Necho that we described in Chapter 3. Trajan develops *Clysma* (Suez) and undertakes repair of the canal.[88] The canal constitutes one of the routes to Arabia, India, and China, without ever really seriously challenging the grand ports of the Red Sea (*Myos Hormos*, on the site called *Qoseir* on Figure 3.1, and *Bérénice*, formerly *Head of Nékheb*).

The port of Carthage, the most important of Africa, is also a point of departure for wheat bound for Rome. We have mentioned the wealth of the Roman *Cartago* earlier, from its refounding by Augustus in 29 BC up until the 7th century AD when it is destined to be eclipsed by the modern Tunis. The port has two basins (Figure 6.37), one for military uses and the other for commercial traffic.

It would be difficult to mention all the other Roman Mediterranean ports. Some are excavated into level ground or in natural creeks, like Carthage and Marseille; others are developed in the mouth of watercourses, as is partially the case of the complex of Ostia, often with the same accompanying problems of sediment management. The port of *Leptis Magna*, the native city of Septimus Severus, is situated in the mouth of the wadi Labda. This emperor develops the port (Figure 6.38) and erects a lighthouse there. As a typical Mediterranean watercourse, the wadi Labda is subject to rapid and violent floods, and these can endanger vessels or cause other significant damage. As we have

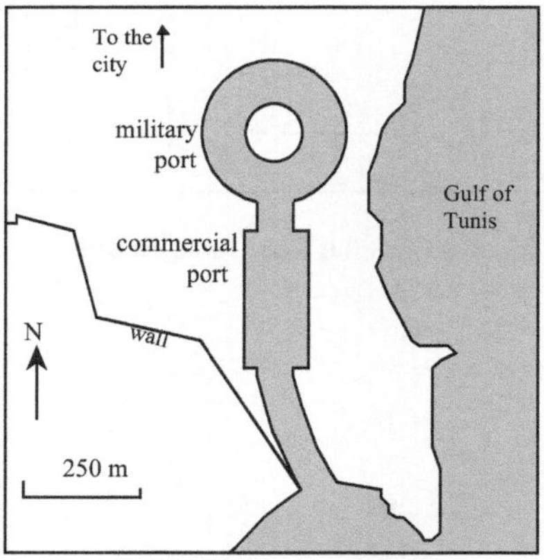

Figure 6.37 The port of Carthage (after Scarre, 1995).

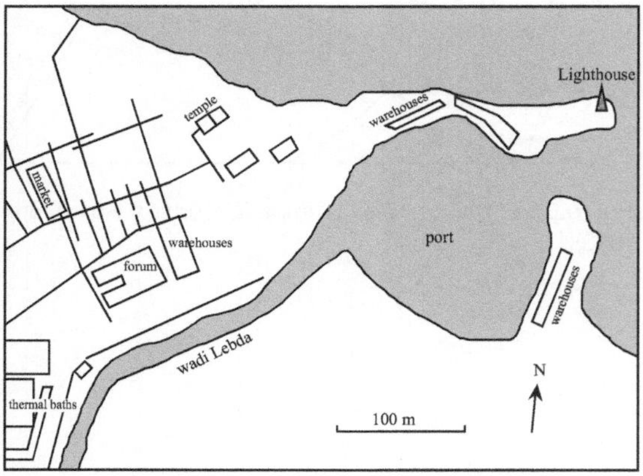

Figure 6.38 The port zone of Leptis Magna (after Scarre, 1995).

88 Mayerson (1996).

seen for some other African dams, a flood diversion structure is built upstream on the wadi Labda. A similar scheme is used by the Romans for Seleucid, the port of Antioch (with a 130-m long tunnel, followed by a 700-m long open cut, 6 to 7 m wide).

The Romans are not great canal builders – even though they maintain and use the Necho canal, or *canal of two seas*, and rename it the *canal of Trajan* after this emperor's work on it. There is some fainthearted thought given to the construction of a canal at Corinth, cutting across the isthmus to enable ships to avoid having to sail to the south of Peloponnese. Caesar, Caligula, and Nero are tempted by this project, and initial work is begun.[89] But there is concern that the sea level may be different from one end of the canal to the other,[90] and that the opening of the canal would provoke a catastrophe. Or perhaps it was simply that the economic stakes were not high enough, for Greece is, under the Empire, more of a symbol than a significant economic player. In any case, the canal was not built in Antiquity.

Roman hydraulic knowledge and knowledge transfer

An observer of the vast array of Roman technical achievements can only be surprised and disappointed at the lack of technical documentation left by the Romans. With the exception of Vitruvius, whose technical descriptions are sometimes precise (the water mill), and sometimes extremely vague (the aqueducts), there is simply no body of technical literature as we know it. We have extensively cited Frontinus' book on the aqueducts of Rome, a remarkable work. Yet it is much more of a precise and documented "audit" report than a manual for the guidance of future builders. Pliny's *The Natural History* is precious and perhaps an exception, but it has little to do with hydraulic works. It seems that the attitude of the educated Roman leisure class insofar as applied sciences were concerned, tends toward that of the ancient Greeks in retreating from the attitude of the Hellenistic world: the application of knowledge is culturally devalued. Therefore treatises are not written about techniques. It is hard to imagine, for example, such a dearth of technical documentation after the works of Archimedes.

As a consequence, one finds great diversity in the technical solutions found in Roman projects. There are as many exceptionally well conceived works as there are very poorly conceived ones, whether it be in the domain of aqueducts or dams. It is true that in the three largest dams of Spain (Figure 6.29) one sees a certain technical progression, if indeed these dams were built in the order that is thought to be correct. But in the Orient, for example, one finds as many solid and long-lasting retention dams (for example the dams of Homs, and of Harbaqa) as dams whose stability is very problematic, because of an excessive height-to-width ratio.

Another anomaly in the transmission of Roman techniques is visible in the aqueducts that have very large variations of slope from one point to another along their length. The canals in these aqueducts are constructed in anticipation of a constant depth,

---

89 Sartre (1991), p. 230, after Suétone and Pausanias.

90 This was a frequent argument in Antiquity. It was used at the time of the Pharaoh Necho in the context of the canal of two seas.

and yet in steeper segments, the water velocity is greater and, as a consequence, the actual depth of water is less.[91] On the other hand one can see on several aqueducts (at Nîmes, for example, but it is not the only one) that, along segments of small slope, the canal walls had to be raised after the initial construction, to prevent overflow. With so many successful projects behind them, how could the Romans not have already experienced this problem and drawn conclusions from it?

Similarly, another problem that has been intriguing to researchers is the Roman evaluation of the discharge of their aqueducts. The *quinaria*, as we have seen earlier, is a unit of area. It is used by Frontinus as the only reference to determine the quantity of water being delivered. And in Table 6.2 we have, like Frontinus, used a unique equivalence between quinariae and discharges in m^3/day. In reality, this equivalence has to be extremely variable. If we consider only several aqueducts of Rome, all having the same slope and the same wall roughness, Table 6.5 shows an important variation, from one aqueduct to the other, of the equivalence between sections (*quinariae*) and discharges.

Table 6.5 Calculation of the discharge of three aqueducts of Rome.[92]
Variation of discharge corresponding to one quinaria.

Aqueduct	Canal width (m)	Section after Frontinus (quinariae)	Water depth calculated from the area (m)	Calculated velocity (m/s)	Calculated discharge (m3/day)	Discharge correspon-ding to 1 quinaria (m3/day)
Tepula	0.8	445	0.23	0.69	11,214	25.2
Juilia	0.7	1,206	0.72	0.95	41,567	34.5
Anio Novus	1	4,738	1.99	1.35	232,059	49.0

Were the Romans ignorant of this fact? In reading Frontinus, and in particular the following significant extracts, it would seem that they were not ignorant of it:

> "Let us remember that every stream of water, whenever it comes from a higher point and flows into a reservoir after a short run, not only comes up to its measure, but actually yields a surplus; but whenever it comes from a lower point, that is, under less pressure, and is conducted a longer distance, it shrinks in volume, owing to the resistance of its conduit; and that, therefore, on this principle it needs either a check or a help in its discharge.
>
> "But the position of the *calix* (intake) is also a factor. Placed at right angles and level, it maintains the normal quantity. Set against the current of the water, and sloping downward, it will take in more. If it slopes to one side, so that the water flows by, and if it is inclined with

91 For example, in the aqueduct of Nimes with a discharge of 66,000 m3/day, if one takes as a simplification a constant slope of 0.38 m/km upstream of the Pont du Gard, and a smaller slope of 0.18 m/km between the bridge and the city of Nîmes, one finds that the water depth should be approximately 0.96 m along the first segment and 1.3 m along the second one, these values representing the useful depth of the canal (calculations done with a wall roughness of 5 mm and a canal width of 1.22 m).
92 After Viollet, Chabard, Laurence, and Esposito (1998); the calculated velocities and discharges correspond to a slope of 1.3 m/km and a wall roughness of mean height 3 mm.

the current, that is, is less favorably placed for taking in water, it will receive the water slowly and in scant quantity."[93]

There is both right and wrong in this observation, but in any case there is an intuitive understanding of the importance of the velocity, and even of what hydraulicians today call the "head". The importance of the velocity in evaluating the discharge appears even more clearly in this additional extract:

"Whence it appears that the total found by me is none too large. The explanation of this is, that the swifter current of water, coming as it does from a large and rapidly flowing river, increases the volume by its very velocity."[94]

This intuition as to the importance of the velocity in determining the discharge was not sufficiently strong to be put to use. What has always seemed incomprehensible is the lack of any connection between the knowledge that Frontinus demonstrated – or did not demonstrate – and the work of Heron of Alexandria. We described this work in Chapter 5, and it has long been assumed to have come before Frontinus' work by at least a century. We now know with certainty that Heron could not have done his writing before 60 AD. Frontinus, while he was studying the aqueducts of Rome, could therefore have been perfectly ignorant of the work of Heron, if indeed that work predated his. It

Figure 6.39 Remains of the arches of an aqueduct near Rome, between the via Appia Antica and the via Appia Nova (photo by the author).

is perhaps even possible that it was after having read Frontinus that Heron was led to write down the correct expression for discharge, as the product of the velocity and the cross-sectional area.

One technology that clearly and uniquely represents Roman know-how is the inverse siphon, such as the one that can be seen in the aqueduct of Gier at Lyon. The number of pipes in each siphon is in effect fixed *a priori* by the size of the head tank,

93 Frontinus, "The Aqueducts of Rome", 35, 36, Translation of Charles E. Bennett in the Loeb edition, 1925 http://penelope.uchicago.edu/Thayer/E/Roman/Texts/Frontinus/De_Aquis/text*.html
94 Ibid., 73.

and by the number of outlet openings in the downstream wall (Figure 6.13). Lacking valves for control of the discharge of the pressurized pipes, one could compensate for dimensioning errors only by plugging one or several of these openings – an expedient visibly employed for certain siphons of the aqueduct of Gier. If the discharge of the pressurized pipes turns out to be too large, causing emptying of the head tank and consequently a dangerous aspiration of air into the siphon, it is thus possible to reduce the discharge. But one has to wonder what expedients were used when, on the contrary, the capacity of the siphon was insufficient, causing the head tank to overflow.

In assessing the technical contributions of the Romans, one must also remember the

Figure 6.40. Aqueduct of Gier for Lyon: seen from the upstream end of the siphon of Yzeron, at the place called "le plat de l'air". From left to right: hidden behind the trees, the arches that support the canal, the remains of the head tank, and the ramp, the inclined plain that carried the lead pipes descending into the valley to the bridge-siphon (photo by the author).

heritage they received from their predecessors, professors and enemies - the Etruscans. The arch, as well as water supply and drainage works, are part of this heritage. We must also not forget to add to the list of Roman shortcomings their inability to maintain the Etruscan systems for land drainage and maintenance built before them in Italy. In their own country, the Romans allowed so many fertile lands to revert back to swamps, soon infested with malaria – for example the famous Pontin marshes. The lazy urbanites of Rome could have produced wheat on these lands, and relieved somewhat the desperate state of the Egyptian peasants.

7. Beyond Rome, The East And The Arab World

Beyond the Roman Empire – Persia and India

Between the Tigris, the Ganges, and the Oxus: multicultural influences

The Indus valley had harbored the great "hydraulic" civilization of Harappa between the third and second millenia BC. This civilization would develop to have exchanges with the Mesopotamian millenia, and had extended its influence along the "lapis-lazuli route", to the north of the Hindu Kush mountains in Bactria (Figures 1.3, 7.1). After the collapse of this civilization only the trading posts to the south remained, at and around the mouth of the Indus at Lothal. Development later continues in the Bactria – a prosperous civilization grows on the banks of the Oxus and its tributaries, a civilization that uses gravity canal irrigation in the cultivation of terraces overlooking the rivers (Figure 7.2).

The ancient Indian civilization grew from Indo-European (Aryan) migration from the northwest in the middle of the IInd millennium BC. The earliest Vedic texts, written in Sanskrit around the 6th century BC, give us our earliest distinct portrait of the development of this civilization. The birth of Buddha is placed in this century at about 560 BC.

After the fall of the Persian Empire and the death of Alexander, all of the region from Mesopotamia to the borders of India becomes the domain of Seleucos, one of Alexander's generals, and then of his descendents the Seleucids. In India itself the Maurya Dynasty, 313 to 180 BC, includes a period of unification from the Ganges to the Indus under the grand sovereign Açoka (about 269 to 232 BC). The development of writing first appeared in his reign, as did the humanistic principles inspired by Buddhism. Only the extreme south of India and Ceylon (modern-day Sri Lanka) are not included in this unification.

Around 250 BC the Parthians from the north of Iran push the Seleucids back toward Syria and settle in Mesopotamia. In so doing, they isolate Bactria and Sogdiana from the rest of the Hellenistic world. This is the beginning of the Greek kingdom of Bactria, destined to spread Hellenistic culture toward India. Around 200 BC, Bactria's king Euthydemus and his son Demetrius set out to conquer large regions of India. These Indo-Greeks were subsequently pushed back out of Bactria by a people who are known to us through Chinese history as the Yuehzi. The Chinese, pushed to the north by the Xiongnu (fellow nomads of unsavory reputation and who were likely the ancestors of those whom we now call the Huns), try to form alliances with the Yuehzi. In the 1st century BC, the Yuehzi found the empire of the Kuchans, occupying all the high valleys of the Ganges and the Indus up until the 2nd century AD. The stability of the great empires across these centuries – Rome to the west, the Parthians and the Kuchans in Asia, the Chinese empire of the Han Dynasty to the east – favors development of the Silk Road.

Meanwhile in Ceylon there are many signs of active commerce with Roman merchants.

The power of the Arabs rises in the near east during this time. In 640 AD the Arabs take Alexandria, occupying Egypt and destroying the Sassanide Persian Empire that had supplanted the Parthians from the 4th century BC. Having been confined to the Indus for a long time, the Arabs occupy Sind as well as Samarcand in 712, and in 751 they affront the Chinese armies on the Talas River to the northeast of Samarcand, in the loop of the Iaxartes (today the Syrdarya). This was an Arab victory in principle, but in reality it marked the end of their expansion toward the east. With the decline of authority of the caliphs of Baghdad, the Ghaznavid Turks become the masters of Persia at the end of the 10th century AD. India collapses under their blows around 1000 AD, and all the north of the country is pillaged. Then successive waves of Mongols sack Mesopotamia in 1258 AD, and continue to ravage the north of India as far as Delhi in 1398 AD. Turkish-Mongol regimes control the sultanate of Delhi from the 13th century AD, and control the entire northern half of India until the 17th century, including the Ganges and Indus valleys.

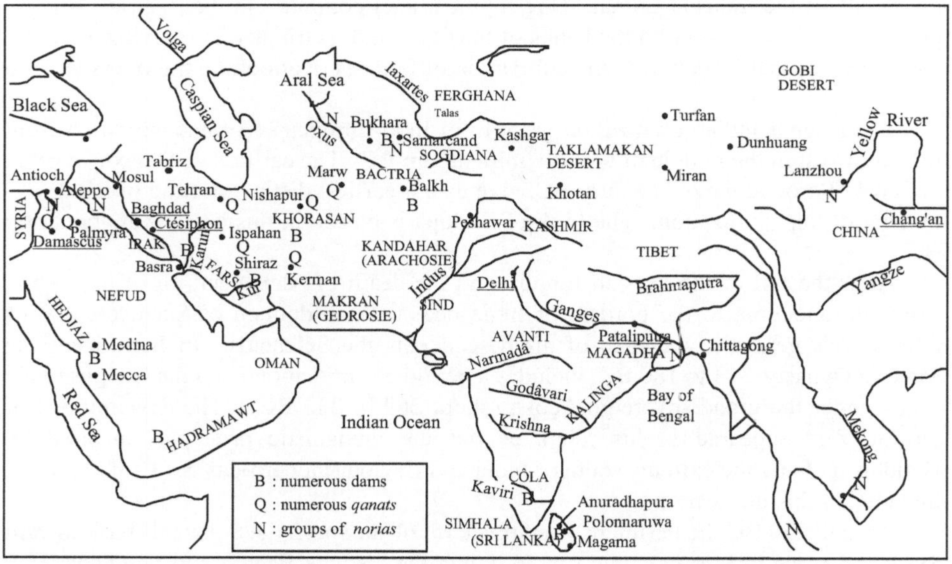

Figure 7.1 From the Syrian coast to the heart of China, in the ancient and medieval eras. Cities whose names are underlined are the great capitals: Ctesiphon for the Sassanide Persians, Balkh (Baktria) for the Greek kingdom of Bactria, Damascus and Baghdad for the Ummeyyades and the Abbassids, Pataliputra (Patna) for the India of the Mauryans and the Guptas, Delhi for the Turkish sultanate that dominates the north of India in the XIIIth century, Chang'an for China of the Han and Tang dynasties.

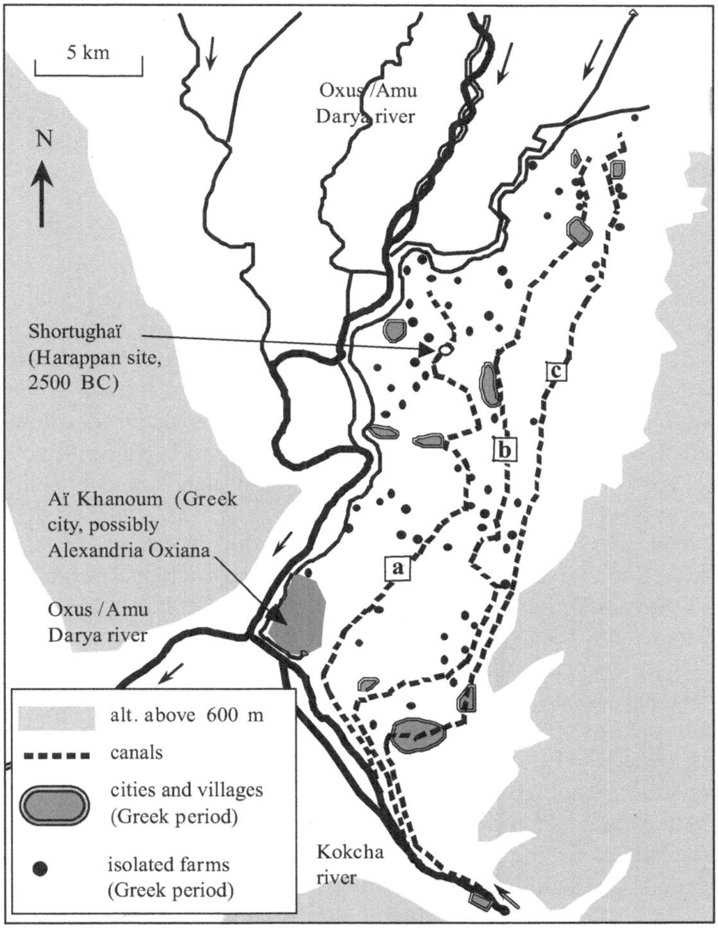

Figure 7.2 Irrigation of the plain of Aï Khanoun in eastern Bactria, at the confluence of the Oxus (Amou Daria) and the Kokcha. Aï Khanoun was probably the Alexandria of the Oxus. The map shows the population patterns and the traces of the principal irrigation canals during the time of the Greek kingdom of Bactria. These principal canals rise from the Kokcha and run along the plateau toward the north. The Oxus flows in an adjacent lower course, 20 m below the irrigated plain, which is why its water could not be used. After 37 centuries of irrigation using this same basic layout of canals, the plain returns to desert after the Mongol invasions of the 13th century AD (Francfort, 1989; Gentelle, 1989; Gardin, 1998). For an overview map, see Figure 1.3.

a. Canal trace dating from the Bronze Age (IIIrd millennium BC) passing by the Harapan site of Shortughai. It supported irrigation of 6,000 hectares of barley, wheat, lentils, and sesame.
b. Canal developed before or during the time of the Persian Empire (Figure 2.20)
c. New canal built in the time of the Greek kingdom of Bactria. It brings the irrigated area to 16,000 hectares. It is abandoned, and then restored with the same general alignment during the Islamic period.

Irrigation in Ancient India

Agriculture is the foundation of the economy of India, and consequently the practice of irrigation is widespread. In the above introduction, we have tried to give some idea of the broad cultural mixing that took place in this country. Because of this mixing, combined with India's traditional lack of interest in its own history and the difficulty of dating Sanskrit texts, it is quite a challenge to find the origins of innovations, and sometimes even to identify references to original hydraulic works in the texts. Nonetheless, ancient documents do mention the existence of canals, reservoirs, gates, and machines for lifting water.[1] The lifting wheel ("rotating wheel fitted with buckets") is mentioned with a date that is perhaps 350 B.C. but impossible to confirm.[2] Moreover, it is impossible to know if this description refers to a simple wheel, a bucket chain (*saqqya*), or perhaps a true hydraulic *noria*, which seems unlikely.

The great reservoirs of Ceylon

About 544 BC the island of Ceylon (Sri Lanka) falls under the domination of princely families of Indo-European origin, coming from Maghada. The Ceylonese civilization develops markedly after the conversion of the island to Buddhism. This conversion is attributed to a certain Mahendra who, according to tradition, is said to be the son or the brother of the great king Açoka. Irrigation is a significant factor in the island's development. Rainfall is abundant, but concentrated during the monsoon period. Since the soil is relatively impervious, very little of the rain is stored as groundwater. The only solution is construction of dam-reservoirs.

We can identify several reservoir-dam installations dating from 370 BC to 540 AD. They range in height from 5 to 20 m, and some are several kilometers long.[3] Eight of them are in the region of the capital Anuradhapura; three in the region of the great southern city, Magama, and four more near Polonnaruwa, which becomes the capital of the island after 781. All of these are earthen dams of trapezoidal section (Figure 7.3), fairly narrow at the top. The talus slope was rather steep for the first of these dams (370 BC), being 1:2.5 on both the upstream and downstream faces. Subsequent dams had somewhat flatter slopes, generally 1:3 on the downstream face (and in unusual cases, 1:5). These rather long dams are not rectilinear, but follow the terrain in such a way as to minimize their height and maximize the volume of stored water – which can be multiples of ten million cubic meters for the larger reservoirs. One dam near Polonnaruwa is an exception to this general configuration. It is relatively short, but is built to a height of 17 m at the time of its initial construction in around 300 BC, and then is raised to 34 m in 460 AD. Depending on the dam, floodwaters are spilled either over a natural rocky sill, or over a thick stone wall covered with a layer of large blocks. Often several dams are built on the same river, or on different tributaries of the same watershed. From the 5th century AD, canals that are 15 to 30 km long transfer water from one valley to another within the same overall watershed.

One of the reservoir-dam installations in the northwest portion of the island is gigantic, having an immense reservoir of 39 million cubic meters. The reservoir is contained by a semi-circular dike that is 9 km long. This project was first undertaken in the 12th century AD by the king Parakrama Bahu, but it remained unfinished, the necessary efforts exceeding the capacities of the population. Eventually Tamil invaders, coming from the south of the Indian subcontinent, replace the authoritarian dynasty that had

1 From Sanskrit texts translated by Louis Renou, The civilization of ancient India, (1950), pp. 148-151
2 Needham and Ling, Science and Civilization in China, IV, II (1965), p. 361
3 Schnitter (1994), p. 34.

developed such vast hydraulic resources for the island of Ceylon. The Tamils neither maintain nor further develop the hydraulic systems, which therefore progressively fall into decay and ruin.

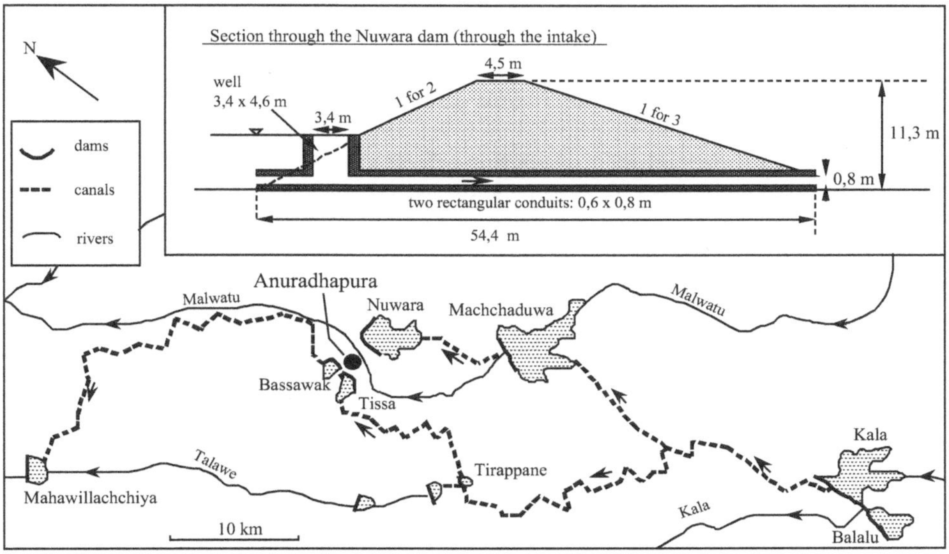

Figure 7.3 Development of the basin of the river Malwatu near the ancient capital of Ceylon, Anuradhapura. The first man-made reservoirs are those of Bassawak and Tissa (about 300 BC), then Nuwara (80 BC), Balalu (40 BC), Mahawilachchiya (70 AD), then Kala and Tirappane (around 470 AD) and Machchaduwa (540 AD). The sketch shows the state of the development at the end of the 6th century AD, and the structural details of the Nuwara dam, 4.8 km long, creating a reservoir of 43 million cubic meters. After Schnitter (1994).

Hydraulic works in medieval India

Dams for control of river flow and for irrigation are also built on the Indian subcontinent itself. One of the oldest of the important dams in India is an earthen structure built about 150 AD on the Kâviri River in the southern region of Côla. This dam, 330 m long and 18 m wide, is rebuilt in the 11th century. River traffic is important along the great rivers of central India – the Ganges and the Indus. Later on, the Hindu civilization takes refuge in the south of India when the Turks occupy the north of the country. From the 9th to the 12th century, in particular, the maritime empire of the Côla unifies all the south of India, even temporarily occupying Ceylon. Numerous dams are built, probably under the influence of the earlier developments in Ceylon. One can identify some fifteen important dams dating from the 11th to the 16th century, generally ranging in height from 9 to 24 m and several hundred meters long; however one of them is no less 16 km

4 Schnitter (1994), p. 98.

in length.[4] As in Ceylon, these are all earthen dams.

The Indians are good mathematicians (they invented algebra). But ancient Indian writings show few innovations in the domain of physics, which remains aligned with the Greek theory of the four elements.[5] Indian intellectual efforts were more focused on medicine than physics.

In the north of India, the 13th and 14th centuries witness the development of the infrastructures of Delhi,[6] capital of Turkish sultanate between 1192 and 1388. The initial site of the capital being far from the river, the city constructs very large open reservoirs, and directs storm runoff to them through canals. The inhabitants then get their water from these reservoirs. A new site closer to the river Jamuna is chosen for the city between 1320 and 1325. Although a dam and dike are constructed to create a reservoir for this new location, the site is abandoned even before being occupied. The original site of Delhi remains occupied, despite the inconvenience of its distance from the river. Therefore the city continues to depend on its reservoirs. The account of Tangerian Ibn Battûta, who lived in Delhi between 1335 and 1341, gives some idea of the scale of these reservoirs, as well as a sense of their fragility:

> "Outside Delhi, one can see an enormous basin that carries the name of the sultan Shams ad-dîn Lalmish and serves a supply of potable water for the inhabitants of the city. [....] This basin is two milles (3.5 km) long and half that distance in width. The western part, next to the musallâ (a place of prayer), is made of stones that are laid in steps like benches at different heights. Under each "bench" are steps that enable one to get to the water. Beside each "bench" one sees an enormous cupola with seats for those taking walks or others who are just relaxing. In the interior of the basin there is a grand cupola two stories in height, made of sculpted stone. [....] When the basin is dry, sugar cane is grown along the edges, as well as cucumbers. [....] Between Delhi and the residence of the caliph is the basin al-Khâsa, even larger than that of the sultan Shams ad-dîn. It has nearly forty cupolas around its edge."[7]

Still later the city of Delhi is moved onto the banks of the Jamuna, under the sultan Firuz Shâh (1351-1388) because the maintenance of the canals and reservoirs was judged to have become too burdensome. Along with the move came construction of a great bridge-dam on one arm of this river, as well as irrigation works. In the end all these efforts are in vain, for Delhi is razed by the Mongols of Tamerlan in 1398, ten years after the death of Firuz Shâh.

Water resources for Persia and the silk road

The traveler coming from Taklamakan or India enroute for the Roman or Arab worlds, whether he crosses the Kush or the high passes of Pamir that lead to Bactria, encounters the vast arid zone of the Persian plateau (or Khorassan). The plateau's sparse and unreliable water resources were exploited by means of several irrigated oases during the

---

5 Basham (1954).

6 Porter (1992).

7 Ibn Battûta, 1995, adapted.

Bronze Age. Then much larger irrigated zones were developed beginning with the period of the Achaeminde Persians. This development was based on the mining of groundwater through *qanats*. The earliest evidence of these projects is the account of the historian Polybius, who describes an expedition led by the Seleucid king Antiochus III against the Parthians in 210 BC. When the army of Antiochus penetrates into the desert, forcing his enemy to retreat, the Parthian sovereign Arsace II has his horsemen destroy the *qanats* of the region:

> "Arsaces had expected Antiochus to advance as far as this region, but he did not think he would venture with such a large force to cross the adjacent desert, chiefly owing to the scarcity of water. For in the region I speak of there is no water visible on the surface, but even in the desert there are a number of underground channels communicating with wells unknown to those not acquainted with the country. About these a true story is told by the inhabitants. They say that at the time when the Persians were the rulers of Asia they gave to those who conveyed a supply of water to places previously unirrigated the right of cultivating the land for five generations, and consequently as the Taurus has many large streams descending from it, people incurred great expense and trouble in making underground channels reaching a long distance, so that at the present day those who make use of the water do not know whence the channels derive their supply. Arsaces, however, when he saw that Antiochus was attempting to march across the desert, endeavored instantly to fill up and destroy the wells. The king when this news reached him sent off Nicomedes with a thousand horse, who, finding that Arsaces had retired with his army, but that some of his cavalry were engaged in destroying the mouths of the channels, attacked and routed these, forcing them to fly, and then returned to Antiochus."[8]

As is noted by Henri Goblot in his study of the *qanats*, the above text shows that the wells were not very well maintained under the Parthian regime, since nobody knows the layout and source of the underground channels. The Parthians began as nomads, caring little for the infrastructure of irrigation. It is certain that the system of *qanats* was once again developed under the Sassanides, and especially under the Arabs and the Turks between the 9th and 11th centuries. The city of Nishapur, in Khorassan, owes its prosperity to these wells from the beginning of the 9th century. A text from 830 AD describes how the judges of all Khorassan and even of Iraq came together to write a book of law regarding use of the *qanats* (the *Kitab al Kani*), given the absence of any prior legal precedents or earlier Muslim law. The Persian mathematician al-Karagi, living in Baghdad, wrote another more technical account of the *qanats* in about 1010 AD.[9]

At Marw (Merv, Antiochia of the Margiana to the Greeks), the river Murgab has a dam whose age is difficult to determine, though we know that it is maintained during the Islamic period. This dam serves to stabilize the upstream progression of settled areas which we have discussed at the end of Chapter 2.[10] It also leads to the regrouping of

---

8 Polybius, Histories, X, 25, Translation of W. R. Paton,
http://penelope.uchicago.edu/Thayer/E/Roman/Texts/Polybius/home.html

9 Goblot (1979); Landry (1990).

10 The Parthian period marks the beginning of the concentration of habitation at Marw, associated, according to Bader et al (1996) with the construction of a dam. Hiebert (1992) places the origin of this dam in the time of the Seljuks.

dwellings inside an enclosure, and provides for a more reliable and regular functioning of the system of irrigation - even though the area also is equipped with *qanats*. Some 10,000 men are employed, under the direction of a superintendent, to maintain the hydraulic system in the 10th century.

At the beginning of the 13th century the Mongols destroy not only the cities, but also the hydraulic systems. Marw and Balkh (Bactra, ancient capital of Bactria) are abandoned, but Nishapur and Harat rise from their ashes, as does Samarcand, the great center of the silk trade in Sogdiana. The *qanats* are rebuilt; Marco Polo notices them to the north of Kerman in about 1272:

> "The fourth day *(of crossing the desert)*, we came upon a fresh-water river that flows mostly underground, but in certain places, there are openings created by the waters, where one can see it flow, but then it immediately returns below the ground. Nonetheless, one can drink to ones full. Not far from there, travelers who are spent by the ardors of the desert they had crossed, rest and refresh themselves and their beasts."[11]

When the traveler Ibn Battûta visits Nishapur about 1335, he writes that this city is called "little Damascus" for the abundance of its running water and the lushness of its gardens. A Persian historian of the 15th century reports the words of another Arab voyager who was not entirely happy with his experience:

> "What a beautiful city Nishapur would be if its canals were above ground and its inhabitants underground!"[12]

The *qanats* are still in operation in this area, but it is not possible to assign dates to them individually. In Iran there are twelve groups of *qanats,* some along an axis parallel to the Caspian Sea, then oriented toward the east, i.e. from Tehran to Nishapur; and others on the eastern foothills of the Zagros mountains (Ispahen) and in the center of the region of Zarand-Kerman (see Figure 7.1). Among the two thousand *qanats* that have been studied the longest is 50 km, in the center of the area where the land is relatively flat. But 81% of the *qanats* are shorter than 5 km, and 36% of them are between 500 m and 2 km. The delivered discharged is normally between 10 and 100 m^3/hour. Among the 180 *qanats* in the Tehran region, the depth of the mother well at the head is usually between 10 and 50 m, but it can be more than 100 m or even 150 m in certain cases.[13]

Important rivers such as the Karun and the Kur rise in the Zagros and Fars mountains of southern Persia. Rather typical hydraulic works are constructed along the Karun, constituting masonry overflow weirs combined with bridges, beginning in the Sassanide period. Foremost among these is the 520-m long weir built by Roman prisoners around 260 AD after the capture of the Emperor Valerius. These projects typically raise the river level to supply irrigation canals. But in addition, beginning in the Arab period, they supply water to batteries of mills, or *norias*. The Amir and Feizabad dams on the river Kur, near Shiraz in Fars, are 9 and 7 meters high and 103 and 222 meters long, respectively. These dams are equipped with an impressive number of water mills:

11 Marco Polo, Le devisement du monde, XXXVIII, adapted.
12 Citation from Goblot (1979)
13 Beaumont (1989)

no less than 30 for the Amir (Figure 7.4) and 22 for the Feizabad. These mills have horizontal wheels on vertical axes, and are driven by water falling vertically onto their blades:[14]

> "Adud al-Dawla closed off the river, between Shiraz and Istakhr, with a great wall, reinforced with lead. The water that accumulated behind this dam formed a great lake. Above this lake, on its two sides, there are hydraulic wheels like those that we have mentioned in Khuzistan. Below each wheel, there is a mill, and today this is one of the marvels of Fars. Later, he constructed a city. Water flows through canals and irrigates 300 villages in the valley."[15]

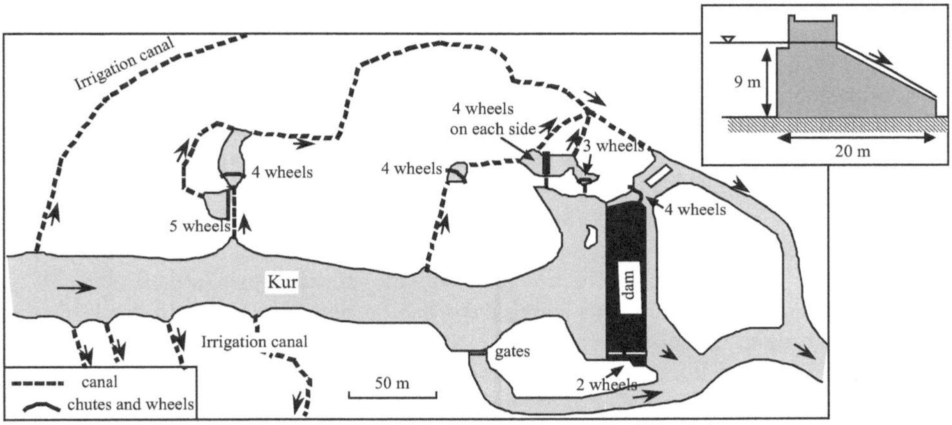

Figure 7.4 The Amir installation on the Kor river, in Fars (10th century) and its thirty water wheels (after Schnitter, 1994).

The regions of Bukhara and Samarcand are to the northeast of Khorassan, between the Oxus and Iaxartes rivers. If Khorassan is the land of *qanats*, Bactria and Sodgiana are the lands of gravity irrigation, through river water diverted into canals by small hydraulic structures. This practice in the Oxus basin stems from the Bronze Age (Figure 2.30), and becomes fully developed in the Greek kingdom of Bactria, then under the Kuchans, between the 3rd century BC and the 2nd century AD. Samarcand solidifies its identity as a great commercial crossroads in the silk trade from this time on. Around Samarcand and Bukhara irrigation networks originating from the Zeravchan River branch out over tens of kilometers.

Ancient Samarcand is located on high ground, and benefits from an advanced water-supply system – a conduit forming a siphon that is destroyed during the siege of the city by the Mongols of Ghengis Khan, in 1219. The city is rebuilt right on the river banks, making it possible to use *norias* to provide water for the city and its gardens. Ibn Battûta visits Samarcand around 1335, a little more than a century after its pillage. Even though the ruins remain visible, it has clearly become a beautiful city once again:

---

14 Schnitter (1994), p. 87.

15 Al-Muqaddasi, Arab geographer, citation after Hill (1997) in History of Arab science, p. 21, adapted.

> "... I reached Samarcand, one of the geatest cities, the most beautiful and the most superb. It is located on the wadi al-Qassârin *(River of the Fullers)* on which there are hydraulic wheels for irrigation of the gardens. The inhabitants get together, after prayers, to stroll and amuse themselves along the banks of the river where one can see benches and seats for resting, and stands that sell fruits and other consumables. Formerly there were, along the banks, imposing palaces and edifices that lead one to imagine the ambition of the inhabitants of Samarcand. But most of these were destroyed, as was a large part of the city [....]. In the city, one can see gardens."[16]

Samarcand is known for the quality of its paper. Fullers (water hammers that shred and mash linen cloth to produce fiber for paper protection) powered by hydraulic force lend their name to the small river that feeds them, *River of the Fullers*, a tributary of the Zeravchan. Strong Turk-Mongol regimes launch their raids on surrounding lands from central Asia, where they also develop some new water resources. In the 11th century, the Ghaznavid Turks build two dams in the Samarcand region, 8 and 15 meters high, and 25 and 52 meters long. They also build another larger structure 23 km to the north of their capital city Ghazni, in the region of present-day Kabul; it is 32 meters high and 220 meters long. The Seljuk Turks who succeed them rebuild a dam at Marw on the Murgab River to provide water for the oasis, a dam that will be rebuilt yet again by the Timurids so the oasis can be repopulated.[17] In the 14th century, Tamerlan and his successors establish Samarcand as the capital of a great central Asian kingdom, and build still more dams in the regions of Teheran, Kashan, Tabas and especially near the new city of Mashadd (not far from Nishapur), present-day capital of Khorasan. Near Kebar and Tabas are three arch dams, among the first known to exist after the few Roman and Byzantine arch structures. The largest, 50 km to the southeast of Tabas, is 28 meters long and 60 meters high, a record that is destined to stand for quite a long time.[18]

In the heart of the Arab world: the splendor of the Umeyyades and the Abbassids

Byzantium, the Sassanides and the new Arab empire

Water supply systems of the Byzantine cities do not measure up to those of the Romans, either in quality or in quantity. In many Byzantine cities, such as Apamea-on-Orontes or even Constantinople itself, aqueducts are abandoned in favor of cisterns, sometimes very large and fed by runoff from rainstorms. Small rural communities located along wadis or on small rivers implement numerous hydraulic developments at their scale[19] –

16 Ibn Battûta, 1995, adapted.

17 Hiebert (1992).

18 Schnitter (1994), pp. 89-92.

19 Ducellier, Kaplan, Martin (1990), p. 132; see Sadler (1990) for a detailed study of such a fertile complex in Syria.

such as modest derivation canals that support gravity irrigation of valley fields, with many small mills.

The Byzantines indulge in large-scale hydraulic activity in Anatolia, where they build several dams. One among them is the Dara dam, constructed under Justinian (527-565) on a tributary of the Khabur. It probably has an arch in its central portion – at least this is what Procopius of Caesarea, in the 6th century, attributes to the dam's architect, Chryses of Alexandria:

> "He did not build the dam in a straight line, but in the form of a crescent, such that this arch, turned against the stream of the water, could better resist its violence."[20]

Figure 7.5 A cistern serving a house of the Byzantine age at Sergilla, in northern Syria (photo by the author).

At about this time, and at the other end of the Syro-Mesopotamian universe, a great misfortune occurred in what had been the land of Sumer: the submergence of the irrigated lands of lower Mesopotamia, with new marshes that rendered the land useless during all of the Middle Ages. The bed of the Persian Gulf had been filling with sediment since the Bronze Age (Figure 2.1). The fields created by alluvial deposits of the great rivers are relatively low in elevation, and only drained thanks to the inhabitants' continuous

20 Citation from Schnitter (1994), p. 80.

efforts.[21] Repeated ruptures of the dikes along the Tigris occur during the time of the Sassanide sovereign Kawadh (488 – 531). His successor Khusraw I (531-579) manages to repair the dikes. But on the eve of the Arab conquest, one hundred fifty years later in 627 or 628, there is a catastrophic flood of the great rivers. Khusraw II is powerless to manage these floods, despite an interesting means of managing human resources. This is told to us by al-Baladhori, one of the most ancient Arab historians (he dies in 892):

> "Then when arrived the year when the Prophet (God bless him and give him peace!) sent as ambassador to Chrosroes-Parviz *(Khusraw II, 590-628)* Abdullah son of Hodhafa as-Sahmi, that is to say in the year 7 or 6 of Hégire the Euphrates and the Tigris had a considerable flood, such as had never been seen before or after: large breaches opened that Parviz tried to close, but the water was stronger and reached the low country, submerging villages and crops and several land districts in this place. Chrosroes came to the site in person to block the breaches: he laid a pile of silver on a leather tablecloth and put to death those workers who did not work hard enough (it is said that on a single dike he put under the cross, in one day, forty of those who worked there), but he could not stop the water. At the same time, the Arabs invaded Iraq and the Persians became henceforth preoccupied by war, to the point that the breaches grew larger without anyone worrying about it: the landowners in the villages were powerless to block them, so large were they, so the marshes grew in extent."[22]

The Umeyyades installed the capital of their empire at Damascus not long after the Arab conquest of Syria, (661 to 750 AD). This empire soon extended from the Atlantic ocean to the Indus. Damascus, a city that had prospered since the Bronze Age thanks to the water taken from Barada where it comes out of the Anti-Lebanon mountains (Figure 7.6), quickly becomes the very image of paradise, with its gardens and orchards. Yazid, the second Umeyyade caliph, builds a new canal coming from the Barada (after which the canal is named) and establishes a pattern for the cultivated zone, the *Ghouta*, that will last for all of the Middle Ages. The Andalusian pilgrim Ibn Jubayr, in the 12th century, tells us of the charms of this region:

> "The gardens surround Damascus as a halo surrounds the moon (….). To the east, the green Ghouta extends as far as the eye can see and no matter which direction one looks, its sparkling splendor transfixes the gaze. How true is what one says of Damascus: If paradise is on the earth, Damascus is it, and if paradise is in heaven, Damascus is its rival and just as wonderful!"[23]

The Umeyyades use Egypt as a granary, much as did the Romans and then the Byzantines, and send wheat to Arabia using the ancient canal of Necho between the Nile and the Red Sea (Figure 3.8). They renovate the canal about 641 and rename it *canal of the Caliphs*.

To foster the economic development of Syria and Iraq, the Umeyyades continue the

---

21 Strabo had already described the difficulty of maintaining the hydraulic system of this region; the reader can refer to the extract cited in Chapter 2.

22 Al-Baladhori, in "Arab Historians" (Sauvaget, 1988).

23 Ibn Jubayr (beginning of the 12th century), Relations de Voyage, adapted from the translation of Paule Charles-Dominique.

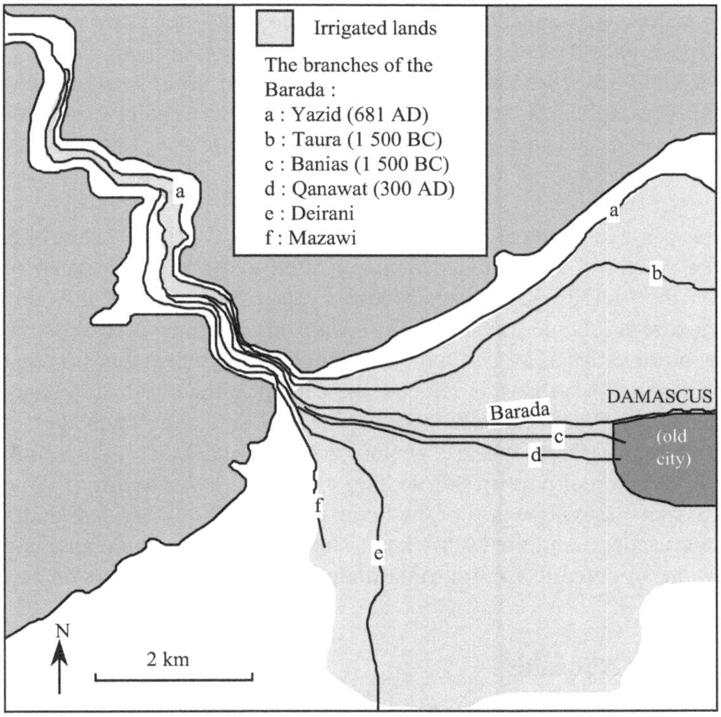

Figure 7.6 The arms of the Barada artificially branched off to irrigate the Ghouta of Damascas (after Kamel, 1990). In 1185, ibn Jubayr observes these seven arms from the top of a hill he climbed as a pilgrim: "This blessed hill (according to the Koran, Jesus and his mother took refuge on it) marks the beginning of the city gardens and the line of separation of the watercourses that divide into seven branches each going in a different direction. The most important of these is the Taura that passes below the hill and that is dug into the rock as an underground canal, as large as a grotto. Now and then an audacious swimmer, a young boy or a man, dives from the heights of the hill into the river and swims underwater to cross the canal below the hill and comes out downstream. But this is a very risky undertaking! From the hill, one can look over all the gardens to the west of the city. No vista rivals this one in beauty, splendor and perspective."[24]

ancient Roman method of conceding undeveloped land to colonists – friends of the powerful, retired soldiers, even entire tribes. Then, when the lands begin to produce a harvest, the colonists pay a tax to the central treasury. Under the first caliph Mu'awiya (661 – 680), and under his successors al-Walid (705 – 715) and Hisham (724 – 743), the marshland reclamation works in lower Mesopotamia are financed by the close colleagues of the caliph, who later take possession of those lands:

> "The breaches having opened during the time of al-Hajjaj *(the governor of Iraq)*, he wrote to al-Walid *(the caliph)* to inform him that the cost of closing them would not be less than three million dirhems. Al-Walid judged this to be excessive, but his brother Maslama offered to

24 Ibid.

undertake it himself, under the condition that he would be given rights to all the lands that would remain submerged, once the three million dirhems had been spent. [....] Al-Walid accepted: Maslama thus obtained several adjoining land districts. Then he had two canals dug from as-Sib, brought in peasants and farmers, and brought these lands under cultivation: these people prospered and formed numerous villages, to take advantage of the protection they would offer."[25]

In 750, the Abassids come to power and massacre all the Umeyyade family save one, who flees to found the caliphate of Cordoue. The definitive closure of the canal of Necho, associated with the troubles in Hedjaz, occurred at the beginning of this new dynasty in 767. The closure is ordered by the caliph al-Mansour, to prevent the shipping of wheat toward the revolting cities of Mecca and Medina and to prevent an invasion that might make use of the canal. The Abbasids relocate the capital to the city of Baghdad, founded in 762 on the Tigris. From 836 to 892 the capital will be moved once again a bit further upstream, to Samarra. From the 11th century this will be the era of the Seljuk Turks, with a general distribution of power. Before long it becomes a *de facto* split. One side of the split is the Syrian-Egyptian world of the Fatimides, the Ayyubids (the Saladin Dynasty, enemy of the crusaders) and the Mamlukes. The other side is the Persian and Iraqi worlds that are more oriented to Central Asia under the influence of Persian and Mongol leaders.

The great libraries of the Abbasids

Two grand libraries flourished under the Abbasids. They were motivated by the same goals as the ancient libraries of Alexandria and Pergamon: the prestige of the sovereign and the attraction to scholars from everywhere. The grand library of Baghdad, the *Bayt al-hikma* was developed under the reign of Haroun al-Rashid (786 - 809). It benefited from the latter's acquisition of ancient Greek works from the court of Constantinople, a practice continued by his successor al-Mamun (813 – 833). A strong memory of the *belle epoch* of these rich libraries persisted in the 15th century:

"The caliphs and the sovereigns had a lively interest in the grand libraries and paid close attention to them, enabling them to acquire beautiful and numerous collections. It is said that the greatest libraries of Islam were the following three: the library of the Abbasid caliphs of Baghdad, [...], the library of the Fatamide caliphs of Cairo [...] and the library of the Umeyyade caliphs of Spain."[26]

The influence of the Baghdad library begins to fade when the capital is relocated to Samarra in 836. But Alexandria, Antioch, Edhessa, Haran, and Nisibia remain great intellectual centers.

---

25 Al-Baladhori, in Arab Historians (Sauvaget, 1988).

26 Al-Qalqashandi, Arab author of the 15th century. Citation after Micheau, in History of the Arab Sciences, III.

Hydraulics and prosperity in the heart of the Arab world

Arab agriculture included exotic crops such as cotton, rice, and sugar cane in addition to traditional grains and fruits. Cotton was known in Mesopotamia since the time of the Assyrians, but was essentially undeveloped. These crops require considerable water, and therefore are grown in the large irrigable zones on the shores of the Khabur, the Euphrates, and the Tigris.[27] The Muslim world uses all known existing techniques to develop irrigation. This includes derivation canals from rivers and wadis, water-lifting machines, and even drip irrigation for young plants, a technique known since the 5th century and wonderfully described in an Arab work of the 12th century.[28] The *shaduf* is used to lift water out of rivers and canals, but on a small scale. The bucket chain or *saqqiya* and the *noria*, are the most widely used devices. The *noria* sees considerable application, especially on rivers of regular flow such as the Orontes and the Khabur, and also on the middle Euphrates. We will come back to this point further on.

Qanats exist also, but it is usually impossible to tell if the origin of a particular installation is Arab, Roman, or even earlier. In Syria (where they are called *foggaras*), there are 45 qanats in the Palmyra region, 50 in the *ghouta* of Damascus, 35 between Damascus and Homs, 20 to the southeast of Homs, 50 to the east of Hama, 15 to the southwest of Aleppo, and 25 to the east of Aleppo.[29] In the 12th century, Ibn Jubayr takes note of them between Homs and Damascus:

> "We camped in a large village of Christians called al-Qâra where no Muslim lived, and that has a caravanserai much like that of a large fortress. In the center, one can see a large basin that is always full of water, supplied by an underground stream coming from a distant source."[30]

This axis Aleppo – Hama – Homs – Damascus, in Syria, is particularly developed and irrigated. We mentioned above the *qanats*, frequently encountered along this corridor, as well as the water taken from the Barada for the *ghouta* of Damascus (Figure 7.6).

Homs is irrigated using canals issuing from the Roman dam of Homs lake (Figure 6.33).

The steep banks of the Orontes at and around the city of Hama are irrigated thanks to large, numerous, and particularly famous *norias* (Figure 7.9). The oldest of these *norias* to which a date can be assigned (for its canal carries an inscription) was constructed in 1361 (Figure 7.7). But one cannot separate the *norias* from the city of Hama; in 1185, Ibn Jubayr tells us they were already there:

> "On the two banks (*of the Orontes*), starting from the hydraulic wheels and extending regularly out, are gardens, whose tree branches hang over the water. [...] On one of the banks that borders the outlying district one finds washing stations arranged like several rooms; the water, raised by a hydraulic wheel, crosses into all the hidden crannies."[31]

---

27 Berthier (1990).
28 Zakri (1990).
29 Safadi (1990).
30 Ibn Jubayr, Accounts of a Voyage, adapted from the translation of Paule Charles-Dominique.
31 Ibid.

Figure 7.7 The noria al-Muhammadiya and its aqueduct at Hama, dated as 1361 from the inscription on a pillar of the aqueduct. As are the other norias of Hama, this one has been maintained and regularly renovated since its initial construction. With a diameter of 21 m, it is the largest known ancient noria (photo by the author).

Figure 7.8 The group called the "four norias" at the entrance of the Orontes into Hama. The aqueduct of the two norias at the right has disappeared (photo by the author).

The remains of some fifteen of these *norias* (or the gardens that they irrigated) can be found in the archives of the 16th century of the tribunal of Hama.[32]

Qanats and *norias* are costly devices that provide copious quantities of water. Therefore social organization is needed to regulate their use. In the Muslim world, periodic hourly time schedules are established at a scale of about ten days. In the Syrian areas irrigated by *qanats* to the north of Damascus, irrigation periods are every twelve days, or rather every 24 half-days (from two hours before dawn to two hours after sunset). The unit of time is one "hour" of a hundred minutes, and through an alternating schedule a user can draw water first for a daytime period, then a nighttime period.[33] At Hama, on the other hand, weekly cycles are used to allocate water from the *norias* among the fields, the mosque fountains, or the public baths.

In some cases, Roman hydraulic works are brought back into service. For example, when the tenth Umeyyade caliph Hisham (724 – 743) decides to build a palace in the

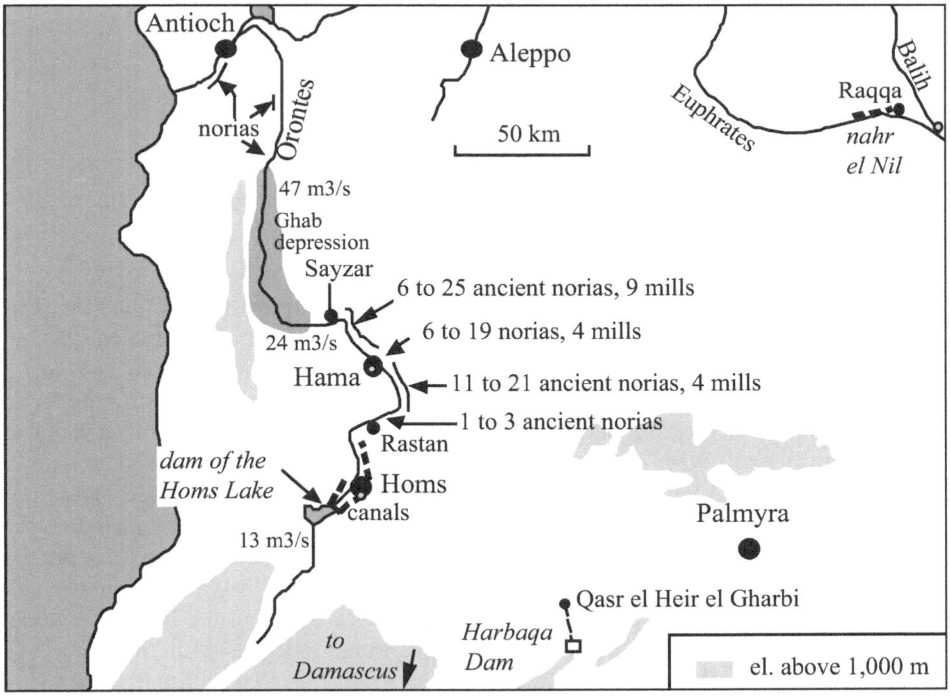

Figure 7.9 The valley of the Orontes, showing the locations of the ancient norias. The lower figure shown corresponds to the number of norias that can be reliably dated from the 16th century (from inscriptions and archive documents analyzed by Zaqzouq, 1990). The implantation of norias is particularly dense between Rastan and Sayzar, where the slope of the Orontes is a fairly regular 1.1 m/km. The indicated discharges are the mean flow of the Orontes at different points (after Delpech, girard, Robine, Roumi, 1997).

32 Zaqzouk (1990).
33 Haj Ibrahim (1990).

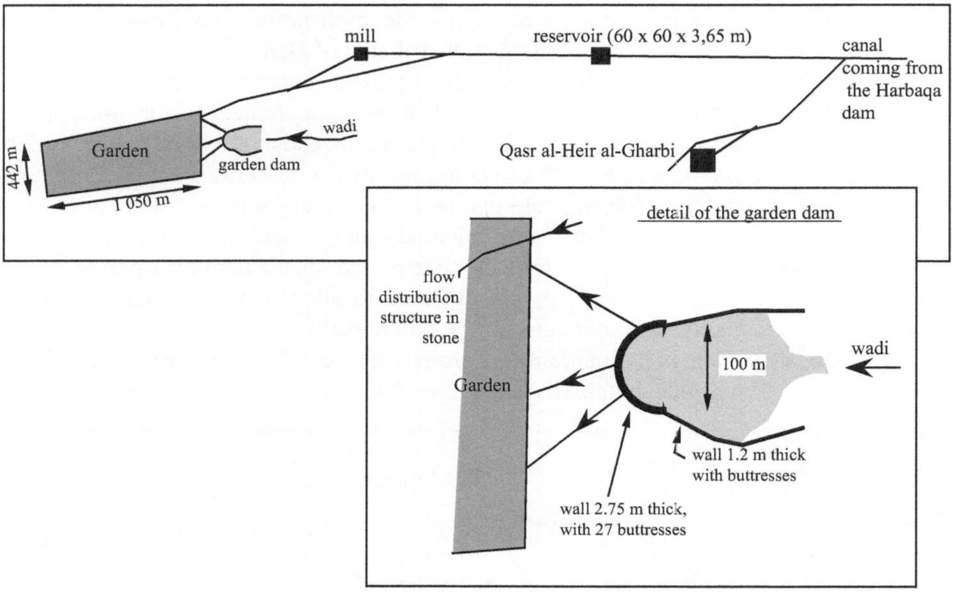

Figure 7.10 Layout of Umeyyade hydraulic installations at Qasr al-Heir al-Gharbi, and of the garden dam (after Sailby, 1990; Calvet and Geyer, 1992).

desert of Palmyra in 727 (at the intersection of the caravan routes, on the road between the capital Damascus and the middle Euphrates), he makes use of the ancient Roman dam of Harbaqa (Figures 6.34, 7.9).[34] This palace, known as *Qasr el-Heir el-Gharbi*, is supplied by an underground canal issuing from this dam that is 16.5 km long. The aqueduct supplies a reservoir that is 60 m on a side and 3.65 m deep. The aqueduct also supplies a mill, and further downstream, a nearly rectangular large cultivated area of 45 hectares, or *garden*. A rather strange dam (the *garden dam*) is built to capture water from the wadis that discharge downstream of the Harbaqa dam and to create a supplementary reservoir near the garden. Since the valley of the wadi is ill defined at this location, the ends of the dam extend in an unusual fashion along the edges of the valley (Figure 7.10).

The open canals downstream of the reservoir are earthen, but the hydraulic works and diversion gates are of stone, following the traditional practices of the region.

Mesopotamia becomes the granary of the great Arab cities. The entire irrigation system inherited from the ancient civilizations is carefully conserved, further developed, or put back into service if previously abandoned. The 35-km long grand canal called *nahr Saïd* (Figure 7.11) is built on the middle Euphrates, apparently starting from the intake of the very ancient bronze-age canal *Isim Yahdun Lim* (Chapter 2). It brings water to the city of Rahba, founded in 820 on the riverbank by the Abbasids and subsequently relocated to the edge of the escarpment after an earthquake in 1157. This canal, along

34 Saliby (1990), Calvet and Geyer (1992).

which there are several offtakes, is excavated into the plain. Therefore it cannot support gravity irrigation of large areas, as was done in the same region by the bronze-age canals.

This canal is instead used as a permanent source of water to be lifted into the gardens using *saqqyas*, remains of jars have been found there.[35] This practice explains why the region's villages, all Islamic, are located directly on the canal and its offtakes. The *nahr Saïd* is destined to be abandoned in the 13th or 14th century during the desertifica-

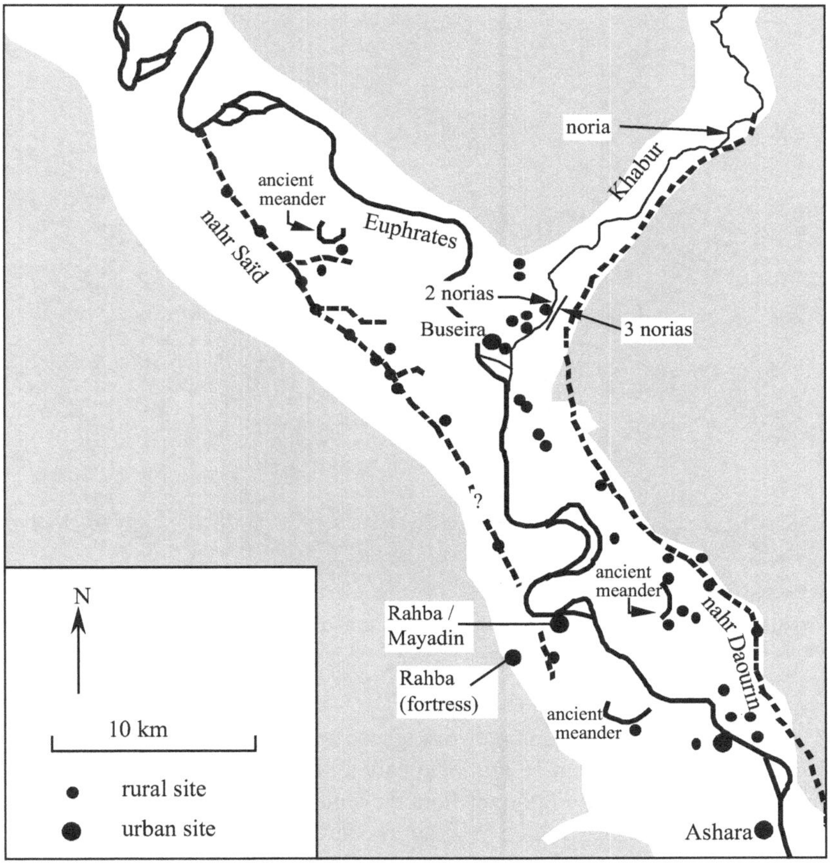

Figure 7.11 The confluence of the Khabur and the Euphrates – an area that is very developed from the 8th to the 12th centuries with the nahr Saïd for irrigation on the right bank and the grand navigation canal on the left bank (nahr Daourin) likely dating from the Bronze Age (Figure 2.7). The most downstream zone (where the ancient Mari was located) is almost desert in this period. Ashara is built on the site of the ancient Terqa. After Geyer (1990), Berthier and d'Ont (1994).

35 The dating of the nahr Saïd and the nature of its use are documented thanks to the investigations of Berthier and d'Ont (1994). The first canal, from at least the 10th century, is the one whose use is described here. A second canal along the same alignment but more recent, may have supported gravity irrigation.

tion of the region after the Mongol invasion (Rahba is abandoned about 1400).

The Raqqa region is further upstream on the Euphrates at the confluence of the Balih, and irrigated on the right bank from the Umeyyade period. The city of Raqqa itself, on the left bank, is an ancient Hellenistic implantation that was refounded by the Abbasid al-Mansour in 772. The caliph Haroun al-Rachid (the caliph of *A Thousand and One Nights)* especially develops the city during his residence there from 797 to 808. The

Figure 7.12 Vestiges of a noria and its mill at al-Lawriyé on the lower Khabur. The photo is taken from the left bank, downstream; to the left, one can see the remains of the dam (photo by the author).

city benefits from an extended irrigation network whose spine is a grand canal from the Euphrates, 10 m wide and more than 16 km long, called the *nahr el-Nil.*[36]

Navigation on the Euphrates is important from the Umeyyade and Abassid period up until the 11th or 12th centuries.[37] Rahba (Mayadin) controls an important fluvial port. The existence of Abbassid sites all along the ancient *nahr Daourin* (Figure 2.12) shows that this navigation canal is to all appearances brought back into service at this period.

On the Tigris, the *nahr Awan* canal services the new city of Baghdad, whose population in the 13th century is nearly a million and a half inhabitants. This grand canal, parallel to the Tigris, was built by the Sassinids, then enlarged and extended to capture water from the Diyala. In the 9th century a dam is built on the Adheim (or Uzaym) river, 150 km to the north of Baghdad. This dam, 15 m high and nearly 200 m long, feeds two new irrigation canals, the *nahr Batt* and the *nahr Rathan*. Like the other dams built by

36 After Kassem Toueir (1990).
37 Bianquis (1986).

Figure 7.13 Remains of a noria at Rweshed on the Khabur, immediately upstream of its confluence with the Euphrates. This noria had three identical wheels. At the left one can see the three canals and at the right, the remains of the aqueduct (photo by the author).

the Arabs (those of Fars, for example), this structure is built of stones interlocked with lead seals.

During the 7th and 8th centuries the simple military camp of Basra becomes a true city with a new irrigation system fed by the Chatt al-Arab waterway, the common course of the Euphrates and the Tigris. According to al-Baladhori, a certain Hassan the Nabatian directs the drainage and irrigation works in this region. The "Hassan reservoir" at Basra is attributed to him. The first tidal mill is built at Basra in the 10th century to operate during the falling tide.[38] But despite these hydraulic works, the swamps of lower Mesopotamia cause Basra to be known for its unhealthy air and the "yellow tint of its inhabitants", as described by Ibn Juzavy, the editor of the memoirs of Ibn Battûta.

In Arabia itself, particularly in the regions of Mecca and Medina where pilgrims congregate, numerous small dams are built on the wadis to provide water reserves through diversion of flood waters into basins and cisterns. There are some fourteen of these near Mecca and four in the region of Medina. The dams are from 2 to 11 m high and 25 to 225 m long. The largest is the Qusaybah dam near Median, notable for its height of 30 m.[39] Certain dams in this region have inscriptions that would date them from the Umeyyade era. But some well-known travellers suggest otherwise. They attribute the water-resource development, needed for the caravans of pilgrims who cross the desert between the holy cities of Mesopotamia, to the queen Zubayda, cousin and

38 Hill (1997), in History of Arab Science, III, pp. 14 and 47.
39 Schnitter (1994), p. 82.

wife of the caliph Haroun al-Rachid. Some descriptions of these watering places mention elaborate structures, not only the dams cited above but also the *qanats* when a permanent water source is found:

> "Friday morning, we camped in a place called Birkat al-Marjum, where there is a basin for which they built, on the hill overlooking it, a pipeline bringing water from far away. The installation is perfect and shows how impressive human resources are and the great things that can be done with them. [….]. All of these basins, all the reservoir, wells, and rest stops between Baghdad and Mecca were developed by Zubayda, daughter of Ja'far ben Abi Ja'far al-Mansur, wife and first cousin of Harun alRashid. She devoted all of her life to this project."[40]

The Arab cities, like the Roman cities, are major consumers of water – for baths, mosque hydrants, and caravanserais. According to Ibn Jubayr, Damascus has "nearly a hundred baths and nearly four hundred washing sites with running water everywhere" in the 12th century. Whether it be from their Roman heritage or from their Arab roots (like Rahba), these cities also have sewers.

Figure 7.14 Remains of an ancient noria on the right bank of the Euphrates, downstream of the Doura Europos escarpment, about 40 km downstream of Rahba (photo by the author).

40 Ibn Jubayr, Accounts of a voyage, adapted from the translation of Paule Charles-Dominique.

The Decline

Catastrophes strike the Arab Orient from the 11th century on, and especially in the 13th century. Its population declines and farmland reverts back to desert or, even worse, changes into swamps like those of the lower Mesopotamian lands. In Syria the crusades cause an exodus from rural settlement beginning in the 11th century; however this exodus is partly reversed by 12th-century agrarian reforms that encouraged a return to rural agriculture. But then the great tragedy of the Mongol invasions occurs. We have already seen how Ghengis Khan had ravaged Samarcand, Bukhara, Marw and the cities of Khorassan. And now in 1258 his grandson Hulagü razes Baghdad (later rebuilt) and irreversibly damages the Mesopotamian irrigation system. The Ottoman era begins in the 16th century; Constantinople becomes Istanbul and is endowed with new water supply systems such as the Maglova aqueduct, constructed in 1564.[41]

The Arab Occident

Water in *al-Andalus*

The Arabs came not only from North Africa, but also from Arabia, Syria, Iraq, Yemen, and Egypt. In 711 they conquered Spain, a country that had been occupied by the Visigoths since the fall of the Roman Empire. They brought with them all the Oriental technologies for water management: the ancient *shaduf*, the bucket chain or *saqqya*, the *noria*, and *qanats*. The Arabs preferred developments of more modest scale in the Mesopotamian tradition compared to the grand Roman hydraulic works. The great Roman dam-reservoirs like those of Proserpina and Cornalvo were apparently not brought back into service.

It is thought that the oldest Arab project is an overflow dam built at Cordova,[42] the capital of the Umeyyade caliphate. Its total length of 425 m exceeds the width of the river Guadalquivir, since it comprises multiple independent segments to increase the effective length of the weir and thus to limit the rise in water level during floods. We have seen this technique earlier in Assyria. The dam is equipped on one side with a large *noria* which lifts water for the l'Alcazar of Cordova. And at the three angles downstream of its broken line it has water mills, each equipped with four wheels. Many other mills existed along the course of the Guadalquivir, from Cordova down to below Sevilla:

> "He who wishes to travel by water from Sevilla to Cordova can embark on the river and travel upstream, passing by the mills of al-Zrâda, by the Mannzil Abân bend, [....] the Nâsih mills, to arrive in Cordova. [....] At Cordova one sees a bridge that surpasses all others in reputation and solidity. [....] Downstream of the bridge, and across the entire width of the river, there is a dike that is built of stones called "coptes". The columns are of unpolished marble. On this dike one can see three buildings, each containing four mills."[43]

---

41 Özis (1999).
42 Smith (1992).
43 Al-Idrissi (12th century), IV, 1, Translation of Jaubert.

But it is agriculture, and its need for irrigation, that is the driving force for the main hydraulic projects. The Arabs establish the cultivation of cotton, sugar cane, and rice in Spain. Agronomy manuals and plants themselves circulate between al-Andalus and the Arab Orient. The Andalusian school of agronomy flourishes, as seen in the treatises of Ibn Wafid and Ibn Bassal at Toledo; Al-Khayr, Ibn Hadjdjadj and Ibn al-Awwan at Sevilla; Ibn Beithar at Malaga.[44] This agricultural development supports significant population growth in al-Andalus between the 9th and 12th centuries.

The most celebrated of the irrigation projects is on the river Turia, to the north of Valencia (Figure 7.15).[45] It is possibly the descendant of an ancient irrigation network of the Roman period. Between 911 and 976 the Arabs built masonry outflow weirs covered with large stones sometimes interlocked with iron pins, as we have seen on numerous Arab or Persian dams. Among the nine sills of this system, from 1.4 to 7 m in height, five very likely date from the Arab period – they are mentioned in an edict of the king Jaime II in 1321 after the Reconquest of Valencia. Almost all of these sills have a downstream face in the form of stairsteps to serve as an overflow energy dissipater. Each sill feeds an irrigation canal, alternately on the right and left banks of the Turia river, and all of these canals subsequently branch out downstream.

From 961 any conflicts arising from water distribution, spillage, and the maintenance of the sills and canals are handled by the *water tribunal* of Valencia. This tribunal, whose deliberations are strictly oral, meets each Thursday in the mosque. After the Reconquest, during which the entire irrigation system is conserved and maintained, this same tribunal meets in the square of the cathedral of Valencia. This tradition continues unchanged to the present day.

Many other projects were developed on virtually all the rivers of the south and east of Spain. Upstream of Murcia there is a rather large dam 7.5 m high and 305 m long on the Segura river[46], built in the reign of al-Hakem (961-976) in a region that had been scarcely cultivated beforehand. It feeds two irrigation canals, one on each bank, with provisions for removal of sediment from the reservoir. These canals branch out extensively, supplying an irrigated domain of 5,000 hectares (and this area triples over the course of the centuries). They extend to downstream of the city of Murcia where another irrigation system, that of Orihuela, takes over. This system makes use of multiple-section sills, as at Valencia, as well as *norias*. Of course there are also mills, and even boat-mounted mills on the Seguro River near Murcia:

> "There are (at Murcia) mills constructed on boats, like the mills of Saragossa, and which, being on the boats, can be transported from one place to another. There one can see gardens, orchards and crops of incalculable number."[47]

All of these irrigation networks are still in service today, having been repaired and maintained over the centuries.

Mills - both water-driven (horizontal wheels up until the 12th century, then occa-

44 Zakri (1990).
45 Fernandez Ordonez (1984).
46 Fernandez Ordonez (1984), Smith (1970).
47 Al-Idrissi, IV, 1, Translation of Jaubert.

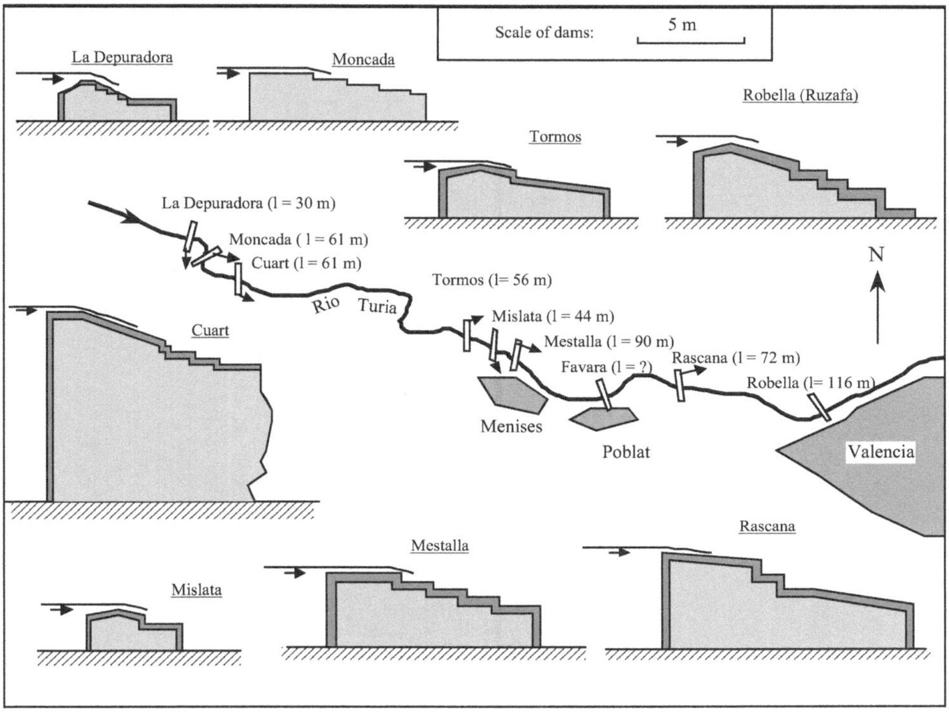

Figure 7.15 The sills of the Turia river, permitting irrigation of the huerta of Valencia from the 10th century. The sections across the structures are shown to the same scale; their lengths are indicated in parentheses. The azudes of Moncada, Mestalla, Favara, Rascana and Ruzafa are surely the oldest; the sills of Cuart and Mislata could be from after the Reconquest of the 13th century. From Fernandez Ordonez (1984).

sional vertical wheels) and wind-driven (such as at Malaga) - were ubiquitous in Muslim Spain. It is said that there were 5,000 mills in the region of Cordova, and 130 within the walls of Grenada alone. The mills belong to private individuals or groups of individuals, as in the Orient. It is only after the Reconquest that they become privileged feudal and ecclesiastical property, as in the Christian Occident. There was no lack of conflicts for the use of water, given the needs of the flour trade and irrigation[48].

One can find *qanats* on the grand plateau of Castilla, the most important being those of Madrid. From the very founding of the city by Mohamed Ist (825-886), networks of *qanats* are put in place to provide water for the city. They are from 7 to 10 km long, with drops of from 80 to 100 m. Indeed, it is perhaps due to the abundance and the quality of water furnished by the *qanats* that Madrid was chosen as capital of Spain by Philippe II, in 1561.[49] According to the geographer al-Idrissi, a large *noria* lifts water from the Tagus River up to the city of Toledo.

48 Lagardère (1991 and 1993).
49 Goblot (1979).

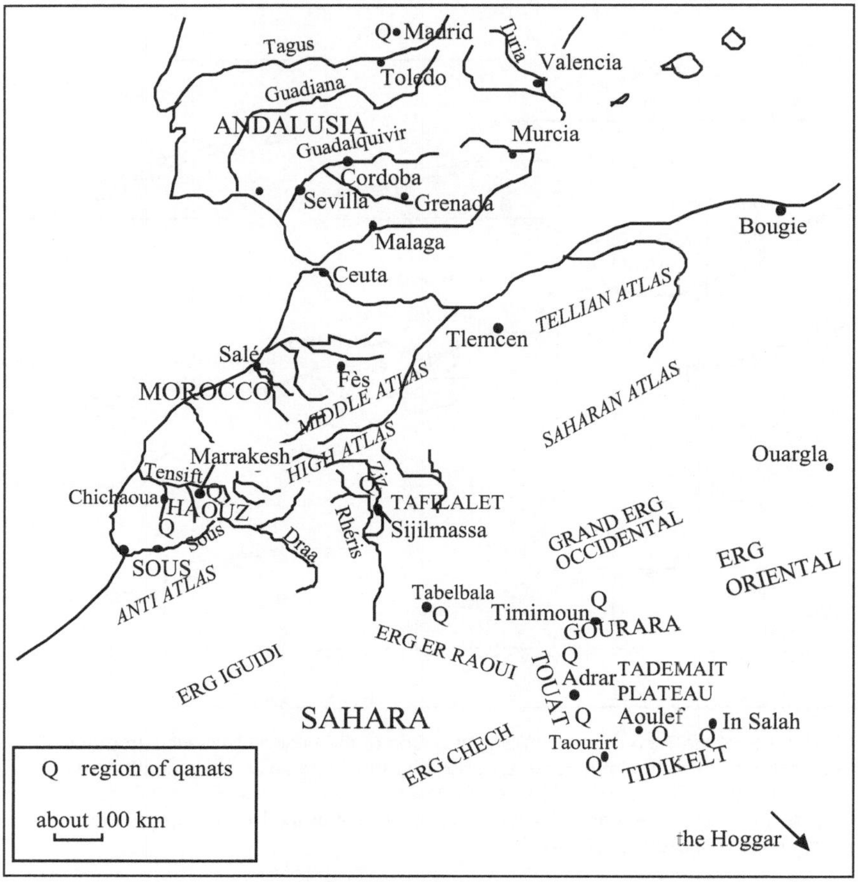

Figure 7.16 The Muslim Occident in the Middle Ages. The regions where one finds qanats (khettaras and foggaras).

The south of Morocco and the Saharan oases

In 1031 the caliphate of Cordova breaks down, fragmenting into small kingdoms. Alphonse VI of Castilla seizes the opportunity to take Madrid in 1083, then Toledo in 1085. The Andalusians call for help from the only powerful Occidental Arab dynasty of the time, the Almoravides of Morocco. Their chief Youssef ben Tachfin puts an end to the advance of Alphonse VI when he passes through Spain, and he brings Morocco and Andalusia under his unified command. His son Ali, raised in an Andalusian culture, succeeds him in 1106. Andalusian scholars and literati follow him to Marrakesh, capital of the Almoravides. This contact probably explains the appearance of the *qanat* technology in Morocco.

The Almoravides, Berbers from the south, had occupied Sijilmassa, the great caravan center of Tafilalet, in 1055, and then conquered all of Morocco and founded Marrakesh in 1060. The construction of the first *qanat* of Marrakesh is documented thanks to al-Idrissi. Ali decided to build it in 1107, and the project was completed by a certain Obeyd Allah ibn Younous (that is, *son of Jonas*) al-Muhandes (which means *the engineer*).

> "At the time there was only one garden [....]. Obeyd Allah went to the highest point of land overlooking this garden. There, he had a very large rectangular pit dug, from which he had dug a single outlet that gradually descended (...) to the garden for which it supplied water in a continuous fashion. With the naked eye, one cannot detect, on the ground, any slope that would enable the water to flow from the bottom of the pit to the surface of the garden. To understand this, one must understand the clever trick that was used to supply the water. This trick consisted in evaluating the difference in ground level *(from the bottom of the pit to the garden)*. The Emir of the Muslims, who very much appreciated the work of the engineer Obeyd Allah, paid him with silver and clothing. [....] After that, inspired by the example provided by this man of art, the inhabitants of Marrakesh set about capturing water and bringing it to their gardens, to the point that many of them could only increase in size, buildings growing up around them, embellishing the skyline of Marrakesh." [50]

According to the text, this Obveyd Allah ibn Younous well understood the technique of the *qanats* when he came to install them in Marrakesh. If one juxtaposes the dates (the event was 80 years after the taking of Madrid, one year after the birth of Ali, who was raised in contact with the Andalusian literati) and the facts (according to al-Idrissi, it was Andalusian engineers that Ali called upon to build the first bridge on the wadi Tensift), it seems reasonable to presume (as did Henri Goblot) that this engineer came from Spain.

Over a thirty year period some fifty *qanats* are built on the plain of Marrakesh (the *Haouz*), where they are called *khettaras*, bringing 5,000 hectares under irrigation. The Almohades took over from the Almoravides after 1160. They built new *qanats* and constructed an extended network of canals supplied by the rivers to increase this irrigated land area to 15,000 hectares. After some retrenchment in the 16th century, the irrigated area grows to about 20,000 hectares in modern times. There are some 600 identified *qanats*, 500 of which were still in service in the middle of the 20th century. These *qanats* are generally quite shallow, for the water table issuing from the Upper Atlas mountains is only about 20 meters below the ground surface. The *qanats* are limited by the land slope to only from 500 m to several kilometers long, and the wells are closer together than those of Iran. The average discharge of a *qanat* at Marrakesh is of the order of 40 m^3/hour.[51]

Sugar cane is grown on the plain of Sous and in the *haouz* of Marrakesh. This crop requires considerable water. Sugar processing is also water-hungry, both in its need for hydraulic energy (mills to grind the cane) and for preparation of sugar bread. This water is most often taken from the rivers, but it can also come from the *qanats* – both those of

---

50 Al-Idrissi (12th century), 31, Translation of Hadj Sadok; a revised translation and analysis of this citation is given in El Faïz (2005).
51 Joffe (1989); see also El Faïz (2005).

Marrakesh as well as others on the plain of Sous, particularly in the region of Aoulouz and the surroundings of Agadir. The sugar industry quickly grows to play a significant role in the economy of Morocco, a role it maintains up until the 17th century. Important hydraulic infrastructure is developed to support this industry, particularly by the Merinides at the end of the 14th century and then by the Saadians in the 16th century, on the plain of Sous and at Chichaoua in the *Haouz*.[52]

Fez was established toward the end of the 8th century on a site that was at the center of an agricultural plain and naturally well supplied with water, the terminus of the route that crosses the Atlas near Sijilmassa. In the 11th century the Almoravides build canals to irrigate the gardens of Fez, and also build water mills. Al-Idrissi observed in the 12th century that in the district with the best water supply, al-Qarawiyyin, the delivery and drainage networks were particularly well developed:

> "Al-Qarawiyyin has an abundance of water that circulates through all the streets and allyways, in conduits that the inhabitants can open when they wish to wash the neighborhoods during the night and have them perfectly clean in the morning. In each house, be it large or small, there is a pipe for both clear or dirty water."[53]

Norias appear in Fez in the 14th century, and numerous water-based activities develop in the city. In the 16th century Leon the African enumerates over a hundred public baths and nearly 400 water mills supporting all sorts of industrial uses.[54]

Sijilmassa is, with Marrakesh and Fez, another grand center of activity in medieval Morocco up to the 14th century:

> "Sijilmassa is an important capital, located some distance from a watercourse that disappears to the south of the city. [....] This city is rich in dates, fresh and dry grapes, fruits, cereals, pomegranates and diverse other agricultural products; the city pleases foreigners who come from all directions in large number. [....] The canton possesses mines of gold and silver. [....] Sijilmassa is surrounded by the sands of the desert, its inhabitants use water holes."[55]

This ancient city was founded in the middle of the 8th century. Recently discovered vestiges have been the focus of American-Moroccan research efforts between 1988 and 1996. The city is located in the Tafilalet, one or two kilometers to the west of the present Rissani (which is some 200 km east of Ouarzazate) near the wadis Ziz and Rheris that descend from the High Atlas. Sijilmassa is a node of communication with the Orient (via Ouargla) and especially with Sudan, a source of gold. Ancient texts present an image of grand agricultural prosperity, the remains being seen in present-day palm groves:

> "She *(Sijilmassa)* has a series of castles, houses and buildings along an abundant water course that comes from the east, i.e. from the desert, and whose flow increases in the summer *(with the snowmelt of the High Atlas)*, much like the Nile. Its waters provide for irrigation of crops,

52 Brignon, Amine, Boutaleb, Martinet, Rosenberger (12967), p. 187.

53 Al-Idrissi, 49, Translation of Hadj Sadok.

54 Brignon, Amine, Boutaleb, Martinet, Rosenberger (1967), pp. 90 and 202; Madani (1999) for the hydraulic network of Fez.

55 Al-Muqaddasi (10th century); citation from Brignon et al. (1967).

which along with the use of Egyptian peasants leads, as everyone knows, to very good harvests. In certain years, following floods of this river, water is so abundant that the grains harvested in the previous year grow again without any need to replant the fields."[56]

This text suggests a practice of flood-recession agriculture, certainly as in Egypt, but also in conditions that would appear to be close to those that we have described in Chapter 3 for the east of Arabia Felix (Yemen). Similarities in architecture between the ancient cities of Yemen and the *ksar* (fortified villages) of Tafilalet suggest another possibility -that the hydro-agricultural techniques used at Sijilmassa have their origin in a Yemeni migration. Field studies [57] indeed show that small dams were constructed on the wadi Rheris, to be destroyed several times by floods but then rebuilt in other locations. But it is especially the wadi Ziz that is developed and managed. Initially, the Ziz ran much more to the east, along a course that is today called the wadi Amerbou. At some undetermined time, the course of the Ziz was changed by a dam situated some 15 km to the north of Sijilmassa (opposite the present-day Erfoud, where traces of a stone structure still exist). The wadi was diverted into a canal that runs along the western side of the city of Sijilmassa (Figure 7.17). This canal is in its own right the source of sec-

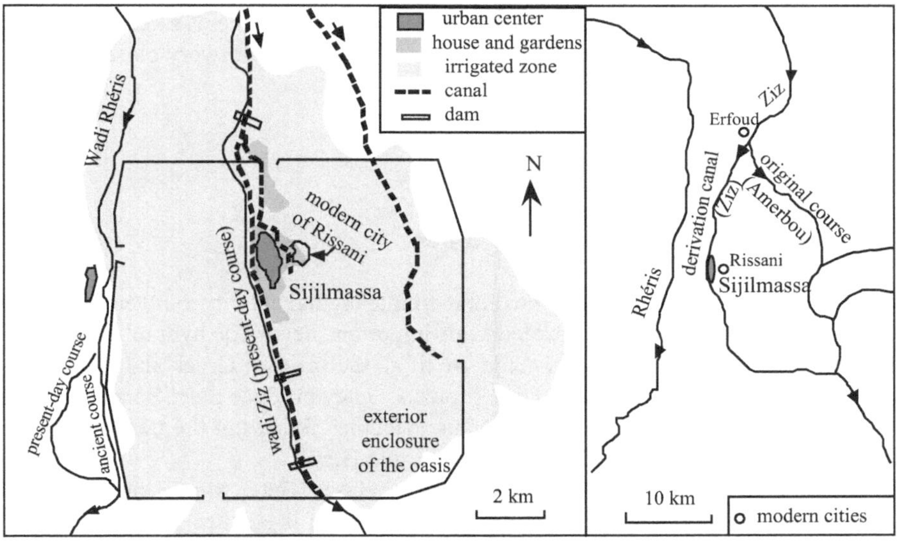

Figure 7.17 The principal hydraulic developments of Sijilmassa (after Messier, 1997).

ondary derivations for irrigation of the oasis and subsequent return flow to the original course further to the south. This original course (the Amerbou of today) handles the drainage of excess flood waters. Over the centuries this canal becomes the source of the

56 Al-Idrissi, 22, Translation of Hadj Sadok.
57 Messier (1997).

modern watercourse of Ziz.

To complete this broad-brush painting of the water resources of Tafilalet, we should note that along with Marrakesh, this region constitutes the second flowering of *qanats* of Morocco. Remains of 300 *qanats* are identifiable today, half of them still in service. Their dates of construction are unknown.

Another very important blossoming of *qanats* is located in the oases of Gourara, Touat and Tidikelt in present-day Algerian Sahara, south of the western Grand Erg. This is a chain of palm groves clustered around the foot of the Tademaït plateau (Figure 7.16).

These oases were undoubtedly populated by migrations of Jewish Berbers from Cyrenaic, the Zenata, fleeing Roman colonialism of the 2nd century AD. But in general we assume that the history of the Saharan *qanats* reaches back only to the 9th or 10th centuries AD, and that their history is unconnected to that of the old Roman *qanats* of Libya and Tunisia (Chapter 6), at this time a forgotten technique.[58] The introduction of *qanats* grew out of new immigration waves from the Orient - we must remember that at this period, the Sahara was not nearly as water-starved as it is today. But without these thousands of *qanats* the grand oases would not have survived to the present. At the beginning of the 20th century some 400 active ones were known in Gourara (especially on either side of Timimoun); 440 in the Touat chain of oases between Adrar and Tauourirt, where they are the sole source of water; and 125 in the Tidikelt, especially around Aoulef and In Salah. They are from 2 to 15 km long, with very closely spaced wells.[59]

Arab Science and Hydraulic Machinery

Hydrodynamics and Marvelous Machines

We have seen that a political climate favorable to intellectual activity evolved under the Abbassides – and this climate coincided with important needs for hydraulic development. Scientists and often engineers made use of all the fruits of Greek and Hellenistic science in parallel with major construction projects. They produce precise mechanisms, water clocks (clepsydres) and other marvelous machines following the tradition of the ancient scholars of Alexandria, Philon, Ctesibios, and Heron.

Three brothers – Muhammad, Ahmad, and al-Hasan Banu Musa, known to the caliph al-Mamun around 820, wrote numerous treatises on these mechanisms. Of particular note is the *Kital al-Hiyal, Book of Ingenious Mechanisms*, written in Baghdad about 850. This book contains descriptions of many devices that reveal a perfect mastery of hydrostatics.

The fact is that Archimedes, as well as other authors such as Aristotle, Euclid, and Heron, had been translated into Arabic by that time. Al-Khazini, a scientist of the 12th century, put together a synthesis of the hydrostatics of Archimedes and the premises of dynamics of Aristotle and his commentators. In his book, called *Kitab mizan al-hikma*

---

58 Lambton (1989).
59 Goblot (1979), pp. 163-164; Bisson (1989).

(book of the balance of wisdom), he differentiates between the two types of action that can be exerted on mobile bodies in water – hydrostatic forces, and hydrodynamic forces arising from the movement of the body. These latter forces "differ because the shapes of the bodies differ". He also extends the theories of Archimedes to hollow bodies, as well as to hollow bodies carrying a load. Here the theory of ship buoyancy makes its appearance.

A characteristically Aristotlean trait appears in the vision of groundwater developed in the 11th century by the mathematician al-Karagi, a vision that we cited earlier in the context of his treatise on *qanats*:

> "God – may he be blessed and exalted – created a compact universe without empty space, and attributed to each element – the celestial sphere, stars, fire, air, water, and earth – its own place, a place to which it tries to return if separated from it. Dense bodies like earth and water (...) seek the center of the universe, and the most dense arrive there first; from which it follows that the earth is at the center of the water that surrounds it."[60]

And from this comes the explanation of groundwater movement:

> "It is in the nature of water to seek, by its movement, the center of the earth, and not to rise."

Springs are then explained by the slope of the impermeable layers over which groundwater flows:

> "When the groundwater has its bed upon a hard surface, and this hardness, lying next to a fissure where the water flows, extends to the summit of a mountain, the water emerges and can be tapped at this summit, if it is nearer the center of the earth than the place from which the fissure is supplied."[61]

Other authors also reveal a great deal of refinement in Arab research into these mechanisms. One is al-Muradi, who in the 11th century describes automatic controls that are powered by waterwheels. Another is al-Jazari who, at the very beginning of the 13th century, brought the art of the *clepsydre* (water clock) to its pinnacle.

We have been emphasizing the influence of the Greek and Hellenistic scientists on Arab thought. But we must not ignore the Chinese influence insofar as technology is concerned. After the battle of Talas, Chinese prisoners introduced the Chinese technology of the hydraulic pestle to the nascent paper industry at Samarcand; we will come back to this in Chapter 8.[62] This technology subsequently spreads to all the Arab world, then into the Occident. A century later, in 850, the historian al-Jahiz assembles an inventory of "products" imported from China to Iraq. In this inventory he naturally mentions silk, but also includes a curious list that contains, in order, "female slaves" and "hydraulic engineers."[63]

---

60 Kitab inbat al-miyah al-hafiyya (The art of extracting hidden water), citation from Landry (1990).

61 Ibid.

62 Hill (1997), p. 45.

63 Al-Jahiz, in Arab Historians (Sauvaget, 1988).

The norias

The *hydraulic noria*, sometimes called the *current noria*, first appeared in the period of the Roman Empire. Herein we simply call this the *noria*. The very first evidence is found in the description of Vitruvius (around 25 BC) as we have cited in Chapter 6. Vitruvius describes the machine very clearly, but says nothing about the location of the particular device. Moreover we are not aware of any other mention of the *noria* by Roman writers. The second piece of evidence is found in the very heart of the land that will become the location of the largest *norias*: it is in a mosaic discovered at Apamea, a Hellenistic and Roman city located some fifty kilometers to the north of Hama opposite the depression of Gharb where the Orontes flows. We know from a signature that this mosaic dates from 469 AD. It is unusual to find machines represented in mosaics, suggesting that during this period the *noria* was considered to be an important element of the patrimony of the region. The tradition persists to this day, since even now the city of Hama maintains and renovates its *norias*. The accounts of Arab travelers in the 12th and 13th centuries, cited earlier in this chapter, speak of beautiful *norias* in the same breath as the Orontes and the city of Hama. Ibn Battûta, in the course of his grand voyage, observes *norias* at Amasya in Anatolia, on the Karun and the Fars, at Samarcand and even at the mouth of the Ganges. He mentions that they also can be found in China (but surely did not see them himself). As we have described earlier, *norias* are also

Figure 7.18 Detail of the wheel of the al-Damalik noria at Hama (note in particular the paddles and runnels). In the background is the wheel of the al-Hudura noria. (photo by the author).

Figure 7.19 In the foreground, the axle of the noria al-Gisriyya at Hama; and in the background the noria of al-Mamuriyya, dating from 1453. Prior to its first reconstruction, this latter noria was the second largest in size (after the noria of al-Mohammadiyya). (Photo by the author).

found in Andalusia (Toledo, Cordova) and at Fez. Still, in the eyes of these travelers, none of those *norias* are as remarkable as those found at Hama.

Norias are machines that can lift water to a great height, but they are also very expensive. They are thus best suited for implementation on rivers of fairly regular discharge, flowing in a well-defined course, and having rather steep banks - and at sites where gravity irrigation would be otherwise difficult. The *norias* of the Orontes have been particularly well studied.[64] The typical machine comprises a narrow wheel whose diameter depends on the desired lift. As is the case for its ancestor the lifting wheel, the *noria* has wooden water boxes or jars, on its perimeter to lift the water. The paddles are mounted on the outside circumference of the wheel; on the Orontes, these paddles are spaced at approximately 50 cm, whatever the diameter of the wheel. Rotation of the wheel is induced by the river current, channeled through a canal scarcely wider than the wheel itself and impinging on these paddles. On modest rivers such as the Orontes or the Khabur, an overflow dam raises the water level at the entrance to this canal, thus augmenting the speed of the current that drives the paddles.

The water boxes have an opening in the side (Figure 7.18). The boxes fill with water when they are immersed in the canal; then, when at the top of the rotation of the wheel, they pour their water out into a trough parallel to the wheel, at the top of the structure

64 Delpech, Girard, Robine, Roumi (1997).

Figure 7.20 Spoke structure of the noria al-Gisriyya at Hama (photo by the author).

called the "tower". The trough becomes an aqueduct that delivers the water to the distribution system. The tower and aqueduct are easily visible on Figure 7.7, as well as on the remains of the *norias* of Khabur (Figures 7.12 and 7.13). One end of the wooden axle of the *noria* rotates in a notch that is part of the tower; the other end rotates on a support that is, on the Orontes, a triangular masonry structure (this "triangle" is clearly recognizable in the mosaic at Apamea, see also Figure 7.19).

A single dam can accommodate several norias (Figure 7.8) and a single noria can moreover be fed by several supply canals (Figure 7.13), each dedicated to one wheel. If these wheels are of different diameter then each has its own outlet trough; if not, they can pour their water into the same trough. Often one can take advantage of the dam to install a mill as well (Figure 7.12).

One finds the largest known *norias* at Hama - those called al-Mohammadiyya and al-Mamuriyya, dating from 1361 and 1453. Their wheels are approximately 21 m in diameter. On rivers subject to destructive floods, for example the Euphrates, there is no need to bring excessive care and quality to the construction of the wheels as they are most often destroyed by the annual flood and thus must be rebuilt each year. On the other hand on the Orontes, a river of quite regular flows, the wheels are built with great care. Large and narrow, they are trued to an astonishing precision (several centimeters on wheels from 10 to 20 m in diameter). They are serviced each year at the end of the irrigation season. Their wooden arms are not radial, but are laid out in an arrangement that is common to all the wheels, but particularly those of the Orontes, an arrangement that is clearly visible on Figure 7.20.

Windmills

The origin of the windmill is controversial. In a manuscript on *Pneumatics*, Heron of Alexandria includes the description of a wheel driven by the wind and driving the piston of an organ. But the authenticity of this passage has not been resolved - it could have been added during the Islamic period. In any case, even if this passage is authentic, the link between this invention and the appearance of the true windmill in central Asia seems to be very weak.

The backdrop of Heron's apparent reference to a windmill is the ubiquitous presence of water mills on all of the rivers. On the Tigris at Baghdad, on the Seguro at Murcia, on the Ebre at Saragossa, there are even boat-mounted water mills. Mills are the indispensable tools of an agricultural economy. But on the Persian plateau there are few, if any, watercourses. This undoubtedly led to the invention of the windmill at an indeterminate date but perhaps prior to the Arab conquest. In about 660 a Persian had affirmed to the caliph Omar that a windmill could be built. The windmills of Seistan (to the east of Khorassan) were known to Arab geographers from the 9th century on, in particular to the Banu Musa brothers.[65] These mills turn on a vertical axis; their vanes are inside a chamber at the top of a sort of tower, open to the four cardinal directions with shutters that one can close depending on the direction or the force of the wind. The rotation of the axle directly turns a millwheel that is in the chamber just below. These mills are described somewhat later by Chinese authors who give a precise account of them:

> "In the western countries called Herat and Samarcand, there are many windmills. Brick walls are built to form a sort of house having openings at its summit, facing the four directions, outside of which screens can be placed to direct the wind. In the chamber below is placed a wooden axle, with sails *(literally planks to ride the wind)* fixed to it. Whatever the direction of the wind, the axle always turns, and the stronger the wind, the more work can be accomplished."[66]

65 Hill (1997), p. 46.
66 After a text dated from 1691 (Chhih Pei Ou Than of Wang Shih-Chen, cited by Needham and Ling (1965), p. 560 (adapted).

8. Rivers, canals, and Hydraulic technology in China

In 329 BC, the army of Alexander the Great conquered Bactria and reached Samarcand, more than 4,000 km from Macedonia. At the same time, on the other side of the deserts of Taklamakan 3,000 km to the east, another warrior kingdom by the name of Qin began an astounding expension. A century later, while Alexandria of the Ptolemites was shining its brightest and while Archimedes was discovering the principles of hydrostatics, this powerful of Qin was unifying an empire that spread across an entire continent. When the name of Qin appeared in the Occident, having passed from mouth to mouth across India, it had been transformed to become what we know as China.[1]

Relations between China and the Near East – the Silk Road

The worlds of ancient China and the ancient West were never truly isolated, even if direct contacts between them were rare. Recent archaeological discoveries[2] in the desert of Taklamakan in Xinjiang (Chinese Turkestan) reveal a very old settlement of oases through which the Silk Road will later pass. This area had widespread irrigation in the first millennium BC. It is possible that wheat was brought to the Near East and as far as China by this route in the Neolithic period. The same may also be true for the technology of bronze, as it appeared very suddenly in China about 1600 BC.

In 135 BC an official Chinese envoy named Zhang Qian reaches, with difficulty, the territory of the Yuehzi at the borders of a country called *Daxia* by the Chinese, thought today to be Bactria.[3] (The Yuehzi later found the Kuchan Empire). There he discovers, among other things, products of South China that reach this country after passing through a region called *Shendu* (India).[4] In 104 BC a Chinese military expedition establishes effective control over this route, destined soon to become the eastern portion of the Silk Road. The Romans discover silk in the 1st century AD and become infatuated with it. The Parthians begin to serve as intermediaries between the two great empires and before long, caravans begin to link them, through many additional points of contact. These included Palmyra or Antioch to the west, and Chang'an (today Xi'an) in China to the east. Starting in the 2nd century AD, a maritime route passing to the south of India (exploiting the seasonal monsoon winds) establishes a more direct link from east to west. This maritime route is especially used by merchants of Alexandria. It is said that a

---

1 In Antiquity, China is known, in the West, by two principle names: Seres, derived from the Chinese Si signifying silk, and Sina, that is thought to have come from the name of the Qin Dynasty. The first name arrived in the Occident through the intermediary of the Greek world; the second seems to have come by the route of the Indies (Needham, 1978).

2 These are the excavations of Kjoumboulak Koum, where the remains of a vast irrigation network have been found in an ancient delta of the Keriya river, in the region of Khotan, after Corinne Debaine-Francfort (personal communication).

3 According to the Chinese historian Sima Qian, who lived around 100 AD, Historical Memoires (Shi Ji), 123.

4 Sima Qian, Shi Ji, 116.

Roman diplomatic expedition, likely made up of Syrian merchants, called at ports in South China during the Han Dynasty, successor to the Qin Dynasty.[5] Later on the Arabs serve as intermediaries between the West and China; eventually in the 15th century the Ming Dynasty launches its own junks on grand expeditions.

From the beginnings of agriculture to the legendary founder of Chinese civilization

Archaeology teaches us that grain cultivation, namely millet, appears in the middle basin of the Yellow River around 6000 BC.[6] Two cultures develop successively in this region: one is called the Yangshao at the end of the VIth millennium BC, during which small-scale farming develops (pork, poultry); the other is the Longshan, at the end of the IIIrd millennium BC, during which wheat and barley develop in addition to millet, and the first fortified villages appear. As a point of reference, recall that this is also the period of the grand civilization of Harappa on the Indus and its extensions in Bactria. The Chinese regions involved are Shaanxi, Shandong and especially Henan and Shanxi. The foundation of Chinese civilization is rooted in the continuity of this culture, localized in the region of the confluence of the Luo[7] and the Wei rivers with the Yellow River.[8] The legendary hero is Yu the Great who is said to have tamed the unpredictable course of the Yellow River through the construction of canals and deepening of its bed:

> "Before him, the overflow waters flowed wantonly"[9]

It is of course, impossible to know if all the accomplishments attributed to Yu resulted from his efforts, or indeed from human intervention at all. But, given what we know and taking into account the seriousness of the ancient Chinese sources, there is no reason to doubt *a priori* the existence of the personage himself. He would be at the origin of the Xia Dynasty, whose domain scarcely extended beyond the terraces and valleys that mark the terminus of the middle course of the Yellow River. Taking the suggestions of the ancient Chinese historians to their logical conclusions, one ends up dating the legendary reign of Yu the Great to around 2200 or 2000 BC.

Civilization in China apparently begins two millennia after it began in Mesopotamia. On the other hand this Chinese civilization is remarkable for its durability, from its earliest beginnings up until the present day.

---

5 For example, Robert (1997).

6 see Blunden & Elvin (1983).

7 There are two important rivers called Luo; one in Shaanxi, the other in Henan. To avoid confusion, we call the first Luo and the second Lo.

8 This does not mean there was nothing further to the south. Archaeology has indeed shown the existence of a culture of rice in the basin of the Yangtze (the Blue river) from 6000 BC. But it is clearly in the north that organized civilization, destined to spread, was established (see for example Debaine-Francfort, 1998).

9 Mencius, citation from Granet (1929), p. 89 in the edition of 1994.

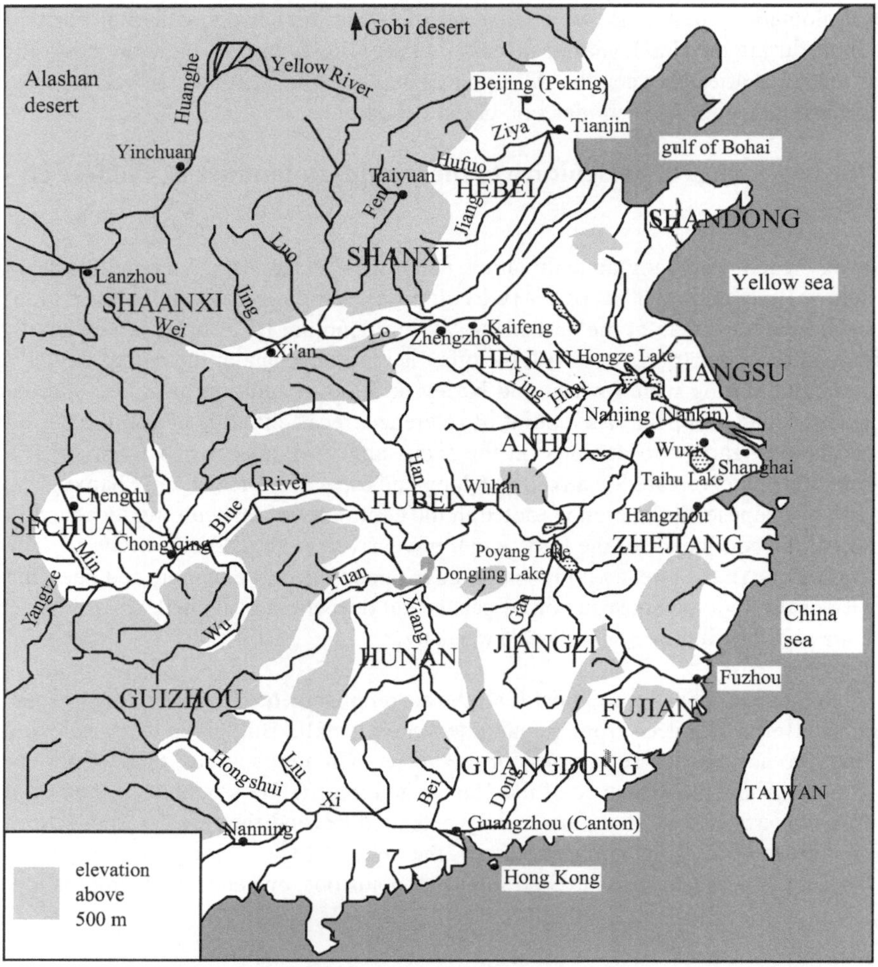

Figure 8.1. The eastern portion of China, showing the middle and lower courses of the Yellow River (Huanghe) and the Blue River (Yangtze). On this map, the rivercourses, coastlines, and city and province names are those of today. The Grand Canal is not shown.

Historical and cultural landmarks

The Bronze Age begins with the Shang Dynasty, succeeding the Xia Dynasty about 1600 BC. This aristocratic and cruel regime is well known in archaeology; the first texts scratched in bone come from it. The influence of this dynasty is limited to the valley of the Yellow river to Shandong. Then about 1100 BC the Zhou, coming from the valley of the Wei (a tributary of the Yellow River in Shaanxi) supplant the Shangs. Under their more humane domination the Chinese civilization reaches not only further to the north, but also and especially toward the south as far as the Yangtze valley. The Zhou establish their capital near Luoyang where the river flows out onto the plain. But the Zhou regime does not have a structure that is capable of coping with the growth of their domain. Several centuries after their advent, provincial powers begin to rival the influence of the central power. This marks the onset of the feudal period, beginning with what is called the *Spring and Autumn period* (771 - 480 BC) and the *Warring States period* (480 - 221 BC), a rather evocative name indeed.

The major currents of Chinese thought developed during this troubled period.[10] Kong Fuzi, or Confucius lived from 551 to 479 BC and founded a social philosophy that seeks to establish justice and equity in the framework of traditional structures and customs. Confucianism preaches peace, order, and return to the path of the "wise kings of Antiquity", motivated by the desire to bring well being to the people. Later this becomes the official doctrine of the empire's administrators, mandarins recruited competitively based on their literary knowledge. The principle disciple of Confucius, Meng Ke, or Mencius, was born in 374 BC and becomes counselor to the princes of the kingdoms of Liang and Qi. He brings the humanist aspects of Confucianism to the forefront: traditions and customs are made for men, and not vice-versa. Opposing this trend of thought, Taoism appears at an indeterminate date during this same period. One of its best-known founders is Laozi (Lao-Tse), who lived in the 4th century BC. Rarely has a philosophy been so caricatured. Nonetheless, it contains the real premises of scientific thought. As a naturalist philosophy, based on observation of natural things, Taoism has as its ideal the search for causes, but without any aspiration to discover at all costs a unique model that might explain everything. Taoist thought, based in observation and experimentation, is more descriptive than explanatory:

> "All phenomena have their causes. In the ignorance of these causes, it can happen that one is correct *(in regard to the facts)*, but it is as if one knows nothing, and in the end remains perplexed. [....] The fact that water exists in the mountains and heads toward the sea does not arise from some antipathy for the mountains, nor from a love for the sea, it is simply due to the effect of altitude such as it is."[11]

So, whereas Confucianism is oriented to action, Taoism is the calm search for an interior pathway. Naturally humble, it nonetheless willingly arrays itself against the powerful.

During the feudal period scholars move from one court to another. The more enlightened of the princes of these kingdoms create academies, the most celebrated being at Linzi, the capital of Qi in Shandong.[12] It is from within this academy that Zou Yan (305 – 240 BC), considered by Joseph Needham[13] to be the founder of Chinese scientific thought, formulates the theory of the five elements: water, fire, wood, metal, and

10 The reader can find deeper analyses of the evolution of Chinese philosophy and science in, for example, Needham (1978), and Gernet (1990).

11 Extract of a work entitled Lüshi chunqiu, 239 BC. Citation from Needham (1978), p. 117.

12 Needham (1978).

13 Joseph Needham is an indispensable reference for those who are interested in the history of science and technology in China. He is the author of a monumental work, Science and Technology in China, to which we will often refer in this Chapter (Volume IV in particular for matters concerning hydraulics).

earth.

> "Of the five elements, the first is called Water, the second Fire, the third Wood, the fourth Metal, and the fifth Earth. Water *(is the quality in nature)* that moistens and tends to sink; Fire *(is the quality in nature)* that flames and tends to rise. Wood *(is the quality in nature)* that permits curved surfaces or straight edges. Metal *(is the quality in nature)* that can follow *(the shape of a mold)* and can harden. Earth *(is the quality in nature)* that permits sowing, *(growth)*, and harvest."[14]

This text shows how the Chinese view is something quite different from the Greek theory of four elements, popularized at about the same period by Aristotle.[15] The Chinese theory of five elements is not a model intended to explain nature, but rather a classification of processes or physical properties (Table 8.1). The theory is essentially descriptive, consistent with Taoism. It will be used throughout the evolution of Chinese thought, completing the vision well known in the West of competition between the opposing principles *ying* and *yang*.

Table 8.1 Correspondence among the five elements of the Chinese literati and a classification of their physical properties.

Element	**Physical Property**
Water	That which is fluid, can flow, and dissolve
Fire	That which emits heat and can burn
Wood	That which is solid and can be shaped (carved)
Metal	That which is solid and can be melted, and which can take the form of a mold
Earth	That which produces useful plants

The first dynasty of the imperial era, that of the Qin, is followed by the long Han Dynasty. The Romans maintained commercial relations with this dynasty and may have even had diplomatic relations.[16] From 221 BC to 190 AD, China knew four centuries of unity.

The feudal period and this first imperial era experience intense economic and demographic development, both supported by hydraulics as we will see further on. The census of year 2 already reports more than 57 million inhabitants.[17]

The fall of the Han Empire is first followed by a splitting of China into three kingdoms (220 to 310 AD) and then by a period of total anarchy (310 to 589 AD). These long centuries of chaos are often called the "Chinese middle ages". It is in the 3rd century that Taoism, a philosophy but also an inspirational movement of secret societies, becomes a religion as well. In the 3rd and 4th centuries, Buddhism is introduced and develops in China.

---

14 Extract of Book of Documents (Shujing). Citation after Needham.
15 The differences between these theories and the dates make it very unlikely that there was any connection between them.
16 Granet (1929), p. 151.
17 Gernet (1990).

In 589 AD China is reunified under the Sui and Tang dynasties. After a troubled period from 906 to 960 The Song Dynasty is again established, and under it the naturalist Chinese movement reaches its greatest development. But starting in 1127 the power of the Songs recedes in South China following a series of invasions. From 1271 the Mongols occupy all of China and call themselves the Yuan Dynasty. The Mongol Empire extends from Persia to the Sea of China, naturally favoring commercial relations between China and the Middle East. In 1368 the Mongols are chased out by the Mings following peasant insurrections caused by famine. The Mings are eventually succeeded by the Manchu Qing Dynasty from 1644.

Naturalist Chinese thought is the inheritor of Taoism and excels in the observation and classification of things. But thanks to Confucianism, oriented toward action and often impelled to relieve manual labor, the Chinese invent many devices that spread toward the West. These include the axial rudder, the wheelbarrow, the magnetic compass, gunpowder, and the navigation lock. The Chinese excel in hydraulic technology as we will see further on in this chapter. But ancient Chinese science also uses mathematics to describe the laws of nature. The Jesuits, who get established in China during the Qing Dynasty in the 17th century, bring with them the methods of modern science.

We have seen that hydraulics is in the fabric of the founding legend of Chinese civilization. Since the very beginning of Chinese history, hydraulic technologies are employed to make new lands productive through irrigation and drainage of swampy valleys. Hydraulics is also brought to bear on the development of the infrastructure for waterborne transport. But there is no relief from a recurrent curse that returns regularly: floods and the associated changing courses of rivers, killing people and ruining entire regions. The inability to stabilize the large rivers, in particular the Yellow River, is often the cause of popular uprisings that end up bringing down dynasties. We therefore proceed to describe this great river, one of the principle actors on the vast stage of China.

The Yellow River, a terrible friend

The first historical treatise from China dates from around 100 BC. This is the work of Sima Qian,[18] who had an official position in the court of the Emperor Wudi of the Hans at the beginning of the imperial era. Sima Qian revived and perpetuated the legendary attribution of the ancient course of the Yellow River to Yu the Great, undoubtedly with some measure of exaggeration. He gives us a rather precise description of the ancient course (Figures 8.2 and 8.3):

> "The documents of the Xia Dynasty tell us that Emperor Yu spent thirteen years controlling and bringing an end to the floods and during that period, though he passed by the very gate of his own house, he did not take the time to enter.
>
> Of all the rivers, the Yellow River caused the greatest damage to China by overflowing its banks and inundating the land, and therefore he turned all his attention to controlling it. Thus he led the Yellow River in a course from Jishi past Longmen and south to the northern side of mount Hua; from there eastward along the foot of Dizhu mountain past the Meng ford and the

18 Herein we use the official modern transcriptions of Chinese into the Roman alphabet.

> confluence of the Lo River to Dapei. At this point Emperor Yu decided that since the river was descending from high ground and the flow of the water was rapid and fierce, it would be difficult to guide it over level ground without danger of frequent disastrous breakthroughs. He therefore divided the flow into two channels, leading it along the higher ground to the north, past the Jiang River and so to Dalu. There he spread it out to form the Nine Rivers, brought it together again to make the backward flowing river *(this is the lower portion of the river which has tidal influence)*, and thence led it into the gulf of Bohai."[19]

The reader may note that the great historian recognizes the hydraulic consequences of the change of slope where the river flows out onto the plain, as well as the effects of the tide. These were surely personal observations, for Sima Qian had traveled all across China.

The Yellow River owes its name to the color of the sediments it carries. During floods it is nearly a river of mud, having one of the world's highest concentrations of sediment. Nearly 5,500 km long, in its middle reaches it flows for 1,200 km across a plateau of loess, carrying fine sediments deposited by the wind. Once the river arrives on the plain, its bed slope suddenly decreases and the water velocity is consequently reduced, as is noted by our historian.[20] As the water velocity decreases, the particles in suspension deposit onto the bed.

Since very early times the Chinese have constructed dikes to protect villages and fertile lands from the floods. The deposited sediments accumulate between these dikes, and this of course raises the bed of the river relative to the plain surrounding it.[21] Quite rightly, Chinese tradition emphasizes both the importance of dredging the riverbed during low-water periods and the importance of building dikes. As we have seen in Chapter 1, it was in dredging the bed of the river that Yu succeeded in conquering the waters; his father Kouen, who only built dikes, had failed in trying to contain the river.

Failure to dredge the riverbed inexorably ends up causing an overflow and dike rupture. The river flows onto the plain with consequences that one can easily imagine, and it may then wander over large distances and establish a new course completely different from that in which it had formerly been contained by dikes. There have been nearly 2,000 dike ruptures recognized in Chinese history up to the present time, during which the course of the Yellow River has undergone 26 significant changes.[22] Over the course of history, the irrigated and fertile plain has become more and more densely populated – and therefore the consequences of inundation become more and more serious.

---

19 Sima Qian, Shi Ji 29, English translation of Burton Watson.
20 The slope of the Yellow River is about 1.1 m/km along its upper course (3,472 km); 0.74 m/km on its middle course (1,200 km); and only 0.11 m/km along the 786 km of its lower course – after Lian Ruiju, Zheng Zhaojin, Hu Jialin (1987).
21 The main bed of the river, the narrow channel in which the river flows during low-water periods, is formed in the bed's sediment, and therefore is not significantly super-elevated compared to the plain. The overflow bed, on the other hand, is much wider and is occupied during periods of high water; and can be more than ten meters above the level of the surrounding plain.
22 After Liang Ruiju, Zheng Zhaojin, Hu Jialin (1987).

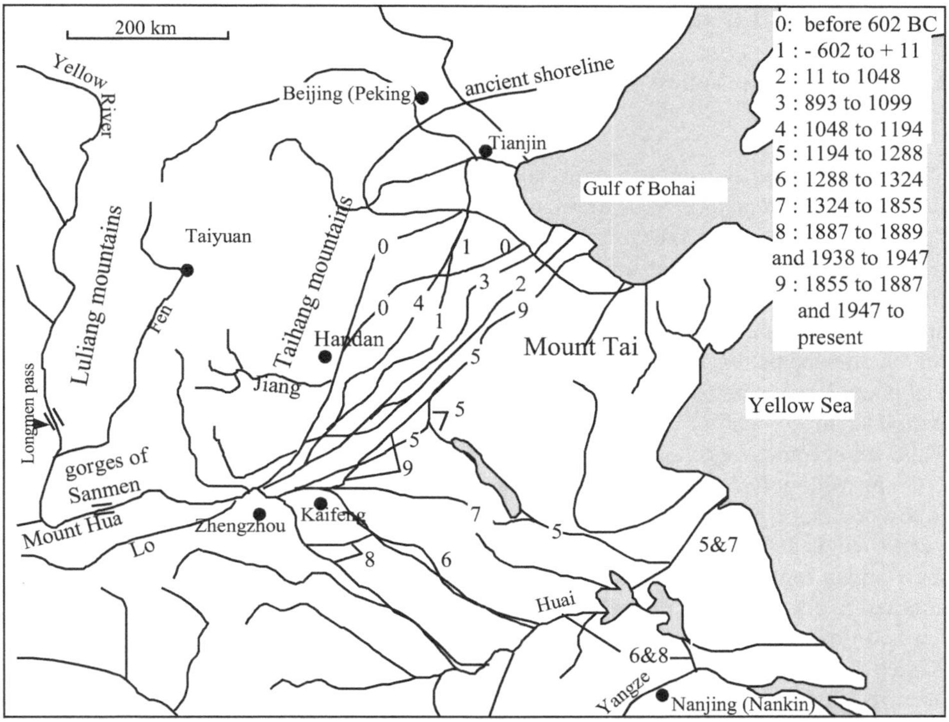

Figure 8.2 The principal historical courses of the Yellow River. N.B. Between 168 and 132 BC, the river flows into the Huai to the south. Moreover, courses 2 and 3 most often comprised two distinct arms each. Finally, it is probable that courses 5 and 7 existed since 1187, and that courses 5 and 6 functioned up until 1495. Other temporary arms are not shown on this map.

The principal flood events took place between the 2nd century BC and the beginning of the Christian era (Figure 8.9), then again between the 11th and the 14th centuries, when the Yellow River flowed to the south of Shandong and joined the Huai. It is not until the 19th century that the river comes back to the north, in the bed of the Ji river. From the beginning of the Qin and Han Empires, Chinese history records multiple examples of populations that are ruined and displaced by floods, eventually to be resettled by the central authority in colonized regions at the edges of their former domains.[23]

23 Granet (1929).

Irrigation and transport works in the feudal period

From fear of the Yellow River to the first grand irrigation projects (7th to 5th centuries BC)

The first large dike construction projects on the lower course of the Yellow River date from the *spring and autumn period*, more precisely the first half of the 7th century BC. The duke Huan de Qi is said to have brought together the "nine rivers" described by Sima Qian into a single course and probably tried to drain the swampy plain.[24] Soon after came the first realignment of the river course mentioned in Chinese history. In 602 BC the river adopted a new course some hundred kilometers to the east (Figure 8.2). Fortunately, the region affected by this first event was probably not very populated. The north of the alluvial plain does not begin to see real development until the beginning of the *warring states period* (in the 5th and 4th centuries BC) when the Qi enters a new era of great prosperity. This will become one of the most active agricultural regions under the Qin and the Han.

The Yellow River is intensely present in Chinese thought during the feudal period; it is worshipped like a god. In some places a young girl is sacrificed to the river each year, such as at Ye, on the ancient course of the river that flows toward the north[25] in the Handan region (Figure 8.3). The girl is chosen by witches of the cult of the river, adorned as for marriage, placed on a wedding bed, then launched onto the river where the bed floats for a brief time, then sinks to drown the sacrificial maiden.[26] A certain Ximen Bao, a disciple of Confucius, was responsible for ending this practice. This end occurred under the enlightened reign of the duke Wen of the Wei kingdom (424 – 387 BC), a period during which this kingdom reached its apogee. We cannot resist the temptation to share the account of this event:

> "Ximen Bao, having arrived in Ye, called together the notables and asked to learn about the custom that was desolating the county; marrying girls to the Lord of the River; the ceremony was described to him; he asked that he be told, without fail, of the day the festival would take place. When the day came, he and his soldiers went to the site of the sacrifice and announced that he came to be sure that a beautiful girl was chosen for the Lord. He looked at the girl chosen for sacrifice and declared to the grand wizard and to the elder of the country that the girl was not at all beautiful: he then sent the wizard to warn the Lord of the River that there was a mistake, that another would be chosen and that the ceremony was put off. The soldiers then threw the grand wizard into the river. Ximen Bao waited for a moment and, when the grand wizard did not come back from his mission, he threw an apprentice wizard into the river, then, when he did not return, a second apprentice, and finally a third. He then had the elder thrown into the river, and he did not return either. [....] Ximen Bao

24 Needham, Ling, Gwei-Djen (1971), p. 232, according to a tradition that comes from the Han era.
25 The hypothesis that Ye is near Handan is taken from the work of Henri Maspero (p. 145) and Jacques Gernet (p. 65). This localization is consistent with the fact that Ye was irrigated by Ximen Bao, using water taken from the Jian, which is precisely what Sima Qian said. It is curious, however, that this city, still active as the center of the cult of the river around 400 BC, was not located on the course of the river itself in this period, since the course had changed in 602 BC. Marcel Granet (Dances and legends, p. 474) situates Ye more to the south, near the place where the ancient course of the Yellow River arrives on the plain and turns toward the north.
26 After Marcel Granet (1926).

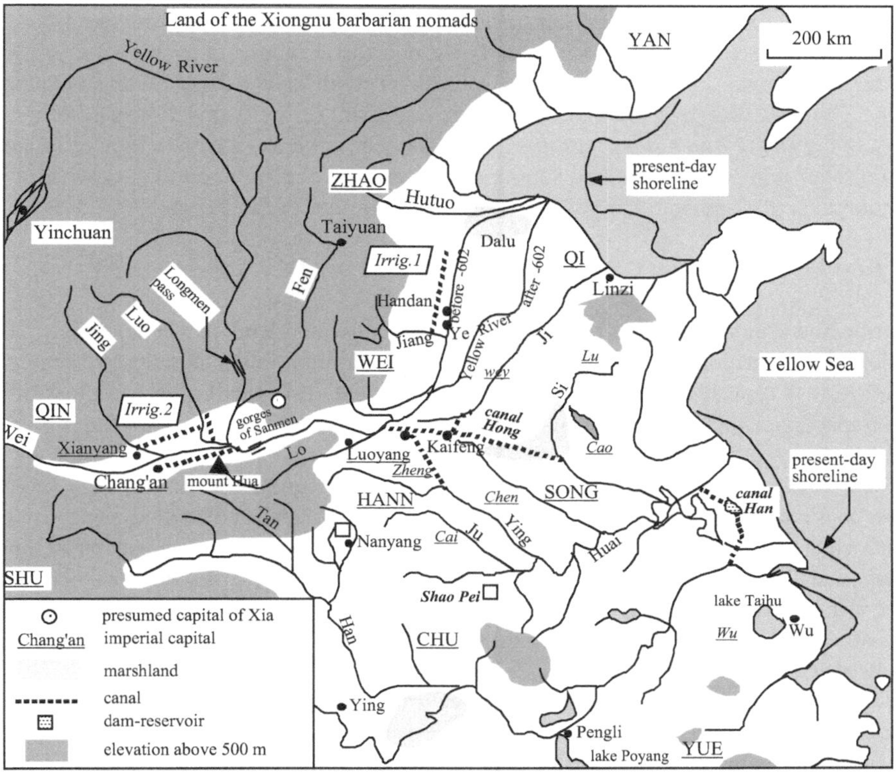

Figure 8.3 The basins of the Yellow River and the Blue River, from the time of the Warring States up to the early Han Empire. The underlined upper-case names refer to these states after 350 BC. The underlined names in italics refer to the regions, more ancient, of the Spring and Autumn period. Grand irrigation projects of the feudal period:

Irrig 1: between the Jiang and the Yellow Rivers (Ximen Bao)
Irrig 2: Zhengguo canal and derivation canals (Zheng Guo)
Irrig 3: Min basin – Figures 8.5 and 8.6 (Li Bing)

then stopped the ceremony and no one dared speak of resuming it."[27]

Ximen Bao is also known for the development of irrigation in the Ye region, on the lower slopes of the Thaihang mountains. Around 410 BC he had dug what amounts to a canal fed by the Jiang (an earlier tributary of the Yellow River), rising toward the north-east on a course parallel to and above that of the ancient bed of the Yellow River. Secondary canals then provided for gravity irrigation of all the region between this derivation and the Yellow River, a region that becomes particularly prosperous according to Sima Qian.[28]

27 Historical memoires (Shi Ji) of Sima Qian, 126, adapted from the translation of E. Chavannes (citation from Granet, 1926, p. 474).
28 Sima Qian, Shi Ji, 29. See also Needham, Ling, Gwei-Djen (1971), p. 271, and Zheng (1991).

Further to the south the first dam-reservoir known in China is constructed around 585 BC, in the state of Chu. It is intended to support irrigation, and is attributed to the minister Sun Shuao. The earthen dam is reinforced with layers of straw and wooden stakes. The dam was originally called the *Shaobei* (dam of the Peony flower), but it is known today as the *Anfengtang*, for it is still in service. It blocks a large valley of rather gradual relief into which flow two southern tributaries of the Huai, coming down from the mountains that separate the valley of the Hua from that of the Yangtze.[29]

The first large transport canals of the 5th century BC

Irrigation and drainage make it possible to develop cultivated land, as we have seen. In addition, the transport of bulk matter (especially grains) relies mainly on canals. Therefore it is typical to find dense networks of irrigation canals branching out from main transport canals during the major kingdoms. The following text of Sima Qian gives us an idea of the scale:

> "Sometimes later *(up to this point the text speaks of the works of Yu the Great)* the Hong Canal was constructed, leading off from the lower reaches of the Yellow River at Xingyang, passing through the states of Song, Zheng, Chen, Cai, Cao, and Wey, and joining up with the Ji, Ru, Huai, and Si rivers. In Chu two canals were built, one in the west from the Han River through the plains of Yunneng, and one in the east to connect the Yangzhe and Huai rivers. In Wu a canal was dug to connect the three mouths of the Yangzhe and the Five Lakes, and in Qi one between the Zi and Ji rivers. (...)
>
> "All of these canals were navigable by boat, and whenever there was an overflow of water it was used for irrigation purposes, so that the people gained great benefit from them. In addition, there were literally millions of smaller canals which led off from the larger ones at numerous points along their courses and were employed to irrigate an increasingly large area of land...."[30]

Let us look more closely at one of these projects. The most impressive transport canal in this period is the Hong canal, or *canal of the wild geese*. This is in fact a system of canals linking the Yellow River, from a city called Xingyang (or Jungyang) near the present-day Kaifeng, to the Ji River that flows to the north of the Shandong mountains but whose origin is quite near, and to the tributaries of the north bank of the Huai river. The canal has two main branches (Figure 8.3). The north branch (or *Bian* canal), the one most used for transport, probably follows the course of the ancient Bian (or Pien) river, which rejoins the Si and the Huai. This is a watercourse some 900 km long, surely artificial along a portion of its course. The south branch (the *Langtanqu* canal) links the Ying, Sui and Kuo rivers at their origins. It is some 400 km long, and constitutes a second fluvial passage between the Yellow River and the Huai,[31] through the Ying River which is navigable. The lengths of these canals are impressive, even though the originally swampy terrain is practically flat between the Yellow River and the Huai. At the

---

29 Needham, Lian, Gwei-Djen (1971), p. 271; Zhang (1991); Schnitter (1994), p. 41.
30 Sima Qian, Shi Ji 29, transl Burton Watson.
31 After Joseph Needham, Wang Ling, Lu Gwei-Djen (1971); p. 270 in the edition of 1987.

beginning, the Hong canal may have simply been a collection of irrigation canals for the north basin of the Huai.

The date of construction of the Hong canal is uncertain. As we have seen, when Sima Qian cites this canal as the very first in his list, he describes the regions linked by this pathway using names that refer to the *period of springs and autumns.* Moreover, Joseph Needham indicates that the Hong canal is mentioned around 330 BC in the account of a diplomat in his discussions of the boundaries between states. So, should we follow those who suppose that the canal dates from the 6th or 5th century BC? The construction of a work of such scale implies a strong central power and an economic motivation for exchanges between the basins of the two rivers. The coexistence of these factors would seem problematic in the troubled period from the 5th to the 3rd centuries BC.

The Hong canal is destined to be well maintained, and remains in use in its original course up until 600 AD. Another important project in Sima Qian's list is the Han canal, linking the Huai and the Yangtze. The king of the southern state of Wu has it built in 486 BC to supply his troops who were on a campaign against his northern neighbors.

Hydraulic development and rise of the Qin kingdom (4th and 3rd centuries BC)

The kingdom of Qin, rising from the western valley of the Wei, begins to grow from 350 BC. Its leaders are tough, uninterested in moderate discourse and Confucian scholarship. In 417 BC they had occupied Lin-Tsin, one of the centers of the cult of the Yellow River at its confluence with the Luo (the other center of the cult was at Ye where Ximen Bao had put an end to the human sacrifices at about the same time). Since the Qin desired the river god's protection for their family, each year they sacrificed a princess in "marriage."[32]

The methods of the Qin were radical, one could even say bloodthirsty. But their leaders well understood the importance, to their own power, of economic development of the regions under their control. They had a marvelous understanding of how to combine development with territorial expansion.

The three projects that we describe below are destined to have an extraordinary future - they remain in service today after more than 2,000 years of uninterrupted use.

The irrigation system in the Min valley, Sechuan

In 316 BC the Qin occupy the land of Shu, in the southwest, as far as the middle course of the Yangtze river. They developed the basin of the Min River, a tributary of the Yangtze in the Sechuan depression. In the region of present-day Chengdu, they implement a gigantic program of irrigation works. Sima Qian mentions this project – all too briefly:

> "In Shu, Li Bing, the governor of Shu, cuts back the Li Escarpment to control the ravages of the Mo *(Min?)* River and also opened up channels for the Two Rivers through the region of Chengdu."[33]

---

32 Granet, 1926.

33 Sima Qian, Shi Ji, 29, Transl. Burton Watson

The crown jewel of this project is a remarkable intake structure on the Min (Figure 8.4),[34] near the city of Dujiangyan (earlier Guanseian). It includes a main dike (the *dike of a thousand feet*) that directs the current toward a structure built of large stone blocks, called the *fish nose*, that divides the river flow into two portions. The resulting two main channels are separated by the *diamond dike* whose crest is above the level of the floods, then by overflow structures that make it possible to spill floodwaters from the left channel (whose bed is higher) toward the right channel, which occupies the ancient river bed.

The left channel, or *interior channel*, is cut into rock across the hill that is some sixty meters in height, and on which the city of Dujiangyan is situated. Sima Qian speaks of the part of this hill that is isolated by the cut called the escarpment of Li. Further downstream, along this same channel, numerous intake structures direct irrigation water toward Chengdu and the plain of Sechuan (Figure 8.5).

It was surely Li Bing, the designer of the project, who also planned for its maintenance. During the low-water period from mid-October to the end of March the canals are cleaned out and the heights of weirs and dikes are brought back to their original levels. This maintenance requires that the two arms of the river be alternately dewatered with portable wood dams. In addition at regular intervals the *fish nose*, the main struc-

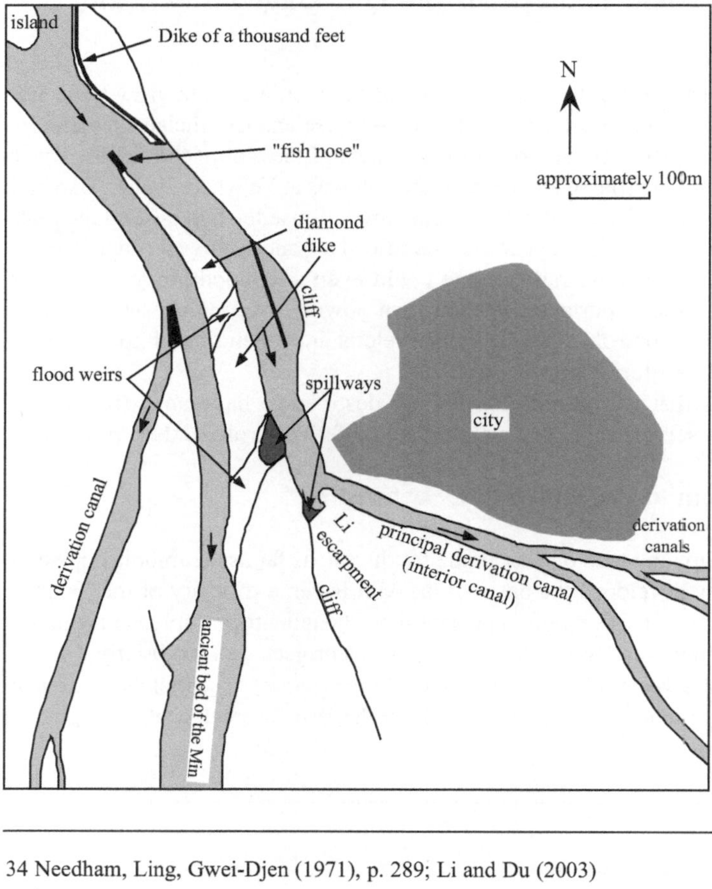

Figure 8.4 The intake structure on the Min River at Dujiangyan, origin of the irrigation system of the Chengdu plain at Sechuan (after Needham, Ling, Gwei-Djen, 1971).

34 Needham, Ling, Gwei-Djen (1971), p. 289; Li and Du (2003)

ture separating the two channels, is maintained.

Li Biing, along with his son who finished these works, is viewed as a benefactor by the inhabitants of the region. He becomes immortal; at the very top of the *escarpment of Li*, a Taoist temple is consecrated to him.

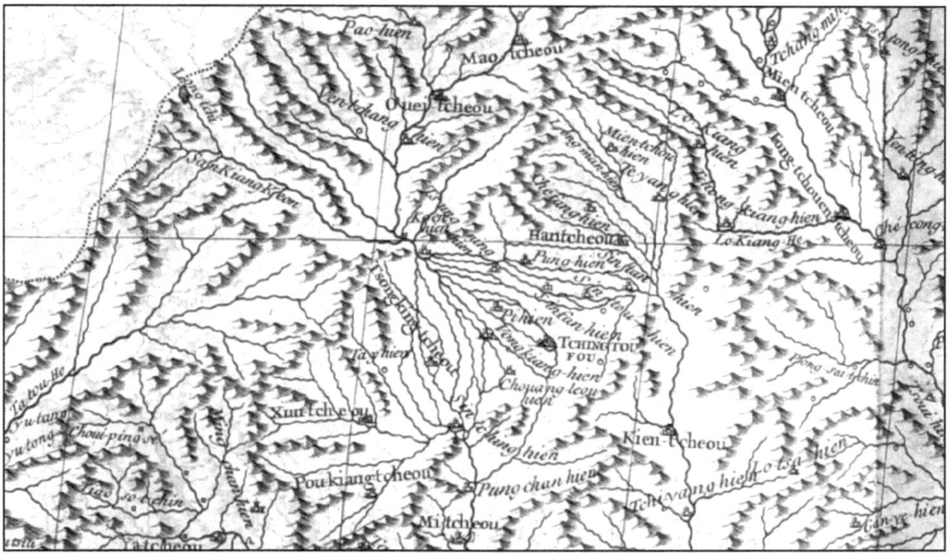

Figure 8.5 The irrigation canals of the Chengdu region, detail from a map established by the Jesuits of the 18th century (Du Halde, 1735 – ancient archives of ENPC).

The Zhengguo irrigation canal, in the basin of the Wei

Later on, probably around 250 BC, the king of the state of Hann felt menaced by the expansion of the Qin. He sought to deflect the warlike ideas of the Qin by turning his dangerous neighbor's energy toward peaceful projects:

> "The state of Hann, learning that the state of Qin was fond of undertaking large projects, dispatched a water engineer named Zheng Guo to go to Qin and persuade the ruler to construct a canal from a point of the Jing river west of Mt Zhong to the pass at Hukou, and from there along the Northern Mountains east into the Luo River, a distance of over 300 *li*. Ostensibly the purpose of the project was to provide irrigation for the fields, though in fact Zheng Guo and the rulers of Hann hoped thereby to wear out the energies of the state of Qin so that it would not march east to attack Hann. Zheng Guo succeeded in getting the project started, but halfway through the real nature of the mission came to light. (...)
>
> "The Qin ruler, deciding that this *(the argumentation of Zheng Guo that the canal would benefit to the Qin)* was sensible, in the end allowed him to go ahead with the canal. When it was finished, it was used to spread muddy, silt-laden water over more than 40,000 *qing* of land which up until this time had been very brackish, bringing the yield of the land up to one *zhong*

per acre (*mu*). As a result the area within the pass was converted into fertile land and no longer suffered from lean years; Qin became rich and powerful and eventually was able to conquer all the other feudal lords and unite the empire."[35]

This canal, put into service in 246 BC, is more than 150 km long, and links the Jing to the Luo in Shaanxi (Figure 8.8). It is called the *Zhengguo* canal in honor of its builder, and supplies numerous secondary canals that provide gravity irrigation for the entire lower region. The canal has been rebuilt several times, even recently, with new intakes further up the course of the Jing to account for the progressive degradation of the river bed and the sediment deposits in the canal itself.

The transport canals – *the magic canal (Lingqu)*

In the very same year that the Zhengguo canal came into service, a twelve-year-old child named Zheng ascends to the throne of Qin. Because of all the irrigation works, he soon inherits unprecedented economic power, and he becomes the first emperor.

In 225 BC Zheng uses the Hong canal for the supply of grain to his army, during his gradual advances toward the south.[36] The main grain storage and distribution center becomes established at the junction of this canal and the Yellow River. Later, this virtual nerve center will become the imperial granary.

The victory of Zheng over the Chu ends the *warring states period* in 221 BC. The conqueror takes the imperial name of Shi Huangdi (or Che Houang-ti). His empire includes, in rough terms, the basins of the middle and lower courses of the Yellow River (Huang) and of the Yangtze.

The emperor, seeking to extend the empire even further to the south and conquer the land of Yue (the region of Canton), plans a fluvial assault using oar-powered military junks fitted with attack towers. In 219 BC, he decides to dig a navigation canal both to transport the army and to carry its provisions.

"... the emperor sent the military commander Tu Sui with a force of men in towered ships to sail south and attack the hundred tribes of Yue, and ordered the supervisor Lu to dig a canal to transport supplies for the men so that they could penetrate deep into the region of Yue."[37]

However the land is mountainous between the Yangtze and the Xi, the river that flows into the Canton sea. The chosen passage is the Xiang river, a southern tributary of the Yangtze that is connected to it through the grand lake Dongling. The *Lingqu* canal *(magic canal)* is therefore opened between the Xiang, flowing toward the north, and the Gui (or Kwei), a tributary of the Xi that flows southerly. The canal follows the course

---

35 Sima Qian, Shi Ji, 29, transl Burton Watson.

36 Granet (1929), p. 119. Marcel Granet attributes the construction of the Hong canal to Zheng. We prefer the hypotheses of Joseph Needham who believes this canal is older.

37 Sima Qian, Shi Ji, 112, transl B. Watson.

38 Needham, Lin, Gwei-Djen (1971), p. 300 and following in the 1987 edition. See also Granet (1929), p. 140, Zheng (1991), and Schnitter (1994), p. 45.

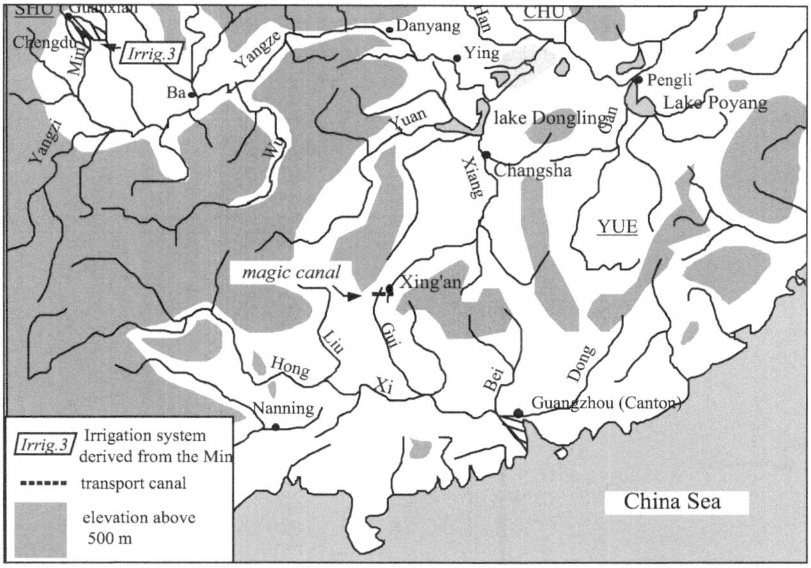

Figure 8.6 The hydraulic works realized by the Qin in south China. The layout of the branched canals in the Chengdu region is taken from the map of von Richthofen (1877).

of the Li, a minor tributary of the Gui (Figure 8.6).[38]

An intake on the Xiang River at Xing'an supplies an artificial canal that has an almost horizontal bed, with just enough slope to convey a discharge that is 30% that of the Xiang. This canal flows for about 5 kilometers, to a point near the source of the Li.[39] The Li is channelized to support navigation along some thirty kilometers of its length, as far as the confluence of this small river with the Gui. Alongside the Xiang there is a lateral canal about 3 kilometers long, and with a very modest section: 1 to 2 m deep, and 5 to 8 m wide.

The Xiang intake structure on the Xing'an is obviously inspired by the one built several decades earlier on the Min River. It includes a separation structure downstream of a dam-spillway in the form of a V. This complex is designed to raise the water level to provide for flow into the canal, and to create a basin in which the current is sufficiently weak to allow boats to be maneuvered - while at the same time providing for the evacuation of flood waters into the ancient bed of the Xiang. This 3.9-m high dam is called the Tianping dam. Several other weirs on the canal itself provide for further regulation of the water level as well as for floodwater overflow, as explained in the following 12th century description:

"The passengers who travel *(on the canal)* are terrified at certain locations for, at about 2 *li (1 km)* from the intake where the "spade head" divides the water and guides one of the branches

---

39 According to Needham, Li mentions the idea of separation. It is therefore possible that the name of the river – the Li – comes from the structure mentioned here, separating the discharge from the Xiang. One could say the same of the Li escarpment, at Dujiangyan: it is separated from the hill by the notch through which the canal passes.

toward the canal, there is another weir *(literally: structure that lets excess water leave)*. Without this weir, the violent force of the springtime flow would damage the canal's support wall, and the water would never get to the south. Thanks to this structure, the violence of the water is calmed, the dike is not broken, and the water in the canal flows gently. [....] This is truly what one can call an ingenious device. The canal waters wind around in the district of Xing'an, and people use it to irrigate their fields."[40]

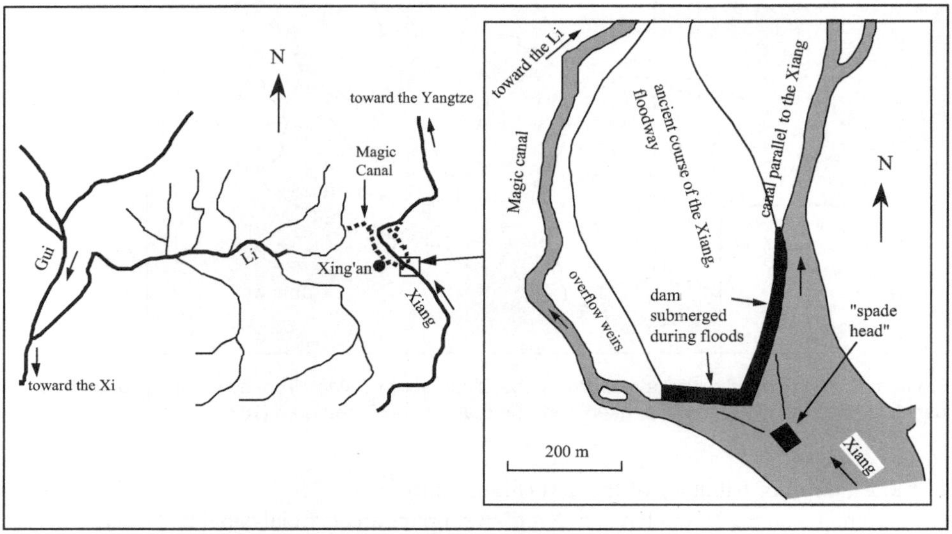

Figure 8.7 The magic canal (Lingqu), communication link between the Yangtze and the Xi (map from Needham, Ling, Gwei-Djen, 1971, detailed plan adapted from Zheng (1991) and Schnitter (1994)). The detail of the installation at the right may not be as it was built under Shi Huangdi, for it underwent important renovations in 825 AD under the Tang.

Later on another navigable channel is dug behind the separation structure to facilitate the turning of barges and their passage from one canal to the other (Figure 8.7).

The *magic canal* is renovated in 825 AD and fitted with single gates (flush locks, a system we discuss further on) on the two canals and possibly also on the channelized Li, to maintain navigability during low-water periods. These flush locks are probably replaced in the 12th century by true chamber locks, with 36 openings in all. The canal is destined to remain in service through all of Chinese history, right up to the present. The completion of this project, along with the canals that had already been created during the feudal period, created a continuous watercourse – though indirect – that links Canton to Chang'an through the Xi, the Li, the Xiang, the Yangtze, the Huai, and then the Yellow River and the Wei canal.

40 Extract from a treatise of 1178 called Ling Wai Tai Ta by Chou Chhü Fei. Adapted from the citation of Needham, Ling, Gwei-Djen (1971), p 304.

The destiny of Shi Huangdi, first emperor

The eulogy of Shi Huangdi (formerly Zheng) is engraved on the gates of the city of Jishi. It recognizes not only his destiny as a conquerer, but also his contributions to hydraulic infrastructure:

> "He took down the inner and outer fortifications,
> "He opened the watercourses and erected dikes,
> "He leveled the dangerous gorges."[41]

His overall influence was clearly civilizing, and he put into place the administrative structures that made the unification of China inevitable. He is harshly treated by Chinese historians, however. Indeed, in a conflict with the scholars in 213 BC he massacres a number of them and burns their writings.

Shi Huangdi's famous army of 6,000 men in terra-cotta is buried in his monumental tomb near his capital Xianyang, about thirty kilometers from the present-day Xi'an, discovered in 1974. A marvelous testimony to the hydraulic developments of Qin is the reproduction of the land's rivers on a scale model in his honor, using mercury as the fluid subject to a flow control system:[42]

> "In the ninth month, Shi Huangdi was buried in the Li mountain. [....] With mercury, the hundred watercourses were made, the Kiang, the Ho, and the vast sea; machines made the mercury flow and transferred it from one to the other. Above were all the signs of the heavens; below all the geographic details. [....] Those who were put to death were very numerous."[43]

The Han Empire: continuing hydraulic development. Awakening of the Yellow River

The First Emperor left the legacy of an energetic, but bloody, monarchy to Chinese historians. And this is no doubt why the Qin Dynasty could not survive it. It is replaced by the long Han Dynasty, dominating China for more than four centuries from 206 BC to 220 AD. The Empire continues to encompass the basins of the two great rivers, and even extends to the south as far as Canton and to the west into the corridor of the Silk Road, nearly to Bactria. But the demographic and economic center of gravity remains in the north. The census of the year 2 AD showed 85% of China's 57 million people to be in the north. The most populated zones[44] are the Wei valley, ancient nursery of the Qin, and the vast alluvial plain of the lower course of the Yellow River down to the sea, between the Jiang River to the north and the Huai to the south. The capital of the early Han, Chang'an, is on the present-day site of Xi'an. Following a temporary usurpation of power by a dignitary named Wang Mang (9 to 23 AD), the capital is relocated to

41 Citation after Blunden and Elvin (1983), p. 81.
42 See for example Debaine-Francfort (1998), p. 93.
43 Sima Qian, Historical memoires (Shiji). Citation from Debaine-Francfort (1998). This account is corroborated by the discovery of traces of mercury in the soil of the tomb.
44 Blunden and Elvin (1983), p. 30.

Luoyang, near the ancient capital of the Zhou (the latter Han), where the Yellow River flows out onto the plain.

Development of the historic heart of China

Development of the Wei basin is actively pursued under the early Han and, particularly under the long reign of the grand Emperor Wudi (141 to 87 BC). This is the region of the capital Chang'an (Figure 8.8). Major extensions to the Zhengguo canal are made in 111 BC, then again in 95 BC. The marshy nature of much of the land led to the development of a special technology for digging a canal deep into very unstable soil. This is the technique called *canals with wells*. The first of such works is the *canal of the dragon's head*. Let us listen again to Sima Qian:

> "The emperor called up a labour force of over 10,000 men and set them to work digging a canal leading off from the Luo River at Zheng and extending to the foot of Mt Shangyan. There, however, it was found that the banks of the canal kept collapsing, so the men dug wells, some of them over forty *zhang* deep, at various points along the course and induced the water to flow from one well to another. Thus the water disappeared from sight at Mt Shanyang and flowed underground to the eastern side of the mountain, a distance of over ten *li*. This was the beginning of the so-called well-canals. In the course of the digging a dragon bone was discovered and the canal was therefore named Dragon Head Canal. It has been over ten years now since it was constructed but, although the water flows through it fairly well, the land has not yet shown much improvement."[45]

One can wonder whether the inspiration for these "*canals with wells*" could not have been traced to *qanats*.

During this period there was apparently a shortage of grain in the historic heart of China. It was therefore necessary to bring grain from the east up the Yellow River and the Wei. It was costly and dangerous to tow boats upriver, especially through the Sanmen pass (the *three gates*). Two projects were undertaken to ameliorate the situation, specifically to reduce the time required for transport of grain from the plain to Chang'an. A proposal to the emperor described construction of a transport canal. Here is the argument developed in 133 BC by Zheng Dangshi, minister of agriculture of Wudi:

> "Up to now grain from east of the Pass has been brought to the capital by being transported up the Wei River. The operation requires six months to complete and the course is over 900 *li* and beset with dangerous places. Now if we were to dig a canal from the Wei River, beginning at Chang'an and following along the Southern Mountains to the Yellow River, the distance could be reduced to something over 300 *li*. We would have a much easier route for transporting grain, and the trip could be accomplished in three months. Moreover, the people living around the canal could utilize the water to irrigate over 10,000 *qing* of farmland."[46]

The emperor gives the go-ahead for this project. An engineer named Xu Bo from the land of the Qi (in present-day Shangdong) is called upon to construct the canal, an

45 Sima Qian, Shi Ji, 29. transl B Watson.
46 Ibid.

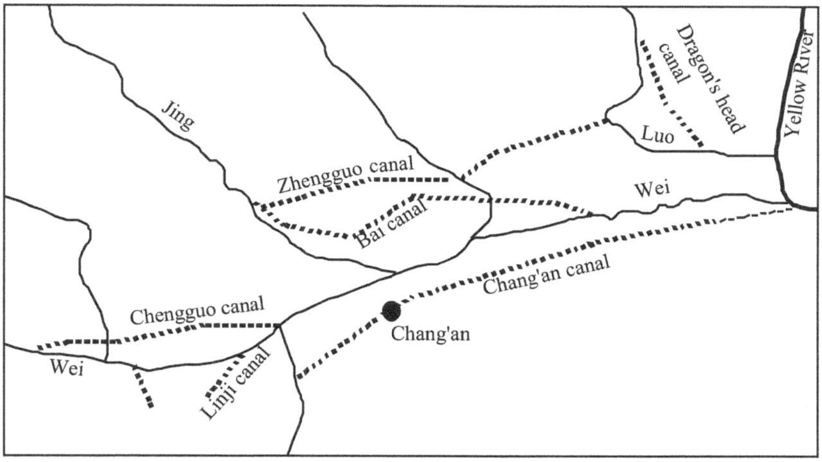

Figure 8.8 The transport and irrigation canals in the Wei basin, during the Han era, adapted from Lian Ruiju, Zheng Zhaojin, Hu Jialin (1987) and other sources.

effort lasting three years and requiring the mobilization of 20,000 to 30,000 peasants.[47] Nonetheless, towage problems remain on the Yellow River. Therefore to further augment production in the region it is decided in 129 BC to construct a canal from the Fen to irrigate the Yellow River valley in the present-day Shanxi. Again, 20,000 to 30,000 peasants work on this project for several years; but in the end the Yellow River changes course and thus nullifies any benefit from this effort. The land, having been prepared for cultivation, must now be abandoned.

Other projects of the early Han

Other canals are constructed in the north in 113 BC. The purpose was to try to bring the region of the grand loop of the Yellow River into cultivation, and thus to settle this area that had served as an ideal corridor for the invasions of the Xiongnu barbarians.[48] These irrigation canals are built on either side of the Yellow River in the region of the present-day Yinchuan. Their construction had in fact been begun around 215 BC under the reign of the first emperor, and saw further new development under the Ming. In the end, they attained a length of about 180 km.

Between 38 and 34 BC a dam-reservoir called the Maren dam was built near the city of Nanyang on one of the principle north tributaries of the Han river. The project is attributed to Zhao Xincheng, who was prefect of Yanyang. This earthen dam is 16 m high and 820 m long, and includes six gates of stone to control flow into the irrigation canals.[49]

47 Granet (1929), p. 142. See also Sima Qian, Shi Ji, 29.
48 Sima Qian, Shi Ji, 110.
49 Needham, Ling, Gwei-Djen (1971), p. 281 and following in the edition of 1987; Zheng (1991); Schnitter (1994), p. 41.

The great Yellow River dike failures in the Han Empire

The Yellow River dikes are continually maintained and raised. In the 2nd century BC, they had already attained a height of some ten meters. The regions they protect along the lower course of the river had become among the most populated of all China, as we have seen earlier. A first dike rupture occurred in 168 BC at a place called Suangao to the northwest of Kaifeng, near the origin of the Ji River (Figure 8.9). This rupture destroys what was called at the time the "metal dike", to the east of Suangao, but in the end the breach is repaired.[50]

Yet again in 132 BC the Yellow River broke through its dikes at Huzi and poured out onto vast agricultural regions to the south. When the flood levels dropped the river continued to occupy its new course, in the beds of the Si and the Huai rivers. The course change that had occurred five centuries earlier, in 602 BC, may have had limited impact, but it is not hard to imagine that the new change had rather dramatic consequences. Dike repairs were attempted, but without success as they failed once again. Finally the struggle was abandoned, perhaps under the influence of a counselor to the emperor whose lands were to the north of the breach, and therefore sheltered from the floods and perhaps increased in value due to the losses elsewhere.[51] But the floods of the Yellow River in its new course could no longer be controlled at all. In 120 BC, some 700,000 victims had to be relocated from this region toward Shaanxi.[52]

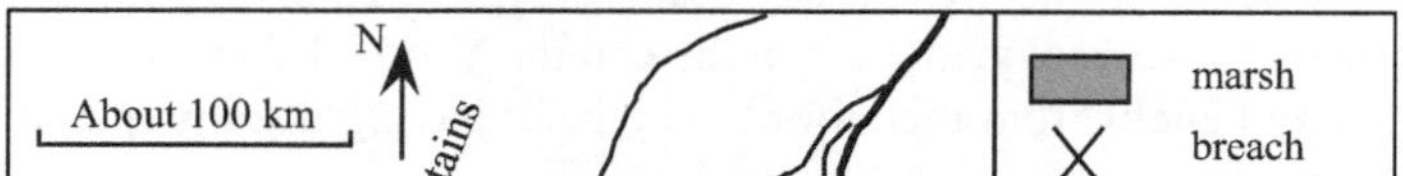

Figure 8.9 Probable locations of the great dike ruptures of the Yellow River under the early Han (adapted from Liang Ruiju, Zheng Zhaojin, Hi Jialin, 1987, and other sources)

50 Sima Qian, Shi Ji, 29.
51 Ibid.
52 Gernet (1990), p. 112.

A particularly dry year occurred in 109 AD, 23 years after the dike rupture. Emperor Wudi took advantage of the situation to bring a considerable armed force to the site, and succeeded in blocking the breach and restoring the river back into its original bed. The blockage was accomplished by throwing bundles of branches into the breach between clusters of bamboo stalks, and quickly covering them with rocks and earth. Sima Qian is, here, no longer a historian but an actual eye witness to the event. He is even an actor in the play, for he tells of having carried bundles of branches on his back, as did everyone in the emperor's entourage. At the outset of his work, the Emperor Wudi threw a jade ring into the river and also sacrificed a horse. Sima Qian faithfully recorded a psalm that the emperor composed on this occasion:

"The river broke through at Huzi;
What could we do?
Beneath its rushing waves,
Villages all became rivers,
The villages have all became rivers
And there is no safety for the land. (...)
The river raged from its boundaries,
It has left its constant course.
Dragons and water monsters leap forth,
Free to wander afar.
Let it return to the old channel
And we will truly bless the gods. (...)
Ask the Lord of the River for me,
"Why are you so cruel?
Your surging inundations will not cease;
You grieve my people!
The city of Niesang is awash;
The Huai and Si brim over,
So long, and yet you will not return
You overstep the watery bounds!" (...)
The Lord of the River hears our plea
But there is not enough brushwood.
There is not enough brushwood -
The fault of the people of Wey.
They have wasted the land with fire –
What can be used to check the waters?
We sink the forest bamboo
And ballast the weir with stones.
We will stem the break at Xuanfang
And bring ten thousand blessings!"[53]

Eighty years later, in 39 BC, new dike ruptures occur further to the north, and the technique for closing them is further improved. Cages of bamboo filled with stones are

---

53 Siman Qian, Shi Ji, 29, transl B Watson .

brought to the site by boat and sunk from the edges of the breach, progressively reducing its width. Finally, boats are sunk at the site of the breach, and then the entire plug is covered with earth.[54]

There is no doubt that the situation on the lower course of the Yellow River continues to preoccupy the leaders. In 6 BC, an engineer called Jia Ran (or Chia Jang) submits a report to the emperor. He points out the potential danger of a poorly monitored dike, and calls attention to population and use of zones that should be kept open as flood plains, zones that represent a dangerous confinement of the main channel.

> "At the present time, the nearer embankments stand at a distance of several hundred paces only from the water, and even the furthest are only several *li* from it. South of Liyang the old "Great Metal Dike" stretched north-westwards from the west bank of the Yellow River to the southern foot of the western mountains. It also ran eastwards to meet the eastern mountains. People built their cottages on the eastern side of the dike. After they have been living here a little over ten years, another dike was thrown out from the eastern mountains southward to connect with the Great Dike. Again, in the prefecture of Neihuang, a swamp with a circumference of several tens of *li* was drained by building a dike round it, and the governor of the district then gave the land within the dike to the people after they had lived here for more than ten years. Now people build cottages in it. These things I have myself seen. In the prefectures of Tungchün (Eastern Commandery) and Paima, the old "Great Embankment" is paralleled by several other embankments (outside it), and people live in between them. From the north of Liyang to the border of (the former state of) Wei, the old "Great Embankment" lies several tens of *li* from the river; but inside it there are also several rows of dykes which were built in earlier generations."

These dikes, built without any planning, provoke the Yellow River into useless and dangerous changes of direction:

> "Thus when the Yellow River flows from Honei north to Leiyang there is a stone embankment forcing it eastwards. When it reaches Tungchün and Phingkang, there is another stone embankment to force it north-west. When it arrives at Leiyang and Kuanhsia it meets a third, changing its flow north-eastwards again. At Tungchün and Chinpei it is diverted north-west, and at Weichün and Chaoyang north-east again – all by stone embankments. Thus in a distance of only a little over a hundred *li*, it is turned westward twice and eastward three times..."[55]

Jia Ran has Taoist instincts. He proposes that the Yellow River be left to occupy its natural flood plains, free of habitations and obstacles. He specifically proposes that the populations in these zones be moved out, and that a natural outlet and floodplain toward the river's ancient bed be opened further to the north (Figure 8.9). In the event that the emperor does not have the will to relocate the population, Jia Ran proposes an alternative. This would be that a dense irrigation canal network be built with solid stone protection, making it possible to irrigate the region during normal periods and to handle the floodwaters in wet periods, thus reducing their threat to the principle course of the river.

A third proposed solution, consisting simply of raising the existing dikes, is clearly

54 Liang Ruiju, Zheng Zhaojin, Hu Jialin (1987).
55 Needham, Ling, Gwei-Djen (1971), p 234.

nothing more than a token one. But the Emperor Aidi is young, the empire no longer having the strength it had under Wudi, and so the energetic measures proposed by Jia Ran are not adopted. This makes catastrophe inevitable.

The catastrophe occurred in 11 AD. The river comes out of its bed never to return, punctuating a three-year period of usurpation of power by a dignitary called Wang Mang who attempts political reform. The new course of the river, destined to be stable for nearly eight centuries, is some fifty kilometers further to the east (Figure 8.2) than the course into which the Wudi had redirected it. The event itself was devastating as can be imagined. And then regional famine ensued, since the crops were destroyed. In subsequent years floods return over and over again, since the river was no longer in its diked bed. The population flees and there are revolts, subversively led by the secret society of the *Red Eyebrows*. All of these troubles likely contribute to the eviction of Wang Mang in 25 AD and the return to power of the Han family (these will become the "latter Han").

The end of the Han and the last great hydraulic projects

Reconstruction efforts mark the beginning of the period of the "latter" Han. The Bian canal had greatly suffered from inundations, and the repair works were directed by an engineer called Wang Ching. In 70 AD the Emperor Mingdi inspects this work:

> "Since the dike ruptures at the outlet of the Bian canal, more than sixty years have passed. [....] The original emplacements of the gates have been lost in the middle of the river. Large expanses of water have formed, to the point that one could no longer recognize the original banklines. But now, the dikes have been reconstructed and the canal has been repaired, the flow stopped, and gates put back into place. The river *(Yellow)* and the canal *(Bian)* flow separately and have returned to the original beds. Because of this, we have sacrificed the most beautiful jade and the purest animals to the spirit of the river."[56]

About 80 AD the *Shaobei* (today the *Afentang* dam) was also rebuilt; it dated from the *period of springs and autumns*.

An agricultural crisis afflicts the empire of the "latter" Han from 170 AD. Major new floods occur on the lower course of the Yellow River. The desperation of the affected population once again creates a fertile ground for agitators; a large revolt is organized in 185 by the secret society *Yellow Turbans*. Although the revolt is quashed the dynasty does not survive it, being abolished in 220 AD.

The result of this collapse is a schism of the empire into three kingdoms, during which new hydraulic projects are undertaken. A vast irrigation system is put into service in 189 in the region of the ancient dam *Shaobei*, with numerous weirs built on small rivers. New transport canals linking the Jiang and the Hutuo in the north are built to support the offensive of the northern sovereign Cao Cao against his rival Yuan Shao in 204 AD. South of the Yangtze, dam-reservoirs are still being built in the 4th century AD; the Han canal, linking the Huai to the Yangtze, is also rebuilt during this period. These new canals already prefigure the future Grand Canal, whose construction accompanies the

56 Adapted from Needham, Ling, Gwei-Djen (1971), p. 346.

rebuilding of the empire. But it will be three long centuries, the chaotic period of the Chinese middle ages, before this construction begins.

Innovations under the Han

The first imperial era was of great importance to the blossoming of China thanks to its cultural unity and construction of hydraulic infrastructure. Before moving on, we need to note the appearance of several other important innovations.

One innovation is the axial ship rudder (see Figure 8.10). We know that it appeared during this period from a terra-cotta scale model of a junk that dates from the 1st century AD and was discovered in a tomb at Canton.[57] At first, this was a movable rudder mounted on the stern of the hull. Later, it is attached to the sternpost of junks. The axial rudder is not adopted in the West until the 11th century.

A second innovation is the water wheel, appearing during the time of Wang Mang at the beginning of the 1st century AD. It is curious that the first literary references to the use of hydraulic energy in China are not about water mills, but about much more complex industrial applications. The first mention of a water wheel appears in 21 AD. This wheel is probably horizontal, and its axle shaft is fitted with cam lobes to drive an assembly of pestles.[58] The system is used to crush grain, and also to power forges. Ten years later (31 AD) a certain Du Shi, the son of Zhao Xincheng who built the dam that we described earlier, introduces the use of hydraulic energy to power piston bellows in the forges at the important metallurgical center Nanyang.[59]

While the axial rudder is clearly a Chinese invention, the hydraulic wheel is more likely an imported technique since there is evidence of its use somewhat earlier in Asia Minor at the beginning of the 1st century BC (see Chapter 5). The spread of this technology into China is more or less contemporary with its appearance across the Roman Empire, albeit for different applications. However the exact origin of this invention remains obscure.

Another innovation that appeared under the "latter" Han appears rudimentary but is extremely effective for raising water. It is a device comprising a wooden chain fitted with rectangular paddles, like quoits, and powered by a sort of chain wheel (Figure 8.18). This square-pallet chain pump, the *dragon backbone machine*, is destined to spread throughout the lands of Chinese culture, and we return to this later on. It likely grew from the need for cities to lift water from cisterns or watercourses. We know from a treatise of Wang Ching in about 80 AD that in the capital Luoyang, men are employed "night and day" to lift water from cisterns to the street level – but he does not describe how it is lifted. A century later, in 186 AD, another treatise mentions the paddle pump very explicitly, again in the context of lifting water to the street level at Luoyang.

These lifting machines probably supplied water distribution networks of some sort. Conduits of stone or terra-cotta have been found dating from the Qin or Han periods.[60]

---

57 Needham (1978), p. 58; Dars (1992).
58 Gernet (1990), p. 128.
59 Needham and Ling (1965), p. 369.
60 Ibid., pp. 344-345.

Bamboo tubes were probably used as well.

Another innovation that appeared in China in the 2nd century AD is the modern sail,[61] that is, a sail carried by a boom or yard that pivots around the mast. This type of sail rig can point into the wind, and it is not necessary to lower the sail when coming about, since the sail pivots on its own when the boat turns across the wind, or tacks. The Chinese sails were made of braided bamboo, stiffened with battens or yards so that they

Figure 8.10 The rudder on Chinese boats (engraving of Chambers, 1757 – ancient archives of ENPC).

do not luff when brought close to the wind. It is not necessary to lower the sails when the boat is moored, and when the wind freshens, it is easy to reduce the sail area.

At the same time as the galley ship is being developed in the West, these Chinese sailing innovations support considerable coastal navigation along the shores and rivers of China, in particular on the Yangtze.

Under the Han, some canals are provided with gated openings. The Bian canal is the principle example; we saw earlier that such gates are mentioned as part of the reconstruction works of 70 AD. They are also mentioned in an older account, again from Jia Ran in 6 BC whose report proposes solutions to prevent flooding from the Yellow River:

"We could construct a rock dike from Chhi-Khou toward the east and build many gates. [....]

61 Steens (1989), p. 422.

> I fear that this proposal will attract criticism saying that the river is too large to be controlled. However, we can evaluate our chances of success from the experience that we have on the Bian canal at Jung-Yang. At this location, the openings or gates were only made of wood, and set directly on the earth of the dike. Therefore, if we constructed these dikes out of rock, with good foundations, the safety of the works would be assured. [....] During the dry season, the lowest gates to the east should be opened to irrigate the countryside of Chichow, and during the flood season, the large gates of the west should be opened to route the river's floodwaters far away."[62]

The existence of gates on small irrigation canals stems from olden times that cannot be precisely pinned down. It is unlikely that they are in general use on large canals

Figure 8.11. A traditional fishing boat on lake Taihu, to the east of Shanghai (photo by the author)

throughout China until the 1st century BC, for Sima Qian does not mention anything like them in his observations. Recall that such gates existed in Egypt during the Ptolemite period, 3rd century BC, at the end of Necho's channel – as well as in the Fayoum depression (Chapter 5). The principle of a portable plank-dam was well known even earlier in Syria (see Chapter 2, the dams around Ugarit). In China, these single gates generally consist of several planks laid horizontally, one on top of the other, between guides set into each bank. On the larger applications, the planks are interlocked, and the assembly is operated using a system of pulleys and counterweights.[63]

On the navigable canals the gates are closed to maintain sufficient water depth when the discharge is low with respect to the slope of the canal. When the gates are opened,

62 Needham, Ling, Gwei-Djen (1971), p. 345, adapted.
63 Ibid., p. 348.

boats moving downstream are carried by the wave that is created. Boats moving upstream, on the other hand, must be winched upstream of the gates using a capstan. After the gate is closed it takes some time before the water depth increases sufficiently for navigation to resume. This clever system constitutes the principle of the *Chinese flush lock*. As we see further on, this type of simple lock sees broad use on the Grand Canal.

Finally, even though this does not directly involve hydraulics, we must also note another Chinese innovation of this period: the humble but ergonomic wheelbarrow.

The Grand Canal

The Grand Canal of the Sui, the Tang, and the Song (6th to 11th century)

In 581 AD, Yang Jian founded the Sui Dynasty at Chang'an. He reunifies China in 589, and in 604 the country sees its new master enthroned as emperor. An imperial necessity appears immediately: to establish a safe communication route between the north of China where the reconstructed capital Chang'an is located, and the Yangtze basin to the south. This need reflects a fundamental change in the relations between north and south

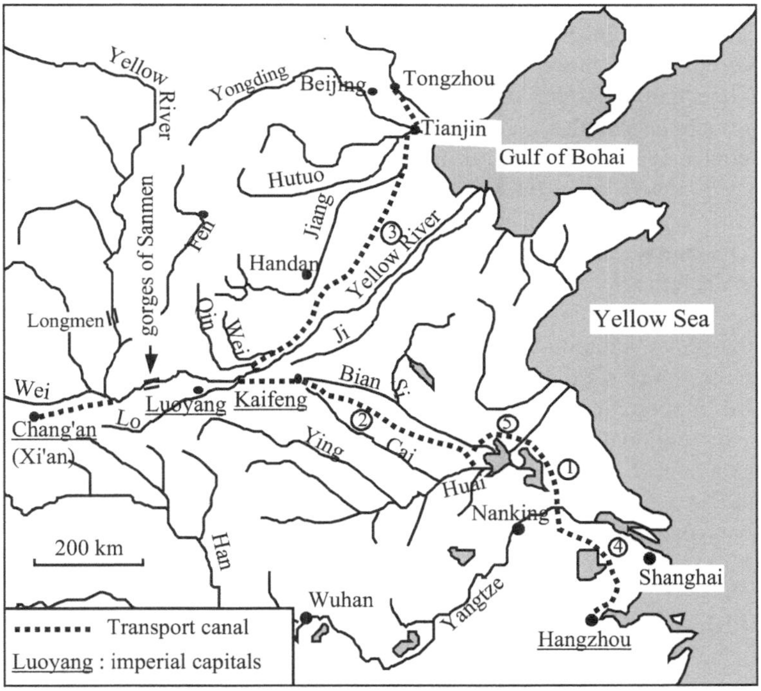

Figure 8.12 The Grand Canal of the Sui, the Tang and the Song.

1. Shanyang traverse (renovated in 587)
2. Tongji canal (605)
3. Yongji canal (608)
4. Jiangnan canal (610)
5. Northern detour of Hongze lake (about 735)

since the Han period. The north was inflicted with a series of wars and barbarian invasions, whereas the lower basin of the Yangtze had begun to develop, propelled by the rapid growth of rice cultivation. The cities of the lower Yangtze, where Chinese intellectuals took refuge, had become the cultural centers of the land. The communication route therefore had high political and economic stakes: cementing the unity of the country and, at the same time, generating tax revenues.

In 587, even before the fall of Nanjing, Yang had restored the canal that connected the Huai and the Yangtze. The ancient Han canal had filled this role in olden times, and its course had already been shortened in 350. The north branch of the old 4th or 5th century Hong canal had linked the Huai to the capital region. But this canal, subsequently called the Bian canal (or Pien), had become clogged with sand. It was decided to conserve the approximate original layout to the west of Kaifeng, but then to depart from the Bian River in cutting more to the south to end up on the Huai to the west of the present-day Hongze lake.[65] Since the emperor wanted to proceed quickly, he acted radically. He mobilized five million people, men and women alike, to dig 1,100 km of this new sixty-meter wide canal in just five months. The new canal is called the Tongji.

The next step involved renovating the communication route toward the north for essentially strategic reasons. Indeed, the threats of barbarian invasions were from the northeast, and there were also plans for a military incursion into Korea. The Yongji canal, some 1,000 km long, is finished in 608. This canal is at first a derivation of water from the Qin (a river that flows into the Yellow River a little downstream of Luoyang), but later ends up connecting to another small river oriented toward the north, and called the Wei (this is not the large Wei that is near Chang'an). This river then joins the Jiang (that occupies the course of the Yellow River prior to 602 BC) near Tianjin. Two years later in 610 the new Jiangnan canal links the Yangtze to the port of Hangzhou to the south. These works are completed with an access canal to the newly reconstructed city of Luoyang (which becomes a second capital) as well as by a complete renovation of the old canal that links Chang'an to a bend of the Yellow River. The ensemble constitutes the first Grand Canal in the form of a gigantic Y with the capital at its base. For four centuries this will be the spinal column of the empire.

But these massive projects exhausted the population, and there are floods in the Shandong. Moreover Emperor Yang suffers several military reverses at the hands of the barbarians. In 618 he is eliminated and immediately replaced by the Tang Dynasty under which China enjoys a sort of golden age for two centuries. Chang'an, the capital and terminus of the Silk Road, becomes a cosmopolitan city in this period. Around 700 the last Sassanide Persian sovereign comes to Chang'an to finish his life in exile. Canton is inhabited by numerous Arab merchants.

The Grand Canal is of course maintained and further developed. A derivation to the north of the present Hongze lake in 735 allows southbound boats to bypass the rapids of the Huai. Management of the canals is facilitated by the construction of gates near their outlets into the large rivers; dikes and rockfill protect the canals at vulnerable locations. In this period some 165,000 tons of grain are carried annually on the Grand Canal; it is

---

65 The Hongze lake is shown on our maps; but is it possible that it did not exist prior to the development works of the 15th century (see the end of this chapter)?

under the Tang that the blossoming of rice cultivation in the Yangtze basin is the most pronounced. Immense granaries are built at the nodal points of the navigable waterway system.

The gorges of Sanmen remain troublesome for navigation as far as Chang'an. The rapids are dangerous and the channel contains dangerous rocks. From 733 a roadway was used to transport merchandise over a land detour of several kilometers. But in 741 a new 300-m long canal was dug through solid rock to cut across the river bend having the most dangerous rapids. In the south, the *magic canal* is improved in 825, as we have seen earlier, and then again in 868.

Between 960 and 1127 the Song set up their capital at Kaifeng (called Bianling at this period). Since ancient times Kaifeng, along with the old Hong canal, had been an important communication node, not far from the junction of the two branches of the Y of the Grand Canal. The Bian, tributary of the Si, flows naturally into the Tongji canal whereas the Cai, flowing toward the south, connects to the Tongji through a canal called the Huimin. In 1071 major work is done to redo the connection between the Yellow River and the Tongji canal.[66] The entire Song period is marked by a very important expansion of navigation on the large rivers and along the coasts.

The appearance of the chamber lock

The chamber lock, with two gates, is invented in the Song Dynasty at the end of the 10th century. In 983, a civil servant named Chiao Wei-Yo is in charge of transport in the Huai region. At this time barges were transferred from the canal to the Huai, at the northern extremity of the Shangyang traverse, by dragging them on an inclined ramp. This operation often damaged the heavily-loaded barges and their cargo. To remedy this difficulty, Chiao Wei-Yo conceived the concept of the chamber lock, the very first such device in the history of man:

> "Chiao Wei-Ho therefore ordered the construction of two gates at the third dam along the west river. The distance between these two gates was a little more than 50 paces *(75 m)* and the entire space was covered with a large roof. [....] When the gates were closed, water accumulated like the tide until the desired level was attained, then, when the right moment arrived, it was allowed to flow out."[67]

Additional evidence of the appearance of the chamber lock comes from the *magic canal*, linking the basins of the Yangtze and the Xi to the south. Joseph Needham notes that up until the 9th century, written accounts mention 18 single gates on this canal, and that from the year 1178 the number increases to 36. He interprets this doubling as the replacement of the single-gate flush locks by chamber locks with two gates. A later account dating from the 16th century also suggests the existence of chamber locks on this canal:

> "On the Ling Chhü (*Lingqu – magic canal*) north to south there are 32 lock-gates, i.e. from the Li to the Thung-Ku Shui. From east to west, entering Yung-fu, there are 6. In the winter (the

66 For the work done under the Song, see the work of Jacques Dars (1992), pp. 149-150.
67 Adapted from Needham, Ling, Gwei-Djen (1971), p. 350.

> canal) dries up and one cannot pass through. But when I made the passage through these lock-gates there was plenty of water, and under the moonlight they looked like steps leading up to some high platform, or like tiers of walls and terraces coming down one behind another from the sky."[68]

The chamber lock is a Chinese invention that does not appear in Europe until the 14th century. In China itself this invention will have a somewhat murky future, since, as we will see, the old principle of the flush lock will remain in broad use on the Grand Canal for some time to come.

The Grand Canal of the Yuan and the Ming Dynasties

In 1126 the Song were forced to abandon the north of China, pushed out by barbarians who had partially adopted Chinese culture (the Jurchen, or Jin). During their retreat the Song destroy the south-bank dike of the Yellow River in the Kaifeng region, and this causes considerable damage to the Tongji canal and destabilizes the river. For more than three centuries after this the Yellow River will tend to form multiple and unstable branches. Pushed to the south, the Song further destroy hydraulic infrastructure and set up their capital at Hangzhou, at the southern extremity of the Grand Canal. During the period of wars that follows, in about 1190,[69] the river starts to migrate toward the south (Figure 8.2). From this date, part of the discharge in effect bifurcates toward the Huai (note this had already occurred between 132 and 109 BC). First the floodwaters, then the Mongols of Ghenghis Khan, sweep over China. From 1214 to 1276 the Mongols destroy the Jurchen powers, then those of the Song. Their first undertaking is to massacre all the inhabitants, as in Mesopotamia. But in 1271 they take on a more noble role and, under the name of the Yuan Dynasty, reign over an immense empire that extends from the Sea of China to the Euphrates.

The Yuan chose Beijing for their capital, this being a more central site for their new empire than the ancient capitals of Chang'an, Luoyang and Kaifeng. Just as the Sui had been impelled to construct the first Grand Canal in the 7th century, the Mongols now saw the need to reestablish a large waterway, ideally as a more direct link from Beijing to the Yangtze basin. In 1275 a Mongol general called Bayan was fighting against the Song of the south. He began studies of a canal that would use the north portion of the Grand Canal of the Sui, but would cut directly toward the southeast from Lingqing. This canal would rejoin the arm of the Yellow River that had bifurcated from there toward the Huai since the 1190s (this is course no. 5 in Figure 8.2). Between 1280 and 1283 the Huangong and Jizhou canals were built more or less along the alignment of this arm.

The Jizhou canal takes a shortcut across the foothills of Shandong, and has 31 gates along a distance of nearly 150 kilometers (250 *li*). These canals make it possible to ship goods from the south up to the bifurcation of the course of the Yellow River, then on the

---

68 Account of a traveller named Kuang Lu from 1585, adapted from Needham, Ling, Gwei-Djen (1971), p 306.
69 1194 according to Joseph Needham, Wang Ling, Lu Gwei-Djen (1971); 1187 according to Liang Ruiju, Zheng Zhaojin, Hu Jialin (1987).

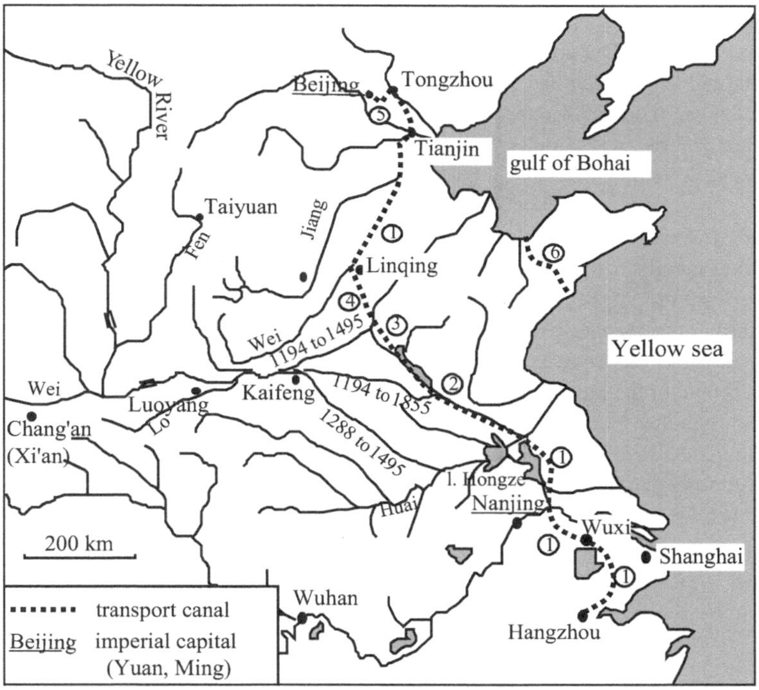

Figure 8.13 The Grand Canal of the Yuan and the Ming.
1: elements of the Grand Canal of the Sui (renovated between 1280 and 1290)
2: Huangong canal (1280)
3: Jizhou canal (1283, 1411)
4: Huitong canal (1289)
5: Tonghui canal (1293)
6: Jiao-Lai canal

northern course to the Gulf of Bohai from which they continue by sea to Tientsin.

Five years later (1288) the Yellow River abandons its principle course to the north of Shandong (though this branch carries some flow until 1945) and flows out to the south into the Huai, even further upstream than earlier. This course change probably motivated and likely facilitated the completion of the waterway project. Indeed just one year later, in 1289, the Huitong canal - whose design had been studied by general Bayan - is completed in its turn. The totality of the project is completed in 1293 by the Tonghui canal linking the capital Beijing to Tongzhou. This canal had to have some 20 gates, for Beijing is at a higher elevation than the rest of the system. The remainder of the waterway network, whose layout follows that of the first Grand Canal of the Sui and the Song, is simply brought back into service and improved (in 1290 for the channelization of the Wei river.).

Marco Polo sojourns in the court of Kubla Khan in 1280, and probably in 1288-1290 is sent on a mission to the south of China (he returns to Italy in 1298). He writes of the city of Guazhou, at the junction of the Yangtze and the Grand Canal, as follows:

"The city is on the river, and it is there that, every year, vast quantities of rice are collected.

And, from this city, they are transported to the large city of Cambaluc *(Beijing)*, to the court of the Great Khan, by water. But understand this is not by sea, but by rivers and lakes. [....] And I tell you that the Great Khan had these waterways between the two cities brought into service, for he had huge trenches dug, very wide and very deep, from one river to another and one lake to another, and had water transported by canals, so that it all became like one large river, and great ships go on it."[70]

Marco Polo also speaks of the city of Jining, which is at the junction of the Huangong and Jizhou canals. He notes how the watercourses coming from the heights of Shandong are captured to supply the highest portions of the Canal:

"And I tell you further that they have a river from which they profit immensely and I will tell you how. The truth is that this large river comes from the midlands to this city of Singi Matu *(Jining)*, and the people of the city, from this large river made two; for they sent half toward the rising sun and the other half toward the setting sun, that is to say that one goes toward Mangi *(the region of the lower Yangtze)* and the other toward the Catai *(the region of Beijing)*. And I tell you truthfully that this city has such grand ships – such a large quantity of boats – that no one could believe it without having seen it. Do not take this to mean that they are large ships: they are of appropriate size for vast rivers. And I tell you that these ships carry to Catai and to Mangi such large masses of merchandise that it is a marvel ...."[71]

Starting in 1327, repeated floods cause serious famines. In 1344, the dikes break downstream of Kaifeng and are not restored until 1349. The plight of the miserable peasants of this region, along with the massive labor conscriptions necessary for the dike repairs, leads to the emergence of the *Red Turbans*. This secret society soon leads an insurrection against the power of the Mongols.[72]

A glance at Figure 8.14 makes it easy to understand the problem of water supply to

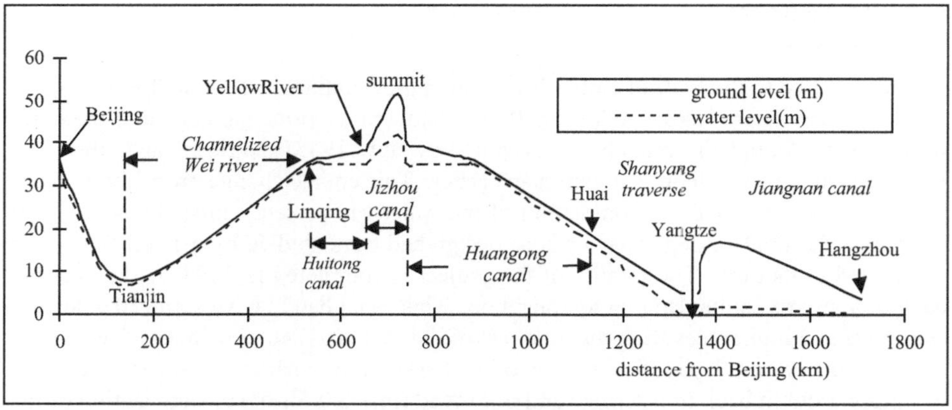

Figure 8.14 Longitudinal profile of the Grand Canal, from a modern survey (Needham et al, 1971).

70 Marco Polo, Le devisement du monde, 149, adapted
71 Ibid., 136.
72 Gernet (1990), p. 340.

the highest portion of the Grand Canal of the Yuans, i.e. the Jizhou canal in Shandong.

This problem is not solved by the Yuan, who continue to use primarily the maritime route and in effect cut off the Shandong peninsula with the Jiao-Lai canal. It is not until the Ming regime, arising from the insurrection that chased out the Mongols in 1368, that an effective solution to this problem is developed using hydraulic works. In 1411, 165,000 people are put to work on the 200-day construction of a dam-reservoir that could assure a sufficient supply of water to maintain the supply to the Jizhou canal throughout the year. Other small reservoirs are constructed near the canal's locks, which were also rebuilt at this time.

Under the Ming the strategy adopted for control of the Yellow River consisted in preventing any return of the course toward the north through construction of a very long dike, stretching for several hundreds of kilometers. Systems of multiple dikes delimiting the flood plain were also constructed. A real strategy for sediment management was put into place. This strategy combined deposition in flood-plain zones managed for this purpose with narrower channels in which the strength of the current was sufficient to prevent deposition. Starting in 1495 the course of the river was finally stabilized, as shown on course no. 7 of Figure 8.2.

Important canal rehabilitation work took place in 1528, still under the Ming, in particular separating the bed of the Grand Canal from that of the Yellow River. Piracy was developing along the shores of the Sea of China at this time, rendering more indispensable than ever the use of the grand artery between the south and north of the country.

It was also during this period that work was done on the Hongze lake. This lake had probably been just a natural flood plain for the southern course of the Yellow River up to then. This zone was transformed into a permanent reservoir of increased volume through construction of a large north-south dike delimiting the eastern bank of the lake over more than 30 km, in 1578.[73] An engineer called Pai Jixun led this work; he was also known for his work on control of Yellow River sediments. At the beginning of the Manchu domination (1660) this dike is extended to a total length of 67 km and its height is raised 1.5 m to attain a maximum of 7 m. The structure itself is a masonry wall whose stones are tied together by steel tendrils. The wall is about 1 m thick and supported by an earthen embankment, and rests on a foundation of piles.

In the 14th and 15th centuries the Grand Canal evolves to a configuration that is essentially the same as today. Its total length of about 1,700 km makes it the largest hydraulic project ever constructed by man. We earlier saw how travelers at the beginning of the 14th century had been impressed by the Grand Canal of the Yuan. Three centuries later another westerner embarks upon the Grand Canal. This is father Matteo Ricci, the founder of the Jesuit missions in China, who traveled from Nanjing to Beijing between September 1597 and February 1598. Being a scientist, he begins by describing the overall fluvial system:

> "This river of Nanjing *(the Yangtze)* goes from Nanjing to the north; then, returning somewhat toward the midlands, flows with great impetuosity into the sea. [....] This is why, to be able to go by water into the royal court of Beijing, the kings of China drew a large canal from this

73 Schnitter (1994), p. 105.

river to another, that is called Yellow. [....] This river [....] as if in revenge for the hate that the Chinese carry to foreigners, very often spoils a large part of the kingdom through its large floods and changes its channel as it pleases, when it is filled with sand that it carries along."[74]

This latter account reveals the isolationism that characterizes the end of the Ming Dynasty. In following sections of his treatise Matteo Ricci gives a very vivid description of the traffic on the upper portions of the Grand Canal, as well as of passages through the flush locks. He also mentions the inclined planes on which boats are dragged using capstans. He does not mention any chamber locks.

"And, however, the multitude of vessels is so excessive that the ships, each blocking the others, are often obliged to wait several days to pass, principally in certain times when there is not enough water in the canals. To solve this, they hold back water in several places with locks of wood, which also, to serve two purposes, are installed as bridges. These locks, when the stream is full, are opened and the boats are carried by the force of the running water. And, thus, the sailors navigate from lock to lock with great difficulty and along a tiresomely long route. The work is made even more difficult since it is very infrequent that, in the narrow strait of the stream, the winds are favorable for the vessels. This is why, ordinarily one uses ropes to advance along the canal and even, it often happens that at the entrance or exit from the locks, when waves rising up like impetuous whirlwinds come to envelope the boats, they are lost in the canals drowning all those who were within. But the ships of magistrates or principals are pulled against the water with machines of wood; and this happens along all the route at the expense of the King."[75]

Reading this account, one can understand what led Chiao Wei-Ho to invent the chamber lock in the 10th century. It is more difficult to understand why this process is not used in the upper reaches of the Grand Canal of the Yuan and the Ming. Does this

Figure 8.15 An ancient stretch of the Grand Canal (Jiangnan canal) at Wuxi (photo by the author)

74 Adapted from History of the Christian expedition to the kingdom of China, IV, 2.
75 Ibid.

perhaps mark the beginning of the decline in the spirit of innovation that will characterize the 17th and 18th centuries?

This is a good time to note the influence of inland water transport on urbanism. Certain cities along the Grand Canal are virtual "Chinese Venices" (Figure 8.16). We can illustrate this through the Venetian Marco Polo's description of the city of Hangzhou (Han-Tchéou). In the 13th century this city becomes the capital of the Song when they retreat to the south. With nearly one and a half million inhabitants,[76] it is also perhaps the largest city in the world (along with Baghdad before the passage of the Mongols):

> "It *(the city of Hangzhou)* is situated in such a manner that it has, on one side, a freshwater lake that is very clear, and, on the other side, an enormous river that, entering in many canals from small to large and flowing through all the districts of the city, carries all the filth, then penetrates into the lake and, from there, flows toward the ocean. This makes the air very healthy. One can visit all the city both on land and on the water. The streets and canals are long and wide, so much so that boats can navigate on them as they please. [....] No one should be surprised to see so many bridges; because I tell you that this city is entirely on the water and surrounded by water. [....] On the other side of the city is a trench that is perhaps forty milles in length, enclosing the city on that side; it is very wide and completely full of water from the aforementioned river. This was done by order of the ancient kings of the province, to be able to redirect the river each time it overflows the dikes."[77]

A wall against the sea - the tidal bore of Hangzhou

The major city of Hangzhou, prominent in the account of Marco Polo as we have seen, is located at the head of an estuary that is nearly 100 km long. The land there is quite flat and thus exposed to the tidal surge and waves associated with strong storms. Moreover in this particular estuary there is an additional extraordinary phenomenon: one of the largest tidal bores in the world.

The tidal bore is a wavelike disturbance, or a series of disturbances, resulting from the progressive steepening of the tidal wave as it propagates into a sufficiently long and shallow estuary. At Hangzhou in the Chien-Thang estuary the mean height of the bore is the order of two meters. But during equinox tides, it can reach 7 or 8 meters. It is very special to be able to come to see this bore during extreme tides, a truly festive occasion. Here is a poetic description dating from the 13th century that gives some idea of the power of this phenomenon:

> "The tidal bore on the Che River is one of the great sights of the world. It reaches its full force from the sixteenth to the eighteenth of the month. When it begins to arise far away at Ocean Gate, it appears but a silver-thread; but as it gradually approaches, it becomes a wall of jade, a snow-laden ridge, bordering the sky on its way. Its gigantic roar is like thunder as it convulses, shakes, dashes, and shoots forth, swallowing up the sky and inundating the sun, for its force is supremely vigorous. "[78]

76 After Pierre-Étienne Will, in Dictionary of Chinese Civilization, p. 344.
77 The Book of Wonders, 153
78 By the traveller and poet Chu Mi, in 1280 – transl by R Strassberg

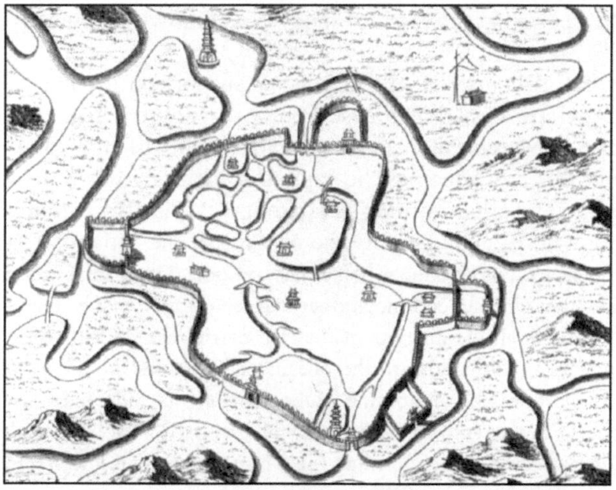

Figure 8.16 The city of Huzhou, near the south bank of Taihu lake. This city is linked to Hangzhou and to the Grand Canal by a network of canals. "It is one of the largest and most considerable cities of China, due to its richness, its commerce, the fertility of its land, and by the beauty of its waters and mountains." (du Halde, 1735 – ancient archives ENPC).

Figure 8.17 Detail of a sketch from 1610, Observation of the tidal bore (Deutsche Staatsbibliothek, Berlin).

The earliest land protection works along this estuary date from the time of the latter Han, around 85 AD. The name of the estuary, *the estuary of the dike of coins*, also dates from this period. The first builder of the dike was the governor of the region called Hua Hsin. He had called upon the people of the region to bring earth and stones for the project, for which they were paid in coins. The dike is exposed to the force of the waves, and the sea apparently destroyed the first few attempted structures. There ensued debates on the mode of construction that should be adopted. Should the dike be built of earth with a rock armor layer, a costly solution? Or should it be built more simply of earth mixed with straw and brush, as were many other dikes and dams? One can also imagine a structure based on rocks contained in bamboo cages, anchored with stakes and stabilized with iron chains, an assembly commonly used to repair fluvial dike breaches. Work was conducted in 910, in 1014, again in 1035, then again in 1169.... Gates were put in place to let the water flow seaward at low tide.

A continuous rock dike supported by an earth fill is eventually achieved in 1368. In the Ming Dynasty, in 1448, a solution is finally found that has remained satisfactory up to the present time. This consists of constructing a wall of cut stone interconnected with steel pins and with a stairstep profile facing the waves, better to break their force. Its foundation is made of wood piles, and the wall leans against old rock outcrops as well as against an earth fill.[79] This type of construction is similar to that used for the large dike of Hongze lake described earlier.

Initially this maritime dike was probably about 80 km long.[80] Today, it is more than 350 km long, and its crest rises more than eight meters above sea level.[81]

Machinery: hydraulic mills and wheels, lifting machines and norias

Lifting machines

Water lifting machines respond to a very basic need of civilization – that of raising water from a river, canal, or cistern for its distribution to agricultural or urban uses. The Near East had used the *shaduf* (Figure 2.4), a balance-beam device, for a very long time - but this device was not used in China. We have seen that more sophisticated devices appeared in Egypt during the period of Alexandria (the end of the 3rd century BC), then in the Roman Empire. Examples include the Ctesibios pump (Figure 5.5), the Archimedes screw (Figure 5.3) and the muscle-powered lifting machine (Figure 6.20).

The first written evidence of lifting machines in China is found in a treatise of Wang Ching during the latter Han period (80 AD). We cited this treatise earlier in summarizing the innovations of the Han period, and here is a specific extract:

> "In the streets of the city of Loyang there was no water. It was therefore pulled up from the River Lo by water-men. If it streamed forth quickly (from the cisterns) day and night, that was their doing".[82]

79 Needham, Ling (1965), pp. 320-323.
80 Will (1998), p. 344.
81 Needham and Ling (1965), p. 323.
82 Ibid., p. 344.

Another text from 189 AD is more explicit, again concerning the question of water supply in the capital Luoyang:

> "He (the minister Chang Jang) further asked Pi Lan to cast bronze statues ... and bronze bells... and also to make (to cast) "Heavenly Pay-off" and "Spread-eagled Toad" (machines) (which would) spout forth water. These were set up to the east of the bridge outside the Phing Men (Peace Gate) where they revolved (continually, sending) water up to the palaces. He also (asked him to) construct square-pallet chain-pumps and "siphons", which were set up to the west of the bridge (outside the same gate) to spray water along the north-south roads of the city, thus saving the expense incurred by the common people (in sprinkling water on these roads, or carrying water to the people living along them)."[83]

This text refers to the simultaneous use of several different machines. According to Joseph Needham and his collaborators, the *siphons* mentioned here are probably simple piston pumps, like those used in China to pump brine from the salt mines. Perhaps other machines in use were manual lifting wheels.

The pump with chain and square paddles (rectangular, in fact), also called the *dragon backbone machine,* is explicitly mentioned in the text. It is typical of Chinese inventions, and Figure 8.18 gives an idea of its workings. It is a wooden chain fitted with rectangular paddles, circulating in a long wooden flume that is open at each end, inclined at 25° or less from the horizontal. Water is lifted by the upwards movement of the paddles inside the flume. The chain to which the paddles are fixed was originally powered by men (or young girls) on a pedalboard. But from the 12th century one encounters descriptions of these machines as powered by animals, and even by hydraulic wheels. The lifting height can be of the order of 5 meters.

Rice cultivation developed rapidly under the Tang and the Song, and this agriculture requires intensive irrigation. The authorities favored the *dragon backbone machine* to lift water into the rice fields. In 828 they even standardized the specifications for this pump, whose great success can be attributed to its ergonomics. In the 12th century its use in Turkestan is noted on the occasion of a voyage of a Taoist wise man, then it appears in Korea in the 15th century. Between the 17th and 19th centuries it is used to various extents throughout the world.[84]

A parallel technology to the *dragon backbone machine* is the *noria.* This hydraulic-powered lifting machine is already well known to us – in Chapter 6 we cited the description given by the Roman Vitruvius in about 25 BC, and we have seen its very broad use in Syria and in the Arab world. But the history of its introduction in China is an enigma. There is clear evidence of the use of hydraulic energy in China ever since the beginning of the Christian era, but the Chinese hydraulic wheels appear to be mostly horizontal, with vertical axes. The *noria* appears in Japan about 800, but one cannot find serious evidence of its use in China before the 10th century, and indeed there is no indisputable evidence prior to 1130. When one encounters the *noria* in Korea in the 15th century, it appears as an alternative lifting technology to the square-pallet chain pump, and it comes from Japan, not China.[85] These observations suggest that there may have been

83 Ibid., p. 345.

84 Bélidor (1737) describes it under the name of "rosary mill", as a machine in widespread use.

Figure 8.18 The square-pallet chain pump, or dragon backbone machine; this is the most wisdspread lifting machine in China. Above, Chinese illustration from 1637 (after Needham and Ling, 1965); below, western engraving from 1757 (Chambers – ancient archives of ENPC).

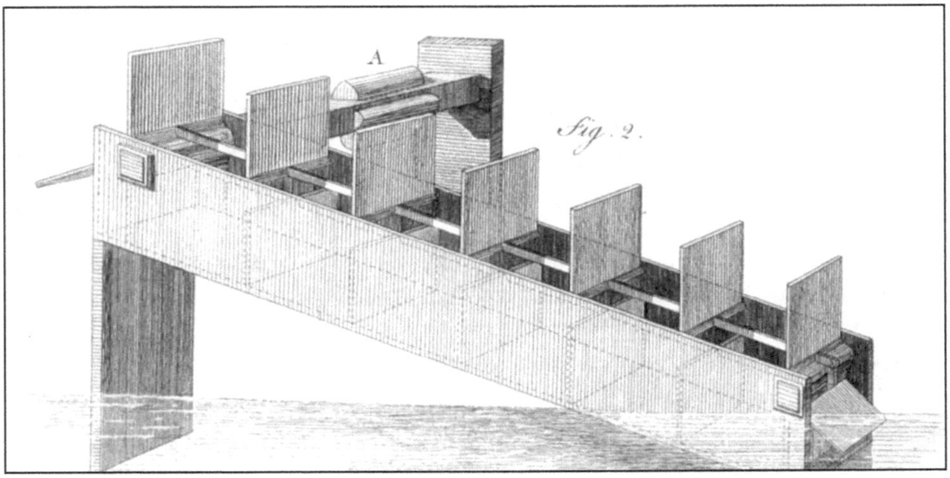

85 These dates and facts are taken from Needham and Ling (1965), pp. 356-362. These authors hypothesize that the noria could have been introduced in China as early as the 2nd century, but we do not see strong evidence of this.

two pathways for the introduction of the *noria* in Asia: by land in the region of the upper course of the Yellow River in China on the Silk Road around the 10th century; and in Japan by sea from India two centuries earlier.

In his account of his *Voyage in China* (1345), Ibn Battûta describes wheels on the Yellow River that appear to be *norias*:

> "*(The Yellow River)* is lined with villages, cultivated fields, orchards, markets, in the manner of the river Nile in Egypt; but here the land is more flourishing, and on the river there are a large number of hydraulic wheels."[86]

The Chinese *noria* is very light, made of wood and bamboo, and therefore it can be turned by a weak current. The most famous of these wheels are those in the region of Lanzhou on the upper course of the Yellow River. Some fifteen meters tall, they are arranged in batteries or sets of up to ten per row.[87]

Figure 8.19 Forge bellows powered by hydraulic force; illustration of 1313 (from Needham and Ling, 1965).

86 Ibn Battûta, Voyages, Translation of C. Defremey and B.R. Sanguinetti, IV, p. 255, adapted.
87 Needham and Ling (1965), p. 356.

We have seen that the bucket chain or *saqqya* was widely used in the Arab world. But it was used only marginally in China, and then probably only in the salt mines of Sechouan at a relatively late date.[88]

Forge bellows and hydraulic energy for metallurgy

According to tradition Tu Shih, the prefect of Nanyang, "loved the common people and wanted to lighten their work". This is why he decided in 31 AD to use hydraulic energy to power forge bellows, as we have seen earlier. Later in about 238 AD it is once again a man of Nanyang who spreads this technique to other metallurgical centers:

> "Han Chi, when Prefect of Lo-Ling, was made Superintendent of Metallurgical Production. The old method was to use horse-power for the blowing-engines, and each picul of refined wrought (iron) took the work of a hundred horses. Man-power was also used, but that too was exceedingly expensive. So Han Chi adapted the furnace bellows to the use of ever-flowing water, and an efficiency three times greater than before was attained. During his seven years of office, (iron) implements became very abundant. Upon receiving his report, the emperor rewarded him and gave him the title of Commander of the Metal-Workers."[89]

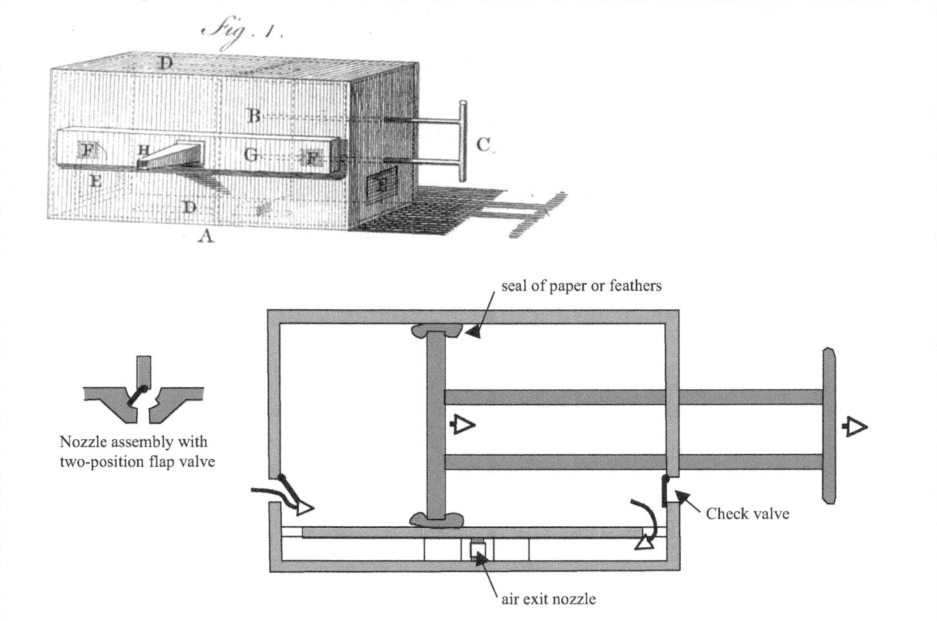

Figure 8.20 The double-action piston bellows: a device to provide a continuous draft for Chinese metallurgy. Above: engraving of Chambers, 1757 (ancient archives of ENPC).

88 Ibid.
89 Extract of a work called San Kuo Chih, Needham and Ling (1965), p. 370.

To understand the importance of this invention, whose usage becomes rapidly established in all of China, one must note how far advanced Chinese metallurgy was compared to that of the West.

Melting and casting were practiced in China very early. The key to this metallurgy is the quality of the furnaces and the draft of forced (blast) air that makes it possible to attain elevated temperatures. An illustration dating from 1313 shows the bellows mechanism powered by hydraulic energy (Figure 8.19). This consists of a horizontal water wheel at the upper end of whose (vertical) axis another horizontal wheel is fixed. Around this latter wheel there passes a belt that drives a third, smaller wheel, which therefore rotates more rapidly. The axis of this small wheel has an eccentric arm on which a cam pivots, thus introducing a rapid alternating movement which in turn activates the bellows.

The bellows itself was initially nothing more than a skin sack fitted with an outlet pipe for the air. Over time, wooden walls are added, and the bellows evolves into an extremely efficient device whose first known description dates from 1280. This was a double-acting bellows, consisting of a piston sliding in a parallelepiped that is divided into two chambers by the piston (Figure 8.20). As the piston moves back and forth, the air in one of the chambers is always being compressed. The evacuated air circulates through a secondary compartment, then into an exit nozzle with a pivoting flap serving as a check valve. Two such check valves also act on each end of the compression chambers. The edges of the piston are covered with feathers or paper to limit air leakage. This remarkably simple device delivers a continuous supply of blast air.

Hydraulic energy - water mills and windmills

We have seen the first use of hydraulic energy in China under the Han Empire, at the beginning of the 1st century AD. The Chinese devices are quite complex from the very beginning, in contrast to the comparable but simpler devices developed at the same time in the Roman Empire. There, simple horizontal-axis mills are used to turn a grindstone. But in China there are complex devices powered by vertical-axis water wheels, according to all evidence. These wheels power batteries of pestles for agricultural or metallurgical use (21 AD). In the industrial center of Nanyang they power batteries of bellows to provide combustion air for the melting and casting of iron (31 AD).

It is probably in the 4th century that the direct use of rotating wheels to grind grain was developed, in response to depopulation resulting from the wars following the fall of the Han Empire. In the 7th century the technology of mills is exported from China toward Korea, Japan, and Tibet. In China of the 8th century one finds many large mills having up to five wheels, the property of rich merchants, Buddhist abbots, imperial concubines and palace eunuchs. Conflicts inevitably arise from competing uses of water: the industrial flour trade, navigation, and agriculture. The Confucian civil servants give priority to traditional uses of water for the general public good, and accordingly they issue edicts to limit the proliferation of mills. In 778, eighty mills are destroyed by order of the administration.[90]

90 Needham and Ling (1965), p. 400.

Hydraulic energy is thoroughly assimilated into Chinese culture at the beginning of the Christian era, when the use of the water wheel is increasingly widespread. Such uses include the powering of batteries of hammers, forge bellows, mills, and also square-pallet chain pumps, and even merry-go-rounds of dolls (in 260), celestial spheres (slowly rotating astronomical models) in 590, and textile spinning machines (the first dating from 1313.[91] The *noria* should, however, be considered separately. As we have said earlier, it is probably a technique imported later from the Near East or India, independently of other uses of hydraulic energy.

Windmills, known in the Islamic world since the 7th century, are apparently introduced into China from Turkestan toward the 12th or 13th centuries:

> "The people of the west use wind mills as the people of the south use water mills."[92]

These are vertical-axis mills, as was the case in Persia. In China, the sails of the mills are improved through adoption of the technology of sails for junks.

The golden age of the Chinese navy, 9th to the 15th century

On the rivers, boats powered by paddle wheels

From the time of the first emperor there was a need for warships in support of military campaigns in the basins of the Yangtze and the Xi (near Canton). These fortified, and sometimes armored, ships were initially powered through oars manned by soldiers. But the specific needs of riverborne military operations led to a very curious invention. It was in 784, under the Tang, that a prince named Li Gao developed warships powered by paddlewheels.[93] This invention may well have come from the 5th century.

This type of boat sees major development on the Yangtze, pushed by a great naval architect named Gao Xuan, at the beginning of the 12th century when the Song retreat into south China. Paddlewheels are mounted on the sides of the vessel; the number of them is variable, but could be as many as eleven on each side, often with a large wheel on the stern in front of the rudder. The wheels are powered by pedalboards, and Chinese authors mention that these ships can attain very high speeds.[94]

The ocean-going junk, precursor of the modern sailboat

We have seen that the technological basis of modern sailboats appeared in China during the Han period, 2nd century AD. The elements of this technology were the axial rudder and the modern sail. Navigation develops widely on the Yangtze, in its immense estuary, and along the southeast coasts of China. The marking of the coasts with beacons

---

91 Needham, 1978.
92 From a text dating from the end of the Songs or from the Yuan Empire (Lao Hsueh Tshung Than by Sheng Jo-Tzu), citation from Needham and Ling (1965), p. 560.
93 Needham (1978), p. 20.
94 Dars (1992), Gernet (1990), p. 273.

develops under the Yuan. Junks having multiple masts appear from the 3rd century.

The large ocean-going junk reaches maturity in the 9th century. This is a very large ship having three or four masts, sometimes even six. In the Song period it can exceed 100 meters in length and can carry several hundred people. One passenger, our Tangiers traveler Ibn Battûta, describes it as follows:

> "The large junk has twelve sails, the others *(i.e. smaller junks)* have up to three. The sails are of bamboo, woven together into mats, they are never lowered but always turn according to the wind direction. When the boat is at anchor, the sails remain hoisted, buffeted by the wind. The crew comprises a thousand men: six hundred sailors and four hundred soldiers: archers, shield carriers, crossbowmen who fire naphtha. [....] On the ship there are four decks with bunks, cabins and salons for the merchants."[95]

In the 11th century the compass appears on Cantonese junks; before it had been used only by astrologists who assured the correct celestial orientation of dwellings. The compass makes seafaring navigation far easier. The large junks navigate on the Sea of China and the Indian Ocean toward Korea, the south of India, and the Persian Gulf. The port of Canton is recognized since the time of the Qin Dynasty; it hosts a cosmopolitan community of sailors and merchants from everywhere. Fuzhou, somewhat further north, is another large port. Hangzhou, a port and city that, in the 13th century, earns the admiration of both Marco Polo and Ibn Battûta when they visit, is exceptionally prosperous in the period after the retreat of the Song into south China.

At the beginning of the Ming Empire, grand maritime expeditions are launched on all seas to celebrate the new rulers and the eviction of the detested Mongols. These expeditions represent the apogee of navigation in imperial China. Between 1405 and 1433 seven expeditions follow one after another, involving an immense fleet of 62 large junks nearly 130 m long and 50 meters wide, in addition to a number of smaller boats. These expeditions are as much diplomatic as commercial, with destinations of Indochina, Java, Sumatra, the south of India, Ceylon, Hormuz in the Persian Gulf, Jeddah in the Red Sea, Aden and the eastern coast of Africa, and perhaps even Mozambique.[96]

But the decline of seafaring navigation in China begins in the 16th century. This decline results from the quasi state monopoly established by the Ming, who authorize only large official expeditions. This decline coincides also with military setbacks in struggles against the Mongols in the north, and with a pullback of the Ming civilization taking refuge in a sort of defensive posture within the borders of China. Western sailors of the 17th and 18th centuries find the Chinese coasts to be infested with pirates, and witness a population turning its back to its coasts.

---

95 Ibn Battûta, 1995, adapted.

96 Dars (1992), p. 62 and following.

9. The Mills of the Middle Ages

The death throes of the aqueducts and the end of the Roman way of life

With the fall of the Roman Empire, the Romano-Hellenistic urban lifestyle in the West begins to retreat rapidly. The roots of this lifestyle are found in the Cretan cities of the Bronze Age. The forums where citizens met to discuss the business of the city are among the victims of the inward-turning that characterized this troubled period. The thermal baths where water so freely flowed disappear; the beautiful public fountains dry up. The cities degrade; new urban streets are unpaved and have neither sewers nor water pipes. Water is no longer delivered, it must be drawn from wells or from the river. This is likely why Lyon, for example, moves down from the heights of Fourvière to the banks of the Saône.

The aqueducts fall into ruin one by one between the 4th and 6th centuries AD, due as much to neglect as to the deliberate action of invaders. And yet these symbols of Roman civilization remain marks of prestige for the powerful ecclesiastics who, one way or another, manage to restore them sufficiently to supply their palaces in the 8th or 9th century. These personages include Aldric, the bishop of Le Mans; Dieudonné the abbot of Saint-Germain d'Auxerre; Rigobert, the archbishop of Reims; and Didier, the bishop of Cahors. Charlemagne himself has the aqueduct of Aix-la-Chapelle[1] restored. In Rome proper, the popes restore several aqueducts to service in the 7th and 8th centuries, to provide water for the mills that they build on the Janicule and for the baptismal fonts of churches.[2] But the people, for their own water needs, must go directly to the Tiber, or dig wells.

It is difficult to determine the number of water mills in use at the end of the Roman Empire. But it is evident that in the 8th and 9th centuries the water mill was widely used in Carolingian agricultural domains in Switzerland and in the northern half of France, whether these domains be lay (the *villas*) or monastic.[3] The Benedictine abbeys that spring up in this period follow the rule of Saint Benoît which dictates that the monasteries be provided with water and a mill:

> "*(the monastery should be)* set up so that everything that is needed can be found there: water, a mill, a garden and shops so that one can practice diverse trades within the enclosure."[4]

Where they exist, these water supply systems are nothing more than modest aqueducts of wood (as at the Swiss monastery of Sant-Gall) or derivations from small rivers (as at the monastery of Fulda in Germany). The importance of these early monasteries to the conservation of hydraulic techniques and the approach to the medieval revival is subject to debate.[5]

---

1 Guillerme (1979).
2 Coates-Stephens, 1998.
3 Champion (1996).
4 Citation from Zettler (1996).
5 Zettler (1996), Hoffmann (1996).

In the 9th century, the Viking invaders from the north travel up the rivers and manage to plunge western Europe into obscurity - it is even forgotten that the earth is round.

With the end of widespread famines and the development of commerce in northern and western Europe, the revival begins in the 10th century, inaugurating a long period of economic and demographic expansion. This expansion will last up until the Hundred Years' War (1337 – 1453) and the great plague of 1348 – 1349. In the 12th century the monasteries rediscover and reproduce certain ancient authors such as Aristotle and Gallien, from Arab translations brought back from the crusades. These Arab documents are brought from al-Andalus or from Sicily where the Norman kingdom founded in 1061 welcomes Arab and Jewish intellectuals.

Figure 9.1 The Borde mill at Saint-Jacquest des Guérets, on a race from the Loir, below the medieval site of Troo in the Sarthe department of France (photo by the author)

The technologies of the medieval revival

The conquest of the waterways: inland water transport and mills

The great technological phenomenon of the Middle Ages is the development of mills, perhaps a natural companion of the revival of interest in watercourses. Since the road network was in bad shape and the countryside was not safe, the development of commerce in the 10th century first relies on the watercourses. Demographic expansion and the nearly exclusive use of cereals to feed the population drive major development of the flour trade. In this period the consumption of bread made from flour develops while consumption of porridge decreases. It was quite natural to build mills along watercourses that were already indispensable for water supply and transport, a heritage from the Roman era. The development of mills is nothing less than extraordinary in all the regions of France, England, and Holland. On the Roussillon canal, mills are already numerous at the end of the 9th century, and 10th century mills with multiple wheels can be found on there.[6]

Twenty years after the Battle of Hastings, in 1086, William the Conqueror put together an inventory of the possessions of his English domains in the "Domesday Book". No less than 5,624 water mills are listed here (but no windmills and only one

Figure 9.2 A tanning mill (to crush bark for its use in a tannery) on a small river, near Luché-Pringé, in the Sarthe department of France (photo by the author)

tidal mill). This amounts to about one mill for every fifty houses, whereas there had only been a total of about a hundred mills a century earlier. At this period all these mills were for grinding wheat. In other regions, such as Picardy, the use of mills grows rapidly – but one must not take all of the data at face value.[7] Boat-mounted mills can also be found under the Grand Pont of Paris.

In the 10th century mills are for the most part the property of lay lords or bishops. This is the "banal" mill, to which vassals, or feudal subjects, are required to bring their grain to be ground and pay an unpopular tax (the "ban"). But the use of hydraulic ener-

6 After Sylvie Caucanas (1995).

7 The rapid growth of references to mills in deeds from Picardy in the 11th and 12th centuries may not be significant, for the number of deeds being rediscovered and archived is, in itself, growing rapidly (see Gies and Gies, 1994; Derville, 1994).

gy goes beyond the grinding of cereals. Industrial uses appear very early on as in Andalusia.[8] The fuller's mill uses a system of cams to beat cloth; it appears in Italy in the 10th century, and in Burgundy (France) in the 12th century. About the same time the use of hydraulic energy to power forgehammers appears using the same principle. The first hydraulic sawmill in the Christian West appears to have been in Normandy (France) in 1204, and further evidence is found here and there in later years. Hydraulic energy has many other such uses: powering bellows for hammer forges, twisting fibers in yarn works, mixing beer, etc.

Horizontal-wheel mills are found in regions under Andalusian influence, for example along the Garonne River in France. Elsewhere the mills are almost always of vertical-wheel design. In the 10th century they are "undershot", driven from below like the Roman mill described by Vitruvius (Figure 6.21). The overshot wheel is driven from above and can provide more power; it appears in the 11th century in mountain monasteries, where it is relatively easy to direct falling water to the device.[9]

The cities that begin to develop in the second half of the 11th century in western Europe establish widespread hydrographic networks, branching out from urban centers on the watercourses. Diverse water-dependent activities such as the flour, tanning, and cloth trades[10] develop along these watercourses. Populations are not very dense, and the running water contributes to the healthy environment of these newly emerging cities.

Ownership of the water mills progressively passes to the large abbeys from the 12th century on – the abbeys clearly wanted to take over both the lands and their associated water resources. Further on we describe an example regarding the very rich abbey of Cîteaux in Burgundy (Figure 9.11). These ambitions of the abbeys obviously can lead to conflicts, especially when the flatness of the river slope limits the number of mills that can be supported on a given watercourse.

On the Somme river, examination of monastic archives reveals that to the west of Ham there was quite a concentration of mills (Figure 9.4) – no less than eight along some ten kilometers, and most of these mills had more than one wheel. In 1160 the monks of Bonneuil, the important parsonage of Prémontré responsible for managing the abbey's resources in this region, raised the height of the dam of the Eppeville mill (acquired before 1138) to increase its capacity. In so doing, they affected the operation of other mills in Ham, a city located several kilometers further upstream. The trial that followed lasted no less than four years, from 1167 to 1171, and involved jurists as well as lay and religious experts. The case was eventually heard in the court of Pope Alexander III. In the end the monks of Prémontré were required to demolish their structure.

We can see that the development of mills necessarily included the transformation of natural rivers into managed watercourses, including overflow weirs to generate the head necessary to power the mills. The mills were sometimes co-located with the dam itself, or sometimes were placed on derivation canals, or races, issuing from the pool upstream of the dam. These races could attain lengths of several kilometers (Figure 9.5). During low-flow periods these installations kept the water level relatively high, thus favoring

8 Hills (1994); Benoît and Berthier (1998).
9 Hoffman (1996).
10 Guillerme (1983).

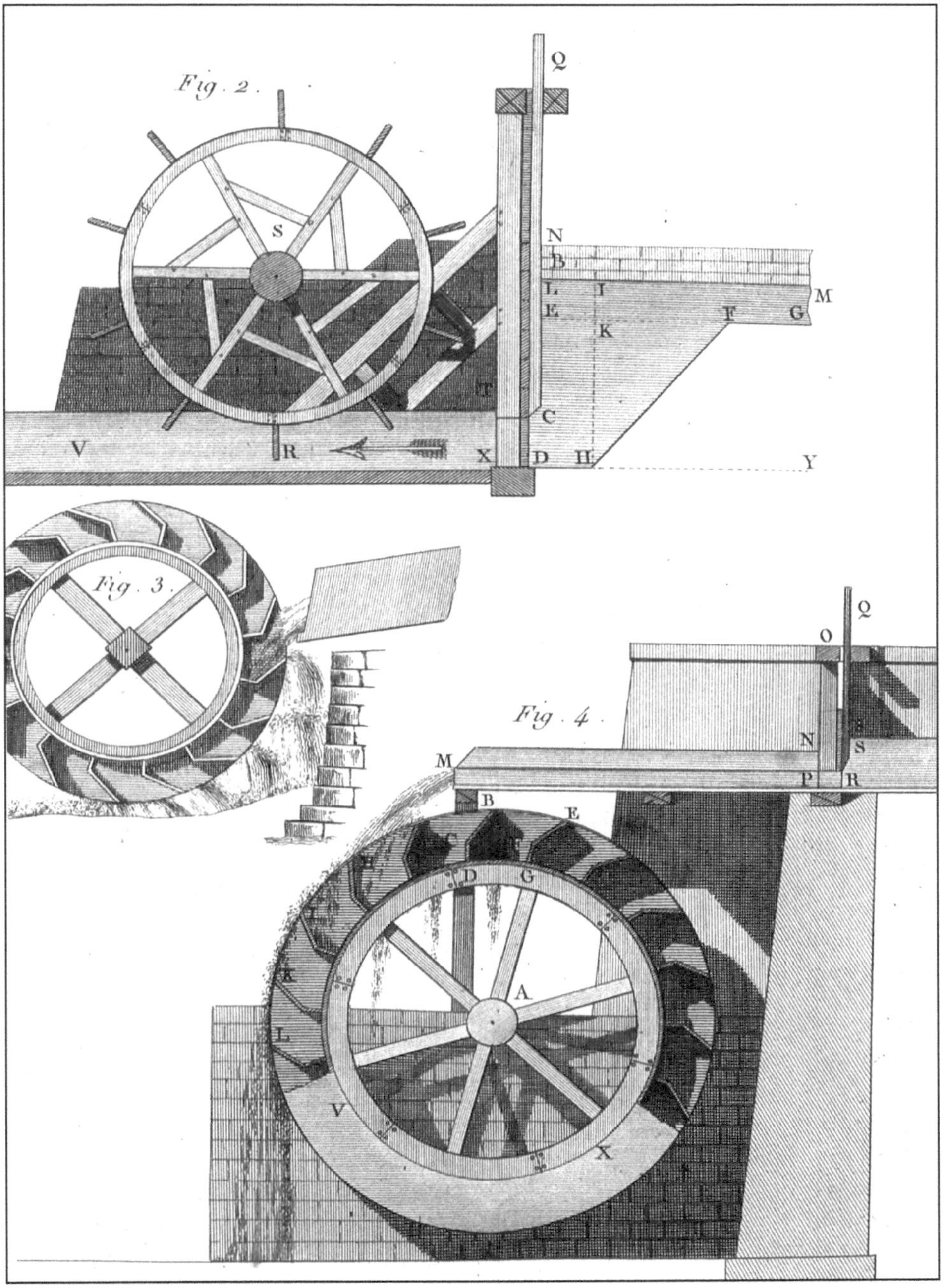

Figure 9.3 Different types of vertical-wheel mills, traditional in France (Belidor, 1737 – ancient archives of ENPC): - above, an undershot wheel, from a river dam (like the mills of Figures 9.1 and 9.2); - below, an overshot wheel, for mountain creeks of low discharge (like the Roman mill of Barbegal, Figures 6.23 and 6.24); - in the middle, an intermediate configuration.

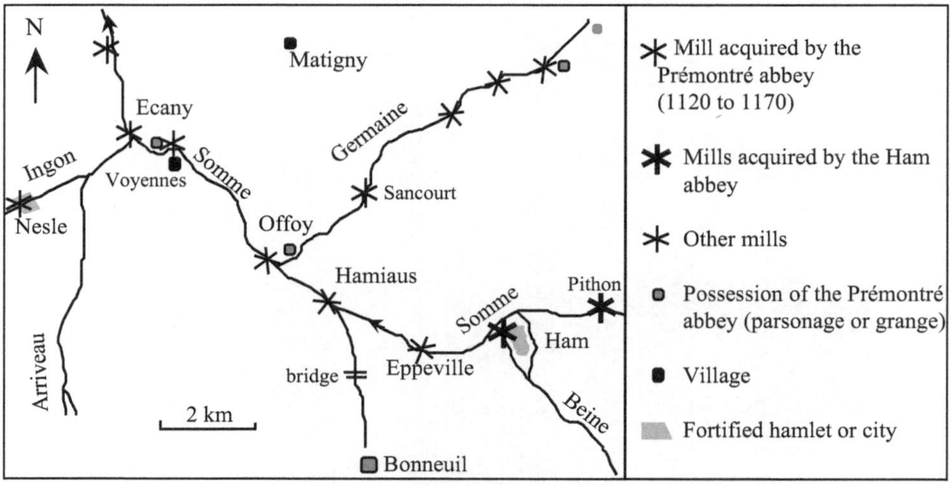

Figure 9.4 The mills on the Somme River (France) west of Ham, in the 12th century. These mills likely existed since the 10th or 11th centuries, and were progressively acquired by the large abbeys like that of Prémontré. From Dietrich Lohrmann (1996).

the development of inland water transport.

The earliest navigation canals may have been in the 11th century, but it is not until the 15th century that locks appear in the West. Prior to this, barges had to be hauled up and down ramps. This inconvenience, along with the improvement of the road network, rapidly reduced inland water transport to a rather modest role in the overall interior commercial activity in Europe. Ferryboats are put into service for river crossings, and where the overbank area (submerged during floods) is much wider than the main channel (in which year-round navigation is possible), transverse levees must be built on the overbank to bring the road to the ferry landing and thus make it possible to cross the river in all seasons[11]. Apparently these transverse levees are of little consequence to the flow of water during floods. They cause the upstream water-surface elevation to be higher, but they attenuate the severity of the flood downstream creating a kind of intermediate storage area.

Up until this time mills had only been built on small or modest rivers. But from the start of the 12th century, the know-how for such installations on large rivers began to cross the Pyrenees. At Toulouse a 400-m long dam was built on the Garonne River. This dam was set at an angle to the river to increase the effective overflow length, a technique that we saw earlier on Andalusian dams. The dam was built of two rows of wooden piles anchored in the riverbed, with rock fill dumped between the two rows to form the structure. Along with two other smaller dams, this installation provided the necessary head to power a total of 45 mills, probably a record for this period (the reader may recall the dams built by the Arabs in the Fars, also designed to provide water for a number of mills

11 Such installations on the Charente (France) are mentioned by Jean Chapelot and Éric Rieth (1995).

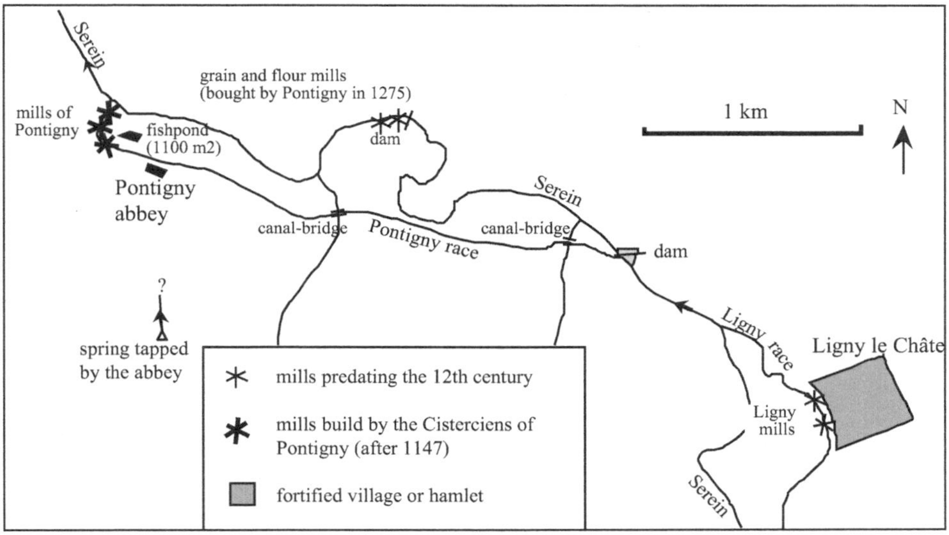

Figure 9.5 Location of mills on the Serein, a tributary of the Yonne, near the abbey of Pontingy in Burgundy (France). The installations at Ligny (dams and race) predate the 12th century. The dam (Figure 9.6), the race and the mills of Pontigny were built after the installation of the Cistercians (1147). Today the three mills of Pontigny have upstream-to-downstream drops of 1.5, 2.6, and 1.3 meters. Numerous conflicts arose between Pontigny and other water users, notably Ligny upstream, and the ancient abbey of Sant-Germain d'Auxerre that had properties downstream on the Serein. From Rouillard (1996) and Kinder 1996).

– Figure 7.4).[12] This dam, whose height is unknown, is mentioned in a text from 1177. To our knowledge, there are no other large structures like this built prior to the 15th century.

Tidal mills

The notion of using tidal energy certainly came naturally to the people on the Atlantic coast, already familiar with both the tide and river mills. The appearances of tidal mills at several different locations would appear to have been essentially independent. We have already mentioned the Bassora mill, in Iraq, built in the 10th century. According to Frances and Joseph Gies there may be some evidence of such a mill in Ireland around the 8th century.[13] But the real development of such mills did not begin until the 12th century, more or less simultaneously at several locations in western Europe: at Bayonne (1120-1125) and in the Basque country; at Wooton in Hampshire (1132); then along all the east coast of England; at La Rochelle where there are the remains of a gift of Alienor of Aquitaine to the Templars (1139); in Suffolk and near London (1170-1180); in the

12 Gies and Gies (1994), p. 117; according to Bélidor (1737), the Garonne mills of the 18th century have horizontal wheels; such is evidently the case for this dam, following the Arab tradition that inspired it.
13 Gies and Gies (1994).

Figure 9.6 The dam on the Serein (renovated in modern times) that forms the race of Pontigny. At the left one can see the start of the race (photo by the author).

land of Guérande (1182); in Brittany at Saint-Coulomb in Ille-et-Vilaine (1181) and at Pencastel in the Gulf of Morbihan (1186); in Normandy at Dieppe (1207) and then Carentan (1277); at Zuicksee in Holland (1220); on the Tagus at Alcantara (1313); at Rupelmonde on the Escaut (1388);[14] and so forth. Mills continue to appear up until the 18th century, becoming particularly common along all the coasts of the British Isles, as well as in Brittany, where some hundred could be found (25 in the gulf of Morbihan, 15 on the Rance). All of the mills in Spain and Portugal had horizontal wheels (a vestige of their Arab heritage); those in England had vertical wheels; and those in Brittany and along the Gulf of Gascogny were both of horizontal and vertical design (but the latter enclosed). These installations most often had a single basin formed by a closure dike into which were installed the headrace, the gate, and the millwheel below it, outside the basin. The mills could begin to operate about three hours after high tide when the difference in water levels between the basin and the sea became sufficiently large. The mills then ran for about six hours, until just after the hour of low tide. This daily operating period could be extended if it was possible to divert the flow of a small river into the basin. But the Middle Ages did not see the development of bidirectional tidal mills; there was no operation during rising tide.

14 Boithias and de la Verne (1989).

Windmills: medieval innovations?

The use of wind energy (over and above the powering of sailboats) had begun on the Persian-Afghan plateau in the 7th century. It then spread to China from the 12th century as we have seen in the preceding chapter. At the beginning of this century, windmills began to see rapid development in Europe as well. Did the idea come from the Crusades, or perhaps from Sicily or Spain? Or did it arise, as we have proposed for tidal mills, quite independently? After two centuries of Muslim occupation, Sicily became Normand from 1061. The geographer al-Idrissi, residing in the court of the Normand king Roger II, gives us a hint of the Sicilian use of the horizontal-wheel Persian mill:

> "One can find there *(in the Sicilian village of Calatubo)* a quarry where stones are cut for both water mills and mills called "Persian"."[15]

Therefore it seems possible that the idea of using wind instead of water came from the Orient. But the medieval European concept of a windmill, with blades in a vertical plane, is quite different from that of the Persian mill.

Figure 9.7 One type of post mill (Bélidor, 1737 – ancient archives of ENPC).

15 Al-Idrissi, IV, 2, Translation of Jaubert.

First of all, in Europe these were post mills. The mechanism of a post mill sits in a wooden enclosure. The mill's four blades are constituted of cloth stretched on a light-weight wooden frame to which the sails are tied, apparently in nautical fashion. At night the cloth sails are removed and folded. The entire assembly, including both the sails and the machinery within the enclosure, can be rotated around a sturdy oak pivot solidly anchored or lashed to the ground, thus making it possible to orient the sail into the wind. A long beam protrudes from the rear of the enclosure, like a tail or lever, to facilitate rotation of the mill. It is said that in Lincolnshire, a horse harnessed to this beam got entangled in the straps and dragged the mill around two full turns before the frantic animal could be calmed down.[16]

This concept had its roots either in the flatlands of Flanders or in England. The known points of reference in Flanders are from 1114 (the Hofland mill) and 1127 (North mill). Those in England are from 1137, when a text mentions the gift of a windmill to the abbey of Reading; then in 1155, Sussex. Four other references to mills between 1180

Figure 9.8 An ancient post mill at the edge of the Loire valley, upstream of Saumur (photo by the author).

and 1190 are from Flanders, at the mouth of the Somme River, and here are some twenty other references from England. This type of mill then spreads rapidly to other regions, since evidence exists from Germany in 1222, from Denmark in 1259, and from Holland in 1274. Some 120 windmills are known to have existed in the Ypres region in the 13th century. Others are found in Russia and at Venice in the 14th century.[17]

A second type of mill with its blades in a vertical plane appears in the Mediterranean area in the 13th century. In this region it was probably very difficult to find the kind of

16 Hills (1994), p. 35.
17 Ibid, p. 36.

Figure 9.9 Remains of a tower mill on a hilltop in the northern extremity of Corsica. In this particular case, the unit cannot be turned into the wind (photo by the author).

Figure 9.9 bis Ancient windmills in Crete, on the Lassithi plateau (photo by the author).

massive wood necessary for fabrication of a pivot. Thus was born the tower mill, in which the mechanism and the millstones are inside a round masonry tower. Only the conical roof of the tower, through which the horizontal axle extends, rotates to orient the sail into the wind. The sail itself also evolves to become more rigid in this kind of mill. Before long, and especially in Mediterranean regions, the technique of a sail stretched over a wooden frame, as for the pivoting mill, begins to disappear. Now each sail is of triangular shape like the lateen sails of Mediterranean sailboats, and is rigged on one spoke of a wheel as it would be on the mast of a ship. The sail is shaped by being sheeted to the neighboring spoke. This new concept for a mill apparently first appears in France or in Portugal in the middle of the 13th century; it then spreads toward the eastern Mediterranean basin during the 14th and 15th centuries.[18] A tower mill dating from 1220 can be seen in Normandy (the *Moulin de Pierre* of Hauville).[19]

The "belle époque" of the Middle Ages, from the dawn of the 12th century to the Hundred Years' War

The hydraulican-monks: Benedictines and Cistercians

We have seen that the earliest Benedictine monasteries at the end of the 8th century already were employing a range of hydraulic techniques to support their activity. Much later, in the 12th century, Saint Bernard founds the abbey of Cîteaux in Bourgogne, and with it the order of the Cistercians. Throughout Europe (France, Germany, Holland, Portugal) the Cistercian abbeys are established near watercourses. But after several years of growth, these abbeys typically found it necessary to augment their water supplies through capturing other rivers or torrents.

The Obazine (Aubazine) abbey was founded in Limousin in 1130 by the hermit Étienne. This abbey quickly entered the fold of the Cistercian order. Here one can identify the trace of the "canal of the monks" dating from around 1150. It is 1.6 km long aqueduct built in truly acrobatic fashion, sometimes laid along a cliff on cantilevered arches, sometimes crossing rocky outcrops. This little canal is only 1.2 m deep and 0.6 m wide; its slope is relatively flat along most of its length (5 m/km), but it ends in a virtual waterfall, descending some 60 m in its last stretch of 230 m. Several centuries later, the ingenuity reflected in this canal is attributed to miracles performed by its founder:

> "It should be further noted that the great Saint Étienne had an aqueduct built to bring water into the abbey, so that all normal needs of hygiene and cleanliness could be satisfied. There has never been seen in France a better built or more expensive canal. It passes through inaccessible places where the Saint broke through the rocks, to the admiration of all those who see it, and, according to legend, we learn that one of these rocks, not having yielded after long and expensive efforts, finally crumbled of its own accord after prayers and the worthiness of the Saint to receive this flow of water."[20]

---

18. Hills (1994).
19. Azema (1995).
20 After Bernadette Barrière (1996).

Cîteaux is the initial site of the Cistercians, and at the end of the 12th century it was a very rich abbey. But the Vouge River on which it was situated did not have a sufficiently ample and regular flow to support the growing population of monks. The abbots covet the water of a gushing torrent called Cent-Fonts, some ten kilometers to the north of the abbey, which powers numerous mills belonging to several local lords. The abbots begin by obtaining a piece of land from one of these lords to build their own mill, between 1175 and 1180. Little by little, in exchange for salvation of the lords' souls, they obtain significant land holdings for their scheme as well as shares in the mills that use the water of this torrent. Adding to their holdings through outright purchases, the abbots eventually own all of the mills. From the Duke of Bourgogne and the Bishop of Langres, the most important personages of the region, they obtain all the authorizations necessary for their project of capturing the Cent-Fonts, in return for which they agree to build and maintain bridges across the canal for the use of the peasants of Noiron, a village on the projected course of the canal (Figure 9.10).

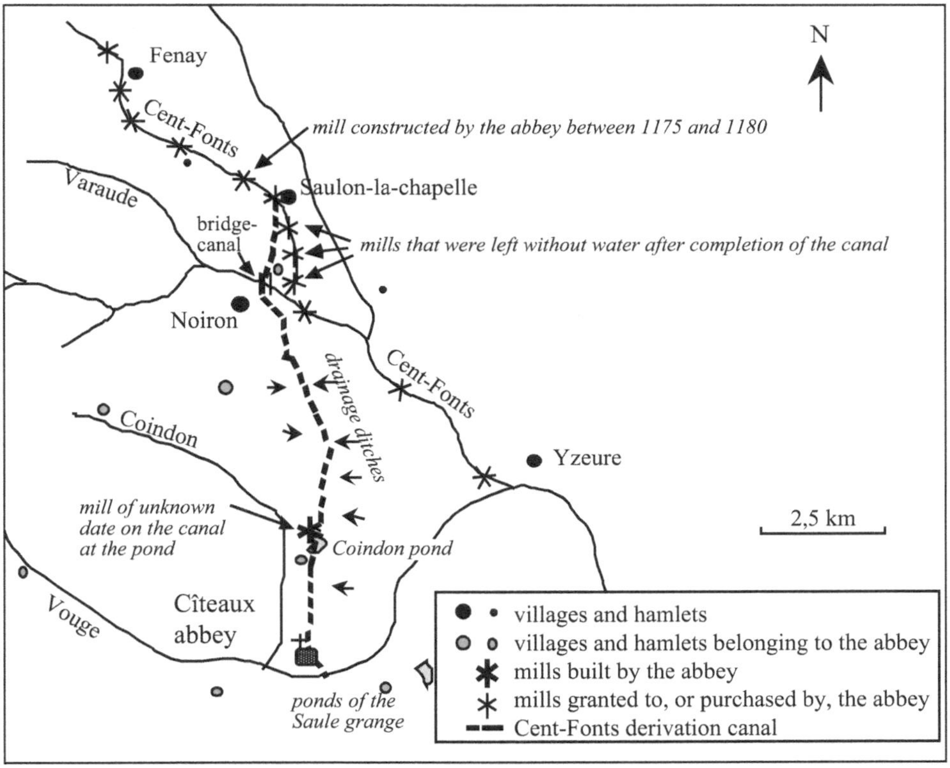

Figure 9.10 Water and land development at the Cistercian abbey of Cîteaux in Bourgogne: derivation canal, mills, pond drainage, between 1175 and 1227 (Berthier, 1996; Sonnet, 1998).

The canal itself is finally built around 1212. Some ten kilometers long and 2 m wide, it crosses the Varaude River, a tributary of the Cent-Fonts, on a 12-meter long canal-bridge (Figure 9.11). Its slope varies from 1.9 to 5.9 m/km, and it probably captures all of the water of the Cent-Fonts (around 320 l/s). Indeed, the mills located between the canal intake and the confluence of the Varaude are left without any water and thus are abandoned. The monks develop the land that they have acquired, draining the fields crossed by the canal by digging ditches and creating ponds.[21]

Figure 9.11 The bridge of Arvaux, carryng the Cent-Fonts canal to the north of Noiron (reconstructed to the original in 1747) (photo by the author).

What did the monasteries do with all this water? In general terms the water was destined for two uses: pure water for human consumption, and abundant and fast-flowing water for hygiene and useful work. Captured source water is decanted in reservoirs, sometimes filtered, and then delivered to kitchens, bathrooms, laundries, and to the apartments of the abbot and his guests. In the monasteries of the order of Chartreux, whose rules require total isolation, each cell has its own supply of water and means of wastewater disposal. The supply pipes are made of clay or lead and even sometimes of wood. Water from a river powers the mills of the abbey, including at least one wheat mill and one fuller's mill, and often several other multiple-purpose mills. Water circulates through the drains and keeps them clean, passes through latrines (the latrine of Royaumont being the most remarkable), and supplies the fish pond. Fish raising is very important, for the monks are forbidden to eat meat by the rules of Saint Benoît. This importance is emphasized by the large size of the fish pond of Obazine, which is 33 m long, 14.5 m wide, and 3 m deep.

21 From Karine Berthier (1996).

Figure 9.12 The Cent-Fonts canal, on the Arvaux bridge (left), and where it arrives at the mill of the Cîteaux abbey (right) (photos by the author).

Land development and management

The minor lay lords of the middle ages were all too often uninterested in the management of the land, or sometimes were simply incapable of applying techniques that went beyond their competence. So it was the monks, and in particular the Benedictines and Cistercians, who became the custodians of land development during the demographic expansion of the 12th and 13th centuries. We have already seen that these monks were quite competent in such matters.

The biggest hydraulic project for reclaiming land from the sea and swamps was in coastal Flanders. Actually there were several distinct operations, but all of them were rather similar and conducted until about the year 1300. These projects were driven by recently founded abbeys, under the authority of the powerful Count of Flanders, with increasing involvement of the rich bourgeois toward the end of the period. The region was undergoing rapid economic and demographic growth during this period; in the 13th century Gand (with 64,000 inhabitants), Bruges (42,000) and Ypres (35,000) were, after Paris, the largest cities to the north of the Alps.

This story begins with the two major incursions of the North Sea in 1014 and 1042.

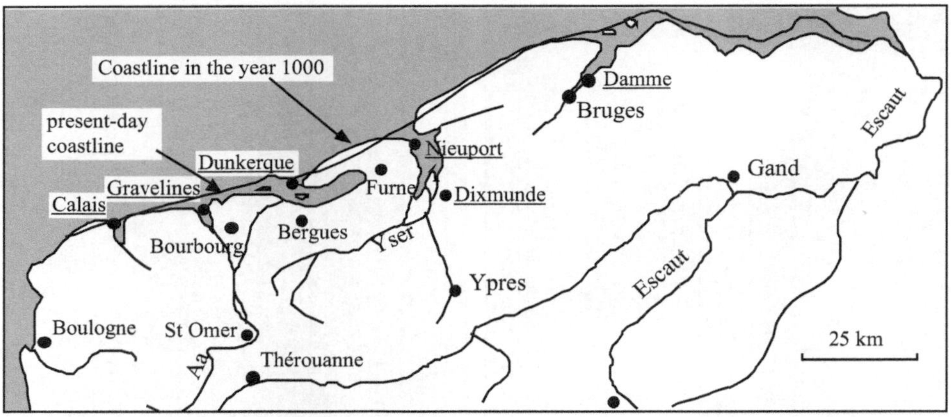

Figure 9.13 Coastal Flanders: shoreline in about the year 1000 (after Parisee, 1994) and after the channelization of rivers and compartmentalization of estuaries, which gave the coast its form from the end of the 13th century up to the present. The cities whose names are underlined are ports founded on reclaimed land in the 11th and 12th centuries.

Sheepraising was widely practiced on the floodable portions of the lowlands. These lowlands were protected by a line of dunes, but this line was broken by the sea incursions. The result was the formation of two large bays, one on the lower course of the Yser, the other to the northeast of Bruges in the region that is today called the Zwin. To protect the adjoining land from these new embayments, the inhabitants built long dikes more or less perpendicular to the coast in the middle of the 11th century. Examples are the 18-km long *Oude Zeedijk* (the "old sea dike"), that protects the area of Furnes from the Yser bay; and the *Blankenberke Dijk* to the northwest of Bruges. During this same period dike construction along the Yser River began downstream of its confluence with the Iepere, the river that flows to Ypres. This resulted in the founding of Dixmunde in 1089 as the seaport for the fast-growing city of Ypres.

Another major event occurred a half-century later, in 1134. This was a new incursion of the sea along all the coast of the North Sea, from Calais as far as the Frise islands to the north of the present-day Holland. This event seems to have marked the beginning of much more aggressive policies to combat the sea, first and foremost by the riparian abbeys. Dikes were progressively extended out from the land in large arcs, capturing the first *polders* (the term first appeared at this time) on the Yser and Zwin gulfs. Ultimately, these gulfs were completely dewatered in the 13th century.[22] The course of the lower Yser River resulted from these projects. The channelization of the river was completed in 1163, at which time the Count of Flanders, Thierry d'Alsace, founded Nieuport at the mouth of the river as the new seaport for the land of Ypres. In this same year he founded Gravelines, at the mouth of the Aa further to the west, and this in turn led to efforts to drain the marshlands of Saint-Omer and channelize the Aa, conducted from 1165 to 1215.[23] All of these projects were extremely difficult, as illustrated by the

22 These projects have been studied by Adriaan Verhulst (1990).

23 Derville (1994).

text of a charter promulgated in 1169 by Philippe d'Alsace, the new Count of Flanders:

> "Between Watten and Bourbourg, a swampland had deposited silt over a vast expanse making it inaccessible and refusing of any human use. At my expense, I drained this muddy sea, at the cost of significant fatigue, and almost violently extracting from it a more favorable natural behavior, I transformed it into fertile land."[24]

In 1180 Philippe founds Dunkirk and Damme, the latter as a port on the Zwim River now conquered and channelized. The south bank of the Escaut River, to the west of Anvers, is similarly diked with the creation of polders. These projects began in the 12th century and without doubt were accelerated after a major flood of the Escaut in 1214; they were completed around 1300. The work was led by the neighboring abbeys, but surely also benefited from initiatives of the rich bourgeoisie of Gand and Anvers.[25]

Public civil institutions called *wateringues* managed the hydraulic systems of Flanders, overseeing maintenance of the canals and gates and regulation of the channelized rivers and polders.

Another remarkable achievement is the development of the Poitevin marsh to the north of La Rochelle.[26] This vast zone of stagnant water surrounded by the waters of the Sèvre, Vendée and Autize rivers, as well as by the bay of Aiguillon, had been the focus of modest developments up until the 12th century. The work included drainage and reclamation of lands bordering higher ground, and salt works. A few fishermen lived in the swamp itself. Villages and abbeys (Benedictine, Clunisian, Augustinian and Templar) are located on buttes that overlook the swamps, or on the edges of plateaus that border the swamp to the north.

In the middle of the 12th century the Cistercians settle on land that was ceded to them, further downstream and thus closer to the ocean than the land developed by their predecessors. These lands were far enough from the coast to be sheltered from dangerously high tides and storm surges. In ten years of work that ends at the turn of the century (around 1200), they manage to drain their marshlands using the technique of the "drying basin". A parcel of land is surrounded by a levee, called the "bot", and then doubly bordered – outside of the levee by a canal connected to a system of "achenaux", or channels, that convey the water toward the sea or into the Sèvre; and on the interior by another canal right up against the levee, and fed by drainage ditches. The system can be controlled by gates in the levee, opened to evacuate the drained water into the hydrographic system, or closed to protect the basin from the inflow of high water. The peasant-fishermen living in the marshland are forced to leave. The principal players in this theater are: south of the Sèvres, the Cistercians of Charron and other monasteries of the region, allies of the Templars of Bernay; and north of the Sèvre, the energetic and dominant figure Ostensius, abbot of Moreilles until 1208.

But this reclamation of the Poitevin marsh has a negative aspect, namely its interference with the flow of water from the upstream Benedictine abbeys and villages to the sea. The first occupants of the threatened area are obliged to undertake additional work

24 Charter relative to draining of the swamp of the Aa; citation from Trenard (1972), p. 99.
25 Verhulst (1990).
26 See the article by Jean-Luc Sarrazin (1996).

to facilitate the seaward flow of floodwaters, while conserving the productive use of their own reclaimed land. From 1217 the "channel of five abbots" (l'achenal des cinq abbés"), a canal some 15 km in length, is constructed thanks to the coordinated efforts of the five large abbeys that occupied the land before the Cistercians arrived (Figure 9.14). Even later, in 1283, another long canal is constructed to connect the Luçon canal to the "channel of the five abbots". This canal makes it possible to drain the edge of the plateau upstream of the land reclaimed for Moreilles by the abbot Ostensius. For a change, this last canal, the "channel of the King" (l'achenal du Roi), does not owe its existence to the monasteries. Since the campaigns of Philippe Auguste and the Treaty of Paris, signed by Saint Louis in 1259, the south of the Poitou is royal land. This canal is thus built under the authority of the King with the financial participation of twelve lay

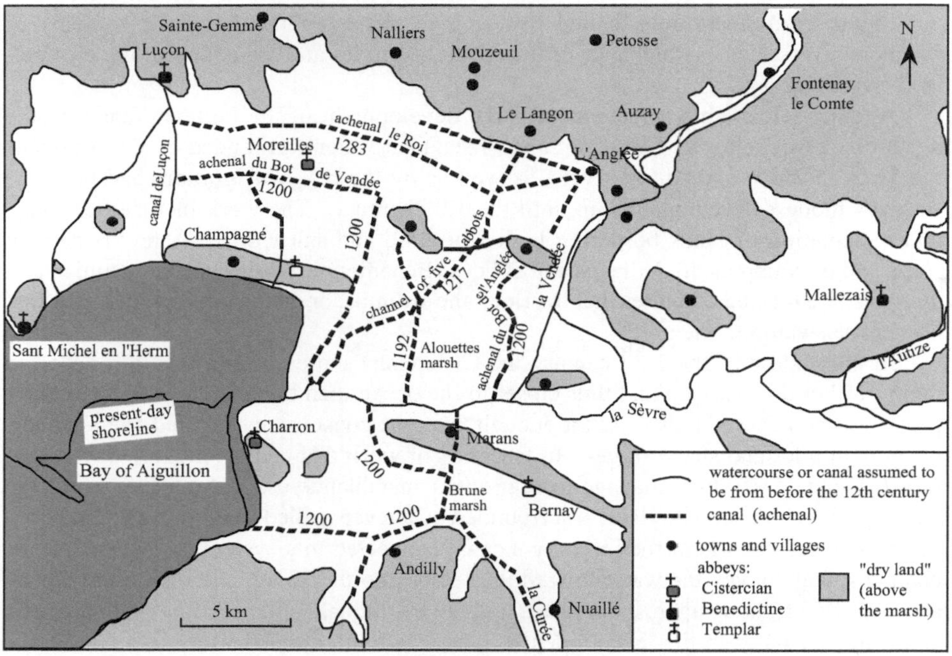

Figure 9.14 Reclamation of the Poitevin marsh, between 1190 and 1283. From Sarrazin (1996).

communities on the plateau.

The energy and hydraulic know-how of the Cistercian monks is seen in many other projects. To the north of Brives the Obazines achieve drainage of the land in the 13th century. They also developed the saltworks of Oléron. The Cistercians of Buzay play an important role in reclamation of marshlands in the Loire River delta.

Again in the Loire valley downstream of Tours, settlements had for ages been built on sandy fluvial deposits, and subsequently kept above the water by the inhabitants through accumulated fragile defenses of earth and turf. In the years 1160 – 1170, undoubtedly at the initiative of the English King Henry II Plantagenêt, the first levee on the Loire is built in the valley of Authion on the north bank of the river, extending some fifty kilometers between Langeais and Saumur. This is a *turcie*, a structure made of driven stakes, branches, sticks, and earth. Henry II settles inhabitants on the structure itself and charges them with the maintenance of the *turcie*; for this service, they are exonerated from military duty. All of this is laid out in a charter dated in 1169. In the 14th century this levee is extended downstream until it is continuous along all the valley of Authion. Large settlements like Tours and Orléans are in their turn protected by similar *turcies*. This leads to a period of calm along the Loire until the middle of the 15th century, perhaps explaining the neglect in maintenance of these levees. But important and repeated floods cause inundations in 1456, 1482, 1494, and again in 1519, 1525, and 1527. Starting in 1482 King Louis XI undertakes to raise the existing *turcies*, to generalize the flood defense system, and even to set standards for the height and the method of construction of the levees. This effort is continued until the reign of Henry IV.[27]

Another enterprise of grand amplitude is the development of the plain of Roussillon – but in a completely different context.[28] Roussillon is in effect part of the Catalan territory. In the 10th and 11th centuries it is part of the county of Barcelona, then a possession of the King of Majorca from 1262 to 1344, and finally it is integrated into the kingdom of Aragon. The land needs irrigation much more than drainage, and it seems logical that the influence of Andalusian techniques is found there. Irrigation networks develop in the basin of the Têt, east of Perpignan, from the 9th century; they particularly flourish between the 12th and 13th centuries (Figure 9.15). The influence of certain abbeys such as those of Lagrass and of Saint-Michel de Cuxa would appear to be decisive insofar as these irrigation works are concerned. As is the case in so many of the fertile regions of Europe, one finds ample evidence of donations of canals and mills from local lords to these abbeys.

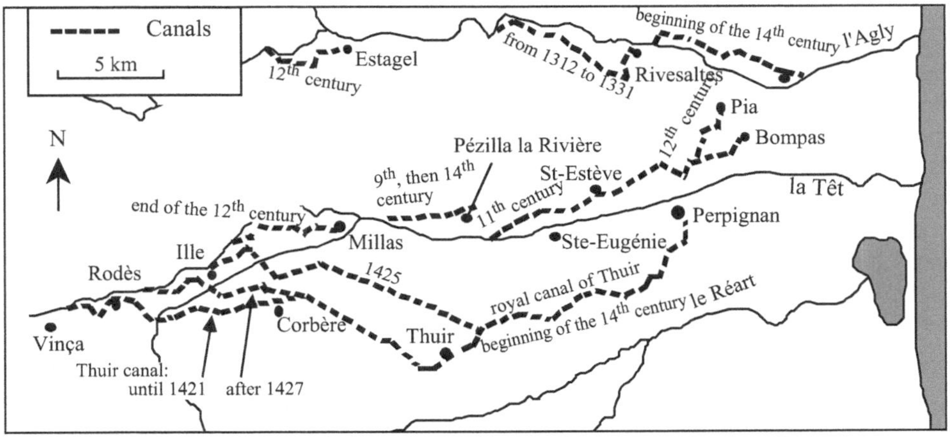

Figure 9.15 The canals in the Roussillon plain: irrigation from the 11th to the 13th century; the "royal Thuir canal" and its wanderings of the 15th century, after Caucanas, 1995.

---

27 See the work of Roger Dion (1961) and the article by Joëlle Burnouf and Nathalie Carcaud (1999).
28 Our source for the canals of Roussillon is the work of Sylvie Caucanas (1995).

At the beginning of the 14th century the development of the cities of Perpignan and Thuir led to a major project to supply water necessary for cultivation, mills, and domestic use. The "royal canal of Thuir" was built by the king of Majorca at the very beginning of this century. The canal is 35 km long, rising in the foothills near Vinça and crossing a hilly countryside. The terrain required that the canal be laid on a slope in some areas, necessitating numerous civil engineering works, and in particular several bridge-canals. The canal supplies six mills at Thuir, and seven at Perpignan where in addition it powers a *noria* that lifts water to the chateau. But the Mediterranean climate, with its violent floods, is a constant threat to this canal. In 1403 flood damages require that the canal be temporarily closed. It was necessary to rebuild the intake, which was accomplished by 1416, but then the canal is again destroyed by the major inundations of 1421. At this point the actions of the Arragon sovereigns are decisive. In 1425 a new canal, the "royal *rech* of Perpignan", rejoins the trace of the earlier canal downstream of Thuir. The sole purpose of this canal, issuing from an intake at Ille further downstream, is to provide water for the city of Perpignan. As for the city of Thuir, the "royal *rech* of Thuir" is built two years later with an intake on the Têt River at an intermediate location between Vinça and Ille (Figure 9.15). Later still, the lord of Corbère has the ancient royal canal of Thuir rebuilt to supply water to his city.

Ships and maritime ports

Initially, the typical vessel in service on the Atlantic coast was the "long boat", the Viking *knorr* whose appearance seems to have sown terror in the 9th century, and enabled the Normands to colonize Iceland and Greenland. This is a light boat, easily pulled onto the beach. In the 11th century quite a different type of boat appeared, one of much greater capacity: the *cog*. The expansion of commerce and the development of more protected harbors led to this evolution in boats. The newer, heavier vessel cannot be hauled up onto the dry beach. Therefore it is not necessary for it to have a flat bottom, and consequently its hull evolves toward a round form with a keel. This keel also gives the ship much better nautical performance, since it drifts less in a crosswind. The boat is equipped with an axial rudder. This device was known in China since the 1st century AD, but it was possibly invented independently on the Atlantic coast of Europe, where the greater freeboard makes it difficult to use the ancient rudder-oar. As was the case for the earlier boats, the *cog* has a single and very large sail rigged on a yard. The sail area could be increased in light weather by the addition of small sails attached to the edge of the mainsail or reduced in heavy weather using reefing straps sewn into the sail. These new boats could attain capacities of 200 tons, even 300 tons in the 13th century.[29] On the Mediterranean Sea, ships having two or three masts and lateen rigging are used by Italian mercantile cities.

Sheltered embayments progressively become ports with wooden quays, especially starting in the middle of the 13th century. Notable among these are Caen and Rouen in Normandy, developed by William the Conqueror; Genoa, Pisa, Venice, Amalfi,

---

29 Gies and Gies (1994), pp. 154-158.

Barcelona in the Mediterranean; La Rochelle, Boulogne, Dunkirk, and the cities of Flanders and of northern Germany, linked to the sea by channelized rivers often with outer harbors such as Nieuport for Ypres. In 1169 navigation is made much safer by the widespread adoption of the magnetic compass from China, and by the advent of navigation buoys and markers, and even beacons - fires burning on towers. But for the most part navigation does not continue through the winter months between December and March, as had been the practice since ancient times.

Navigation markers are set on shoals to warn of danger or mark an access channel. Often made of wood, they proliferate around the ports of the English Channel and the North Sea, as well as in the Mediterranean. For example the turret of *Méloria* was built of cut stone in 1157 in the mouth of the Arno River at the port of Pisa. Lighted beacons had already existed during the time of the Roman Empire, and were not entirely forgotten in the Middle Ages. Indeed in 810 Charlemagne had the *Tower of Order* of Boulogne, originally constructed in 40 AD by the Romans, rebuilt and illuminated.[30] But this development of aids to nighttime navigation accelerates in the 13th and 14th centuries, as shown in table 9.1. However not all of these beacons are permanently illuminated, for wood is an expensive commodity.

Table 9.1 Lighted beacons between the 12th and 16th centuries
(after Fichou, Le Hénaff, Mével, 1999)

Place or Port	Date	Place or Port	Date
Nieuport	1288	Cordouan (entry to the Gironde)	1360
Ostende	1365	Port-Vendres	?
Calais	1290	Aigues-Mortes (tower of Constance)	1246
Dunkirk	1365	Marseille (Île du Planier)	1319
St. Catherine (Isle of Wight)	1320	La Ciotat	1564
Abbey of point St. Mathieu (extremity of Brittany)	1250	Livorno (tower of Meloria)	1158
		Porto Pi (Majorca)	?
Les Sables-d'Olonne	16th	La Rochelle (tower of the Lantern)	1465

The dawn of the 14th century, from the era of abbeys to the era of cities, and the rebirth of central power

We have seen that the role of the abbeys, and in particular the Cistercians, was quite important in the conquest of new agricultural lands in the 12th and 13th centuries. However an assessment of the social success of these monk-hydraulicians is not as clear as it might be. Unless supported by a powerful lord or by royal authority, the villagers and peasants are not in a position of strength vis-à-vis the abbeys, especially since the

30 Fichou, Le Hénaff, Mével (1999), pp. 17-19. According to these authors, the Romans had also built beacons at Dover and at Coruña on the Atlantic coast. We have already described, in Chapter 6, several Roman ports developed on the Mediterranean.

abbeys are themselves favored by the local lords. Often the villagers are chased off by monks who want to develop their land; yet these lands, once developed and intended to serve a single order, almost always engender numerous conflicts with other occupants, whether they be lay or monastic.

The turn of the 13th to the 14th century sees a strengthening role of central power. We have seen how the kings intervened in major projects such as the "King's channel" in Poitou, and the "royal canal of Thuir" in Roussillon. We also saw this on the shores of the North Sea, and in the actions of the count of Flanders (and rich townspeople) in appropriating land on the seashore, developing ports, marking them and constructing beacons.

The middle of the 14th century was a period of catastrophes in the West – from a grave economic and monetary crisis, to the beginning of the Hundred Years' War (1337-1453), to the great plague of 1348-1349. In this period there was also a recurrence of the great famines that had not been seen since the 10th century. There is a general decline in population, the abbeys are decimated by the plague, and the grand era of land conquest is more or less finished. A new power emerged: that of the cities, concentrated and enclosed within walls and behind moats in response to the general insecurity, and through which water no longer flowed freely, but became stagnant and unhealthy.

Over time an awareness of the need for the flow of wastewater and for the management of water supply developed. Wells used by townspeople who were located some distance from the rivers were terribly contaminated by the infiltration of stagnant water polluted by all sorts of wastes. These cities begin to develop sewer networks and to organize freshwater delivery to public hydrants and fountains, although hesitantly at first. This movement had slowly begun in the 12th century in Italy but did not reach the cities of western and northern Europe until the 14th and 15th centuries.[31] Public water fountains, essentially unknown since the end of Roman civilization, timidly reappear. Some are found around 1100 in the port city of Genoa; three are built at Sienna between 1220 and 1227; and large ones appear in Viterbo and Perugia around 1251 and 1277, respectively. But Florence and Milan must wait until the 15th century before seeing serious development of public fountains, and Bordeaux must wait until 1520.

Portions of Roman aqueducts are restored to service, and sometimes new delivery canals are built, as we have seen for Perpignan, under the influence of the king of Majorca. In the cities of northern Europe the movement toward public fountains begins, again very slowly, with the use of water supplies originally developed for religious establishments. Philippe Auguste is credited with establishment of the first public fountain in Paris in 1182 at the grand Halle, through derivation of some of the water of a monastic groundwater supply from Pré-Saint-Gervais. This first fountain is soon followed by others, so that there are ten in Paris by the end of the 14th century, and seventeen a century later.[32] But it is still not until the 20th century that the quantity of water per inhabitant becomes comparable to the abundance seen during the Roman period.

---

31 Guillerme (1983); Heers (1990), pp. 300-325.
32 Lacordaire (1979), pp. 66-81.

Conclusion

From our earliest ancestors who created irrigation canals by scratching the ground to channel the water of a stream, to the appearance of the first technology, from the great engineering projects of Mesopotamia and China to the blossoming of the first great cities - civilization and hydraulics have always advanced hand-in-hand.

Technique, power and society

In numerous ancient civilizations, the legitimacy of those who govern - both in their own eyes and in those of the governed - rests on the social utility of their hydraulic projects. Technique and society are indissolubly linked.

Hammurabi of Babylon, conqueror of all of Mesopotamia, proclaims himself "lord of the city of Uruk" and immediately adds, as if to ensure the legitimacy of this domination, that it was he who "allocated to these people the water of prosperity". He nonetheless owed his power to his weaponry – and to diplomacy. Yahdun Lim at Mari made the same kind of claim, as did many other leaders of the land of Sumer, all of whom associated the legitimacy of their power and the glory of their reign with the hydraulic projects that they had effected.

A thousand years later, when Cyrus the Great entered Babylon, he legitimized his power in "raising the brick banks lining the city's ditches". Alexander, another great historical figure, acts no differently after having occupied Mesopotamia when he ensures that the canals are maintained and tears down the dams the Persians had constructed to block navigation. These technical acts are tantamount to acts of coronation. Was not the legend that grew up around Yu the Great, the pacifier of the Yellow River, the very legitimization of central power in China? We have seen so clearly that in this land it was often hydraulic catastrophes that caused the fall of dynasties. In ancient Egypt, whose cultural and political stability make it somewhat of a special case, the sovereign is, even here, identified with the river from which all blessings come.

Innovation

The legacy of Alexander is somewhat mixed insofar as innovation is concerned. To his credit there is the city of Alexandria, with its cultural diversity and intellectual fertility. Of course there is also Archimedes, founder of hydrostatics and supposed inventor of the "Archimedes screw". There is also Ctesibios, inventor of the fire pump. But other innovations never really emerged from their cocoons to find practical application. These include the aeolipile (wind ball) of Heron, a device whose further development could easily have led to the steam engine. There were many other such inventions, and seemingly useless gadgets, that were destined to be investigated or rediscovered by Arab scientists or, even later, by Leonardo da Vinci and other great thinkers of the Renaissance. Even during the shining period of Alexandria, the greatest innovations seem to have risen from obscurity, from the shadows, from anonymous inventors. The paddle wheel, the water mill, the *noria* are all born somewhere in the Orient and then progress through history silently, leaving traces of their passage only by chance, here in a description by Strabo, there in a Greek poem.... The windmill is not conceived in the great library of Baghdad but in the Persian countryside. Also in China, nursery of innovation from the 3rd century through the end of the Middle Ages, inventions come from a few obscure civil servants such as Tu Shih, who "loved the common people and wanted to lighten their work", and to this end brought hydraulic energy to the forges. Or Chiao Wei Ho who invented the lock to avoid damaging boats that had to be dragged from one section of a channel to another. Or such as the anonymous inventor who developed the axial rudder for ships. These innovations did not come from the minds of scholars working in imperial courts.

From the above observation, one has to ask: of what use are teams of scientists and research institutes? Where these existed in ancient times, they were instrumental in the spread and standardization of useful inventions, fostering the rapid refinement and optimization of new devices, ensuring that optimal configurations and designs were adopted in practice. The dimensions of the Chinese "dragon backbone machine" were standardized from the 9th century on. Without manuals and other technical documentation, the dissemination of technical innovation is an extremely slow process – errors are repeated, and the "optimal" design develops very slowly. The Roman aqueducts are characterized by surprising conceptual flaws given the experience that could and should have been accumulated. Similarly Roman dams are sometimes well conceived, but just as often are very badly designed. The arch dam seems to have been "re-invented" many times over.

Innovations transcend the boundaries of civilizations. But lacking written traditions, their dissemination and spread is extremely slow. Perhaps the most significant example of this is the spread of the *qanat*, the device for tapping groundwater that is so simple in principle, if not in application. Conceived in Armenia or in Persia between the 10th and 8th century BC, it is spread into the Orient by the Persians, and is further developed by the Romans who take it to Lybia and Tunisia. But Morocco does not see the *qanat* until it has made at least two other journeys: one with the westward migration toward the Saharan oases, and another that comes from Muslim Spain to Morocco following the Reconquest.

In the absence of written descriptions, such technical devices tend to be developed differently from one locality to another, reflecting the vicissitudes of oral transmission of know-how rather than the best adaptation of the technology to the local situation. The water mill has a horizontal wheel in the China and the Arab world, but a vertical wheel in the Occident, and the same is true of the windmill at a much later period. The sail rigs on Chinese junks, so perfectly adapted to the needs of coastal commerce in China, do not spread into the West where oarsmen perish in the galleys.

The early technologies

One may be skeptical of the overall contribution of Hellenistic science - its relative disconnection with practical application, or its failure to document the significant revolution represented by hydraulic energy. But the incontestable fact remains that in the study of the science and techniques of Antiquity preceding the Middle Ages, one cannot avoid marking pre- and post-Alexandria. The Hellenistic period represents a watershed, or divide, that is reflected in the two distinct parts of this book.

The earliest technology appeared before the turmoil that followed the epic reign of Alexander the Great. The elements of this technology can be briefly listed in the order of their appearance as follows:

- dams, made of rocks or earth, whose earliest traces are found on the Syro-Mesopotamian steppes at the end of the IVth millennium BC (Jawa, Khirbet el-Umbashi); the characteristics of the oldest of these dams are summarized in Table 10.1.
- derivation canals, sometimes created through cut-and-fill on the floodplain, with gates and guide vanes of stone; the oldest of the major works of this kind are probably those in the ancient land of Sumer, in the IVth millennium BC, but the genesis of this technology really belongs to all of the fertile crescent.
- drainage facilities, sewer systems of stone, bricks, or clay; we have seen them in the ancient cities of the Indus, in Sumerian settlements; they are particularly prominent in Crete.
- navigation canals, necessary adjuncts of the cities of lower Mesopotamia, for each Sumerian city has its own port; on the middle Euphrates, the Semiramis canal and the *nahr* Daourin are surely the oldest; in China, from the 5th century BC, the Hong canal connects the basins of the Yellow River to those of the Yangtze;
- the sailing vessel, first appearing in the Persian gulf, then in Egypt and the Aegean Sea;
- water supply systems, whose technology seems to first appear in Minoan Crete;
- and, somewhat of a special case since it is really an evolutionary technique that is conceived during the iron age: the *qanats*.

From the great cauldron of ideas that was Alexandria emerged the principles of hydrostatics, and the fundamental understanding of the effects of pressure. The inverse siphon and the pressure conduit are used in water supply systems. And hydraulic energy makes its appearance in the form of the watermill and the *noria*, two systems destined to see considerable development in the Middle Ages – the *noria* in the East and Far East, the water mill everywhere. The blossoming of hydraulic technology in the Middle Ages is seen in the mills dotting the landscape from the Atlantic to the Sea of Japan. Later, we can credit the Persians with the idea of the windmill, and the Chinese with the axial rudder and modern sail, as well as the navigation lock.

To design, and then to maintain

Planning for the maintenance of an engineering structure as an integral part of its design has become routine in modern practice. Yet, by necessity, this preoccupation was also present in numerous ancient projects. For example planning of the irrigation system of Sechuan, with its intake at Dujiangyang, obviously took into account the need to clean the intakes and to maintain the dikes and the intake control mechanisms. The designers also anticipate the need for a procedure to dewater the works for maintenance during low-flow periods. Canal cleaning was a continuous activity in the old land of Sumer. From the archives of Mari, we know that these maintenance efforts were tedious and

Table 10.1 The oldest known dams

Name	Region	Probable Date	Ht. (m)	Len. (m)	Type	Water-Course	Purpose
Jawa	Djebel el-Arab (Jordan)	3000 BC or earlier	5.5	80	Fill between stone walls (Fig 2.5)	Derivation canal from the wadi Rajil	Floodwater storage (Fig 2.7). The oldest known dam.
Khirbet el-Umbashi	Djebel el-Arab (Syria)	3000 BC	7	40	Earth dam	Wadi el-Umbashi	Reservoir in the bed of the wadi itself (Fig 2.8)
Sadd el-Kafara	Egypt (near Memphis)	2650 BC	14	113	Fill between two rock faces	Wadi Garawi	Protection against floods of the wadi Garawi. The first known large dam (Fig 3.3, 3.3a)
Weir of Khanouqa	Middle Euphrates (Syria)	1800 BC ?			Rocks (uncut basalt blocks)	Euphrates	Weir on the Euphrates, headworks of the "Semiramis canal" (Fig 2.13)
Boedria	Copaïde (Greece)	1300 BC	2	1250	Fill between two walls	Kephissos	Reservoir (Fig 4.6). See other dams, table 4.1
Kofini	Tiryns (Greece)	1200 BC	10	100	Fill between two rock walls	Lakissa	Rerouting of a river, flood protection (Fig 4.12, 4.13)
Lake Rusa (north)	Urartu (Armenia)	720 BC	15	75	?	Lake Rusa	Reservoir: lake Rusa (Fig 2.18). In service until 1861 AD, then rebuilt in 1952.
Lake Rusa (south)	Urartu (Armenia)	720 BC	7	60	?	Lake Rusa	Reservoir; lake Rusa (Fig 2.18)
Weir of Ajileh	Assyria (Iraq)	694 BC	3	230	Large blocks of cut stone	Khosr	Weir on the Khosr, headworks of a derivation canal of the Khosr to Nineveh
Bavian	Assyria (Iraq)	690 BC	?	?		Gomel	Weir on the Gomel, headworks of the Sennacherib canal
Shaobei, or Anfengtang	Anhui (China)	585 BC	11		Earth, straw and wooden stakes	Tributaries of the Huai	Reservoir; still in service today
Maryab	Yemen	510 BC	15	650	Earthen dike, rock protection	Wadi Dhana	Intake works for two canals conveying flood waters of the wadi (Fig 3.12). Breached in the 7th century AD.
Panda	Sri Lanka (Ceylon)	370 BC	7	2,600	Earthen dike		Seasonal reservoir
Bassawak, Tissa	Sri Lanka (Ceylon)	300 BC	8	1,800 3,300	Earthen dikes		Reservoirs (Fig 7.3)
Paskanda	Sri Lanka (Ceylon)	300 BC	17	?	Earthen dike		Seasonal reservoir
Mala'a (lake Moeris)	Fayoum (Egypt)	250 BC	7	8,000	Masonry	Joseph canal	Reservoir (lake Moeris) fed by floodwaters of the Nile (Fig 3.6). In use until the 18th century AD.

N.B. Two other dams, whose conditions were known in the Roman period, are candidates for being even older: the dam of the wadi el-Souab, a possible headworks on an irrigation canal of the ancient Mari (Fig 2.11), and the dam of the lake of Homs (Fig 6.33).

---

sometimes difficult. The ports of Pylos in the IIIrd millennium BC, and the port of Rome built by Trajan, are examples of projects designed from the beginning to use the flow of the river to keep the entrances open. Of course any project subject to sediment deposition or erosion can quickly be rendered useless due to lack of maintenance. This is why the reclaimed land of Mesopotamia lapsed back into desert so quickly after the Mongol invasions of the 13th century. Another example is that marshes quickly reappear as soon as the Etruscan know-how in lowland drainage is lost in Italy.

City and countryside

The historical record shows how important it was for hydraulic engineering to have social utility in Antiquity. Its effects must be recognized by the beneficiaries – but often these beneficiaries are far from the hydraulic projects themselves. If they are in the countryside, they may easily recognize the utility of large irrigation canals, such as the thirty-kilometer long ones on the Euphrates and the Oxus from the IIIrd millennium BC. In these pages we have not often come across the "paradise lost", the dream of a small community to be able to manage its own technological development at the local level. Such situations probably existed in the very early development of agriculture, and we find it again in the Syrian and Anatolian countryside during the Byzantine Empire. In order to try to survive, to struggle against floods that threatened houses and crops, and to avoid death when the river on which they depended for their livelihood overflowed its banks, civilizations had to assemble and organize significant manpower. This in itself was surely a potent element in the creation of civilizations, as has been proposed by numerous theories and as we have tried to point out in this book.

The cities need raw material and food. The early Sumerian cities had to import wood, rocks, and metals. The Pharaohs import beautiful stone for their Nubian monuments, and wood from the mountains of Lebanon. Rome imports its wheat from Sicily, Tunisia, and Egypt. The successive capitals of the imperial Chinese dynasties import their grains from the alluvial plain of the Yellow River and, later, their rice from the Yangtze basin. Watercourses, their ports and canals, provide the primary support for all these exchanges. The Nahr Daourin, parallel to the Euphrates, flows along an impressive 120 km, likely from the very beginning of the Bronze Age. And in China during the Middle Ages, the Grand Canal stretches from the south to the north of the middle empire, over hundreds of kilometers.

Cities need water. The "so numerous and necessary aqueducts" that the Romans extended over all their empire are works of "great transport", crisscrossing the countryside to meet the urban water needs of Rome, Lyon, Nîmes, Toledo, Carthage, Antioch, Apamea, Jerusalem..... The Roman lifestyle required these aqueducts. And when the barbarian invasions in the West put an end to this lifestyle, they also put an end to the need for these aqueducts, causing their demise just as if the barbarians had destroyed them, though generally they did not do so. But fortunately this destruction did not generally occur. Still, few aqueducts survive the closed mindedness that characterized the Middle Ages in the West. But in the Orient the Arabs perpetuated the Romano-Hellinistic patrician lifestyle to some degree. The pleasures of the city are first and foremost the pleasures of water – baths, ablutions, strolls in gardens or along the banks of rivers. It is water that makes of Damascus, Samarcand and Nishapur the very images of paradise for the Arabs.

Of course there is also a prosaic dimension to water in the city. Wastewater disposal requires its own hydraulic techniques. From the first gutters used to drain wastewater from houses in the Neolithic village of El-Kowm in Syria, this concern for wastewater – that one might think to be only a modern preoccupation – is continuous in the Bronze Age in the cities of the Indus, in the new cities like Habuba Kebira and Mari on the Euphrates, and in Crete where the refinement of urban hydraulics reaches its pinnacle. We also find attention given to wastewater in Roman cities and in many Arab settlements. But during the Middle Ages in the West, and even in our recent Age of Enlightenment, this preoccupation too often falls by the wayside.

Technology for eternity

How many achievements called "eternal" survive the civilization, or even the regime, under which they were created? Perhaps not many, but certain of the hydraulic works of Antiquity have survived their origins. The canal connecting the Nile River and the Red Sea – built by the Pharaoh Necho, finished by the Persian Darius, perfected by Ptolemy, the successor of Alexander, and renovated by the Emperor Trajan, then by the Umeyyade caliphs – functioned, though surely with a few interruptions, for an "eternity" of thirteen centuries. We will have to come back a thousand years from now to see if our Suez Canal, descendent of the Necho canal, is still there…..

The great irrigation systems and several water allocation plans of ancient China, as well as dams in the Armenian kingdom of Urartu, in Roman Spain, and in Arab Andalusia, are still in operation today. Others have disappeared even before seeing practical use - like the Sadd el-Kafara dam project attempted by the Pharaohs of the Ancient Empire south of Memphis, reflecting missed opportunities for innovation and effectively stalling technical progress in Egyptian dams for several centuries. Other such dam projects were useful for a period, then fell into ruin, like the large Maryab dam in the land of the queen of Sheba, today reconstructed to be almost identical to the original by the modern state of Yemen. Other such dams that could have endured were abandoned, like the one built six thousand years ago by anonymous refugees, likely townspeople fleeing some unknown menace in the black basalt desert of Jordan. To survive in this desolate land they built, at Jawa, the first known dam in the history of humanity, before abandoning it to flee to some unknown destination.

Civilizations die. Often their works die with them. But their technologies survive. We are the inheritors of the hydraulic innovations of the millennia, whether they came to us through the work of some scholar who described them in writing that was subsequently recopied and translated, or whether they came to us slowly through the random process of migration and commercial exchanges. Yes, these innovations have come down to us, who are depleting our groundwater resources, who are setting the stage for future water wars. The ancient water technologies are part of the patrimony that our children will so urgently need.

Chronological Table

Dates	History and Civilizations	Hydraulic Science and Technology
9500 BC	**Beginnings of agriculture in the Near-East**	
6500 BC		**First evidence of irrigation** (Choga Mami)
		Earliest evidence of drainage in houses.
6000 BC	Copper metallurgy	
		First wells in Mesopotamia
	Beginnings of agriculture in China	
4000 BC		Development of irrigation in lower Mesopotamia
	First cities in the land of Sumer (Uruk)	
		Development of navigation on the Euphrates.
		First appearance of the sail (Eridu)
3500 BC	**Beginning of the Bronze Age**	Sewers perfected at Habuba Kebira, Sumerian
	Appearance of writing (Uruk, Suse, then Egypt)	Trading post on the Euphrates
		Irrigation at Geoksyr in Margiana (Turkmenistan)
		Oldest known dams (Jawa)
	Invention of the wheel	Earliest water-lifting machine in Mesopotamia (the shaduf)
		First evidence of sailboats on the Nile
3000 BC	Unification of Upper and Lower Egypt,	Irrigation in Margiana and Bactria
	Founding of Memphis	Dams and reservoirs at Khirbet el-Umbashi
	Beginning of the Indus civilization	Irrigated oases in Oman
	Founding of Mari	Irrigation system of Mari, navigation canal
		Dam of Sadd el-Kafara (Egypt)
2500 BC	Maritime civilization of the Cylades	Irrigation at Shorughai in eastern Bactria
	Beginning of the Minoan civilization in Crete	First water supplies in Crete
	Legend of Yu the Great on the Yellow River	(Yu was said to have "dug the river")
	Sargon of Akkad unifies Mesopotamia	Irrigation at Marw(Merv) in Margiana
2000 BC	1st intermediate period (2180–2040) and beginning of the Middle Empire in Egypt	
	Reign of Hammurabi in Babylon (1792-1750)	Hydaulic developmens of Fayoum by Amenemhat III (Moeris)
	Destruction of Mari (1760)	
	End of the Indus civilization	
	Beginning of the Mycenaean civilization in Greece	
	Bronze metallurgy appears in China; begin-	

	ning of the Shang Dynasty, in the basin of the Yellow River Hittite Empire in Anatolia	
1500 BC	2nd intermediate period (1730-1560) and beginning of the New Empire in Egypt End of the Minoan civilization Reign of Ramses II Hebrews arrive in Palestine	Irrigation of the western Bactrian oases Irrigation in the ghouta of Damascus Artificial port of Pylos; drainage of lake Copais
1200 BC	The Trojan War The "Sea People" plunder the Levantine; destruction of Ugarit. End of the Hittites	Catastrophic inundation at Tiryns; construction of a dam and canal
1100 BC	End of the Mycenaean civilization **Beginning of the Iron Age**	
1000 BC	Reign of David in Palestine	
900 BC	**Appearance of the Phoenician alphabet** Beginning of the Assyrian Empire	
800 BC	Founding of Carthage Arrival of the Etruscans in Italy	Canal of Menua in Urartu (Armenia) **Appearance of the qanat** Development of irrigation in Arabia Felix (Yemen)
700 BC	Reign of Sennacherib in Assyria (704-681) Reign of Karib'Il Watar in the land of Sheba	Water supplied to Nineveh; bridge-aqueduct of Jerwan Dams of lake Rusa in Urartu
600 BC	End of the Assyrian Empire (606); reign of Nebuchadnezzar in Babylonia; Saite renaissance in Egypt The Phoenicians found Marseilles	Necho II, pharaoh of the Saite Dynasty, constructs the first canal between the Nile and the Red Sea Thalès of Milet establishes that the earth is round Tarquin the Elder constructs the cloaca maxima at Rome Dam-reservoir Anfengtang, in the Huai basin (China)
550 BC	Birth of Buddha Birth of Confucius Cyrus the Great enters Babylon (539) and founds the Achaemenid Persian Empire	Hong Canal, first great navigation canal in China Polycrate constructs the "tunnel of Samos"
500 BC	Median wars in Greece (490-480 Beginning of the Republic of Rome (509) Voyages of Herodotus in Egypt and Babylon (460) Peloponnesian war Arrival of the Nabatians in Palestine?	Maryab dam in the land of Sheba (Yemen) Irrigation at Djouboulak Koum (Taklamakan desert)
400 BC	Expedition of the Xenophon's "Ten Thousand" in Babylon (401)	Plato adopts the theory of the "four elements", later taken up by Aristotle

350 BC		Aristotle creates the Lyceum at Athens
	Founding of Alexandria (331). The same year, Alexander the Great enters into Babylon	
	Laozi (Lao-Tse) founds Taoism (unknown date)	
	The Qin, coming from the valley of the Wei, occupy Sechuan (316)	Irrigation system of Sechuan, with its intake works at Dujiangyan
		The Aqua Appia, first Roman aqueduct (312)
300 BC	**Ptolemy I founds the Museum and Library of Alexandria**	Euclid founds modern geometry
		Straton of Lampasaque defines a "vacuum"
		Ctesibios invents the fire pump
		First reservoir-dams in Ceylon
		Zou Yan constructs the Chinese theory of "five elements"
250 BC		First lifting wheels (Philon of Byzantium?)
		Irrigation works in the Fayoum; dam of Mala'a
		The Archimedes screw or limaçon
		Zhengguo irrigation canal in China
	Maurya Dynasty in India (313-180)	**Archimedes founds hydrostatics**
	The Parthians evict the Seleucides from Mesopotamia	Eratosthene of Cyrene measures the radius of the earth
	Shi Huangdi, first Emperor of China (221)	"Magic Canal", communication route to southern China
	Beginning of the Han Dynasty in China (206)	
	Fall of Carthage (202)	
200 BC	Apogee of the Greek kingdom of Bactria	First water supplies using the inverse siphon, in the Orient then at Rome (Aqua Marcia)
		Completion of the great siphon of Pergamon
150 BC	Wudi of the Han, Emperor of China (141-87)	Irrigation and transport canals in China
	Rome inherits the Pergamon kingdom (133)	In 109, the Yellow River is restored to the course it had abandoned in 132
100 BC		**Appearance of the water mill** (first evidence in the lands of Mithridate, king of Pontus)
	Death of Cleopatra and annexation of Egypt by Augustus (31)	Hydraulic developments of the Nabatians (Petra, Negev desert)
	Strabo writes Geography, Vitruvius writes On Architecture	Vitruvius describes the water mill and the noria
	Augustus refounds Carthage	The first known arch dam near Glanum in Provence (date uncertain)
	Kouchan Empire in central Asia	Earth dam at Nanyang (China)
1 AD	Jesus Christ in Palestine	Pontius Pilate constructs the "pools of Solomon" and the Jerusalem aqueduct
		The Yellow River breaks through its dikes and

		changes course (11) First mention of the water wheel in China, to power pestles (21), then forge bellows at Nanyang (31)
50 AD	Claudius is Roman Emperor (41-54) Roman Emperors Nero (54), Vespasian (70), Titus (79) Domitian is assassinated (96); Nerva is elected emperor at Rome	Numerous aqueducts at Rome, Lyon, Nimes... The "port of Claudius" at the mouth of the Tiber Development of the water mill in Italy (Pliny) Heron of Alexandria: the aeolipile, discharge calculation in a canal (continuity principle) Frontinus studies the 9 Roman aqueducts and reforms the water distribution system **The axial rudder and the modern sail appear in China**
100 AD	Trajan is Roman Emperor (98-117) End of the Han Dynasty (185). Fragmentation of China	The "port of Trajan" at the mouth of the Tiber. Numerous aqueducts in the Roman provinces: Apamea, Carthage... Invention of the square-pallet chain pump in China Roman dams in Spain and the Orient Roman flour mill at Barbegal (in Provence)
200 AD	 Valerius is captured by the Sassanide Shapur (260)	Dams and Roman qanats in Tunisia, in Cyrenaica (Libya) and in the Orient Mills at Rome (Janicule, Thermes of Caracalla)
400 AD	The Bishop Theophilos destroys the Serapeion at Alexandria (391) **Fall of the Occidental Roman Empire** (410)	
450 AD		Proof of the existence of norias on the Oronte (Apamea mosaic, 469)
500 AD		Repeated dike ruptures on the Tigris. Definitive rupture of the Maryab dam (Sheba) Dara arch dam in Anatolia
600 AD	Yang Jian reunifies China and founds the Sui Dynasty (604) Tang Dynasty (618) The hegira of Mohammed in Arabia (622) Taking of Alexandria by the Arabs (640)	 **The Grand Canal of the Sui and the Tang** Jean Philopon of Alexandria explores resistance and motion through the air
650 AD	Beginning of the Umeyyade Dynasty (661)	The "Persian" **windmill** at Seistan
700 AD	The Arabs in Spain (711)	
750 AD	End of the Umeyyades. Beginning of the Abbasids (750) Founding of Baghdad (762) The Chinese armies are beaten by the Arabs at Talas (Ferghana)	 Chinese prisoners introduce the paper industry to Samarcand (pestle mills)
800 AD	Harun al-Rashid founds the great library of	The Book of Ingenious Mechanisms of the Banu

	Baghdad	Musa brothers in Baghdad
	Founding of Fez	Construction of qanats at Madrid
900 AD	Song Dynasty in China (960)	**Invention of the chamber lock in China**
	The Ghaznavid Turks (977), then Seljuk Turks (1040) in central Asia	Tidal mill at Bassora Major irrigation works in Andalusia
1000 AD	End of the caliphate of Cordoue (1031)	
	The Almoravides in Morocco	
	Founding of Marrakesh	First qanats (khettaras) at Marrakesh Al-Karagi explains the flow of groundwater
	Beginnings of demographic expansion in western Europe	Drainage, drying of polders in Flanders
1100 AD	Beginning of the Crusades	Al-Khazini picks up the work of Archimedes on hydrostatics
	Voyage of the Andalusian Ibn Jubayr in the Orient	Invention of the post windmill (Flanders or England); development of the tidal mill on the Atlantic coast First public fountains in the West since the Roman Empire First levees on the Loire (1169)
	The Song are chased out of northern China by the Jurchen (1126), they destroy the dikes	Course of the Yellow River shifts to the south of Shandong (1194)
1200 AD	The Mongols raze Samarcand (1219), then Baghdad (1258)	Destruction of hydraulic infrastructures in Mesopotamia, Bactria, and Khorassan
	The Mongols occupy Kaifeng (1233), Hangzhou (1276), and in China take the name of the Yuan Dynasty (1271)	Drainage of the poitevin marsh (1190-1283)
	Sojourn in China of the Venetian Marco Polo (1280)	**The Grand Canal of the Yuan**
	The Andalusians lose Cordova (1236), Valencia (1238), Sevilla (1248)	Invention of the double-action piston bellows in China The Yellow River shifts completely to the south (1288)
1300 AD	Voyages of the Tangerian Ibn Battûta (1330-1350)	Dike ruptures on the Yellow River (1327; 1344)
	Beginning of the Hundred Years' War (1337)	
	The great plague in the West (1348-1349)	
	The Mongols are chased out of China; beginning of the Ming Dynasty (1368)	Arch dam of Almansa (1384)
1400 AD	The Mongols of Tamerlan pillage Delhi (1398) then Baghdad (1401	
	Great maritime expeditions of the Ming (1405-1433)	Dam-reservoir for the upper portions of the Grand Canal (1411)
	Taking of Constantinople by the Turks (1453)	
		Louis XI reinforces and extends the levees of the

	Loire (1482)
	The course of the Yellow River stabilizes (1495)
End of the Reconquest of Spain (1492)	Turk-Mongol dams in Persian and in Afghanistan; arch dams
The Occidental Renaissance	**Leonardo da Vinci (1452-1519) rediscovers the principle of continuity**

Units of measure for lengths and areas

Sumerian-Akkadian, Babylonian (sexagesimal system, base 60)

- *lengths*

1 digit = 1.66 cm
1 cubit = 30 digits = 0.5 m
1 cane = 6 cubits = 3 m
1 *nindan* = 360 digits = 6 m
1 chord = 10 *nindan* = 3,600 digits = 60 m
1 *beru* = 180 chords = 10.7 km

- *areas*

1 *iku* = 1 square chord = 3,600 m^2
1 *gur* = 50,000 or 54,000 square cubits (about 13,000 m^2)

N.B.: At Mari, the names of units sometimes are used for small measures. Thus 1 cane at Mari is probably 1.2 m.

Egyptian

- *lengths*

1 digit = 1.87 cm
1 hand = 4 digits = 7.5 cm
1 nilometric cubit = 28 digits = 0.525 m
1 *iterou* = 10.5 km
1 corde = 100 cubits = 52.5 m

1 stadia (Ptolemeic) = 400 cubits = 210 m (but Eratosthene uses a stadia of 300 Egyptians cubits, or 157.5 m)

1 schoene (Ptolemeic) = 30 stadia = 5.4 km, but the schoene is 60 stadia south of Memphis, or even 120 stadia (Herodotus, II, 149; Strabo, XVII, 25)

- *areas*

1 aroure = 10,000 square cubits (i.e. 100 cubits by 100 cubits) = 2,750 m^2

Greek

1 digit = 1.85 cm
1 empan = 12 digits = 0.222 m
1 foot = 16 digits = 0.296 m
1 cubit = 24 digits = 0.444 m
1 *orgye* or fathom = 6 feet = 1.776 m

1 *plethre* = 100 feet = 29.6 m
1 stadia = 100 orgoyes = 400 cubits = 177.6 m
(Herodotus, II, 149)

Persian

1 parasange = 5.94 km

Roman

1 pace = 1.48 m
1 stadia = 185 m
1 mille = 1,000 paces = 1,478.5 m

Chinese

The Chinese *li* is defined from two bases: it is first of all the length of an acoustic pipe tuned to the sound of a particular clock. Under the Tang, it is given by the equivalence:

195 *li* = 1° of arc of latitude, or: 1 *li* = 570 m

Later, the *li* is modified such that 200 *li* corresponds to 1° of latitude.

Sources: Steinkeller (1988), Powell (1988), Yoyotte (notes on *The Journey to Egypt* of Strabo, 1997), Barguet (notes on the *Inquest* of Herodotus, 1985), Needham and Ling, IV, 1 (1962).

Margueron, Jean-Cl., *Mari – Métropole de l'Euphrate*, Picard, 2004)

Wikander, O. (2000), *Handbook of Ancient Water Technology*, Brill

Bibliographic References

Bibliography.

Sources in alphabetical order

Mc Adams R., Nissen H., *The Uruk Countryside,* The University of Chicago Press, 1972.

Agusta-Boularot S., Paillet J.-L., « Le barrage et l'aqueduc de Glanum : le premier barrage-voûte de l'histoire des techniques? », *Revue Archéologique,* 1997-1, pp. 27-78.

Andrieu J. L., « Techniques de construction des aqueducs sur les territoires de Béziers et de Narbonne », in *Les aqueducs de la Gaule romaine,* p. 73-108, PULIM, 1997.

Andrieu J.L., Cazal G., « L'aqueduc gallo-romain de Cabezac près de Narbonne », in *Les aqueducs de la Gaule romaine,* p. 109-132, PULIM, 1997.

Ardhuin M., « L'aqueduc romain de Reims », in *Les aqueducs de la Gaule romaine,* p. 333-352, PULIM, 1997.

Archimedes, *Peri Ton Ochoumenon - Floating Bodies,* French translation by C.Mugler, *Les Belles Lettres*, 1971,

Archimedes, *The Works of Archimedes*, English translation by Reviel Netz, Cambridge University Press, 2004.

Argoud G., « Eau et Agriculture en Grèce », in *L'Homme et l'Eau en Méditerranée et au Proche Orient,* IV, pp. 25-43, Maison de l'Orient Méditerranéen, Lyon, 1987.

Aristotle, *Aristotle's Physics, books I and II* , Oxford ; New York : Clarendon Press, 1983.

Argyropoulos P.A., « Ancient Greek hydraulic engineering experience and technique », *Proc. AIHR Congres*, Cagliari, Sept 1979.

Azema J-P., *Les moulins de France*, Ouest-France ed., 1995.

Bader A., Gaibov V., Gubaev A., Koshelenko G., « The oasis of Merv : the dynamics of its settling and irrigation », *Ancient Civilizations,* vol 3, 1, p. 49-60, 1996.

Bagg, A.M., "Irrigation in Northern Mesopotamia; Water for the Assyrian Capitols (12th - 7th centuries BC)", *Irrigation and Drainage Systems* 14, pp. 301-324, 2000.

Bailhache M., « L'évolution du débit des aqueducs gallo-romains », *Dossiers de l'archéologie,* 38, pp. 82-87, 1979.

Balcer J.M., « The Mycenaean dam at Tiryns », *American Journal of Archeology,* 78, pp 142-149, 1974.

Balty J.C., « Problèmes de l'eau à Apamée de Syrie », in *L'homme et l'eau en Méditerranée et au proche Orient - L'eau dans l'agriculture,* pp. 9-23, Maison de l'Orient, Lyon, 1987.

Barrière B., « Les Cisterciens d'Obazine en Bas Limousin, les transformations du milieu naturel », in *L'hydraulique monastique,* pp 13-33, Créaphis, 1996.

Basham A. L., *The Wonder That Was India,* South Asia Books, 1954 (3rd ed in 2000).

Beaumont P., « The qanat : a means of water provision from groundwater sources », in *Qanat, Kariz and Khattara :Traditionnal Water Systems in the Middle East and North Africa,* pp. 13-31, Middle East & North African Studies Press, 1989.

de Belidor B. F., *Architecture hydraulique ou l'art de conduire, d'élever et de ménager les eaux,* Charles-Antoine Jombert, Libraire, Paris, 1737 (coll. ENPC).

Bellwald, U., *"The hydraulic infrastructure of Petra"*, proc. 12th Int. Congress on the History of Water Management and Hydraulic Engineering in the Mediterranean Region, Ephesus, Oct 2-10, 2004, p. 239-249, Wiplinger ed., Peeters, 2006

Benoit P., Berthier K., « L'innovation dans l'exploitation de l'énergie hydraulique d'après le cas des monastères cisterciens de Bourgogne, Champagne et Franche-Comté. », in *L'innovation technique au Moyen Age*, p. 52-66, Errances, 1998.

Berthier K., « L'hydraulique de l'abbaye de Cîteaux au XIIIème siècle », in *L'hydraulique monastique,* pp 35-43, Créaphis, 1996.

Berthier S., « Recherche en cours sur les aménagements hydro-agricoles dans la vallée de l'Euphrate à l'époque islamique », in *Techniques et pratiques hydro-agricoles traditionnelles en domaine irrigué* (proc. IFAPO conf. of Damas), vol I, pp. 229-234 , Geuthner, Paris, 1990.

Berthier S., d'Hont O., « Le peuplement rural de la moyenne vallée de l'Euphrate à l'époque islamique (VIIème - XXème siècle) : premiers résultats », *Archéologie Islamique,* 4, pp. 153-175, 1994.

Bianquis T., « Quelques problèmes d'hydraulique soulevés lors des fouilles de Rahba-Mayadin », in *L'homme et l'eau en Méditerranée et au Proche Orient III,* pp 121-128, Maison de l'Orient, Lyon, 1986.

Bisson J., « The origin and evolution of foggara oases in the Algerian Sahara », in *Qanat, Kariz and Khattara - Traditional Water Systems in the Middle East and North Africa,* pp. 179-194, P. Beaumont, M. Bonine, K. Mc Lachlan ed., Middle East & North African Studies Press, 1989.

Biswas A.K., *History of hydrology,* North Holland, 1970.

Blunden C. and Elvin M., *Cultural Atlas of China,* Phaidon Press, Oxford, 1983.

Boithias J.L., de la Vernhe A., *Les moulins à mer*, Créer, 1989

Bond C.J., « Les systèmes hydrauliques monastiques dans la Grande Bretagne médiévale », in *L'hydraulique monastique,* pp 457-473, Créaphis, 1996.

Bonneau D., « Le nilomètre : aspects techniques », in *L'homme et l'eau en Méditerranée et au Proche Orient - l'eau dans les techniques,* Maison de l'Orient, Lyon, 1986.

Bonneau D. *Le régime administratif de l'eau du Nil dans l'Egypte grecque, romaine et byzantine,* E.J. Brill, 1993.

Bonnin J., *L'eau dans l'antiquité, l'hydraulique avant notre ère,* Eyrolles, 1984.

Braemer F., Echallier J.C., Taraqji A., « Khirbet el Umbashi (Syrie) : rapport préliminaire sur les campagnes 1993 et 1994 », *Syria*, vol 73, p. 117 - 127, 1996.

Breton J.F., *L'Arabie heureuse au temps de la reine de Saba,* Hachette, 1998.

Brignon J., Amine A., Boutaleb B., Martinet G., Rosenberger B., *Histoire du Maroc,* Hatier, 1967.

Brun J.P., Borréani M., « Deux moulins hydrauliques du haut-empire romain en Narbonnaise », *Gallia,* n°55, p. 279-326, 1998.

Burdy J., « Lugdunum et ses quatre aqueducs », *Dossiers d'Archéologie,* 38, pp. 62-73, 1979.

Burdy J., *L'aqueduc romain de l'Yzeron,* Conseil Général du Rhône, Lyon, 1991.

Burdy J., *L'aqueduc romain de la Brévenne,* Conseil Général du Rhône, Lyon, 1993.

Burdy J., *L'aqueduc romain du Gier*, Conseil général du Rhône, Lyon, 1996.

Burnouf J., Carcaud N., « Le val de Loire en Anjou Touraine : un cours forcé par les sociétés riveraines », *Médiévales,* p. 17-29, vol. 36, 1999.

Burkhalter F., « La maison Egypte », in : *Alexandrie, IIIème siècle av J.-C.,* C. Jacob & F. de Polignac eds., Autrement, 1992.

Butzer K.W., « Late quaternary problems of the Egyptian Nile : stratigraphy, environments, prehistory », *Paléorient,* vol 23/2, p. 151-173, 1998.

Canfora L., *La Biblioteca Scomparsa,* Sellerio, Palermo, 1986.

Calvet Y. Geyer B., *Barrages antiques de Syrie,* Maison de l'Orient Méditerranéen, Lyon, 1992.

Casson L., *Ships and seamanship in the ancient world,* Princeton University Press, 1971.

Castleden R., *Life in bronze age Crete,* Routledge, 1990.

Castleden R., *Atlantis destroyed,* Routledge, 1998.

Caucanas S., *Moulins et irrigation en Roussillon du IXème au XVème siècle,* CNRS ed., Paris, 1995.

Cauvin J., *Naissance des divinités, naissance de l'agriculture - la révolution des symboles au néolithique*, CNRS ed. *Paris,* 1997.

Cauvin J., Cauvin M.C., Helmer D., Willcox G., « L'homme et son environnement an Levant nord entre 30 000 et 7 500 BP », *Paléorient,* vol 23/2, p. 51-68, 1998.

Chambers M., *Dessins des édifices, meubles, habits, machines et ustensiles des Chinois,* 1757 (coll. ENPC).

Champion E., « Les moulins à eau dans les polyptiques carolingiens d'entre Loire et Rhin », in *L'hydraulique monastique,* pp 321-335, Créaphis, 1996.

Chanson H., *Hydraulics of Roman Aqueducs : Steep Chutes, Cascades and Dropshafts,* American Journal of Archeology, 104, p. 47-72, 2000.

Charpin D. « La politique hydraulique des rois paleo-babyloniens ». *Annales d'Histoire et Sciences Sociales,* n° 57-3, p 625-664, may-june 2002.

Chevallier, *Ostie Antique, ville et port*, Les Belles Lettres, Paris, 1986.

Christol M., Nony, D. *Rome et son Empire,* Hachette, 1997.

Clamagirand E., Rais S., Chahed J., Smaoui L., « L'aqueduc de Carthage », *La Houille Blanche*, n°6, p. 423-429, 1990.

Cleuziou S., « Les pays du Golfe il y a 5000 ans », *Les dossiers de l'archéologie,* 122, pp 46-47, 1987.

Cleuziou S., « Construire et protéger son terroir : les oasis d'Oman à l'âge du bronze », *XVIIth international Archeology and History Symposium of Antibes*, p 389-412, APDCA, 1997.

Coates-Stephens R., « The walls and aqueducts of Rome in the early middle ages, A.D. 500-1000 », *The Journal of Roman Studies,* vol 88, p. 166-178, 1998.

Coque Delhuile B., « Les oasis de l'Arabie Heureuse », *Dossiers d'Archéologie,* n° 263, p 18-21, 2001.

Ctesias, « Histories of the Persic », in *Histoires de l'Orient,* trad & ed. by J. Auberger, Les Belles Lettres, 1991.

Dars J., *La marine chinoise du Xème siècle au XIVème siècle,* Economica,1992.

Derville A., « Naissance du capitalisme », in *Le XIIème siècle,* p. 33-60, F. Gasparri ed., Le Léopard d'or, 1994.

Debaine-Francfort C., *La redécouverte de la Chine ancienne,* Gallimard, 1998.

Degeorge G., *Damas, des origines aux Mamelouks,* L'Harmattan, 1997.

Deletie P., Dosseur H., Durouchoux C., Ho Ta Khanh M., Lardin G.N., Levasseur C., Saupin R., *Pétra, projet d'aménagement hydraulique du site,* Electricité de France, Mécénat Technologique et Scientifique, Novembre 1995.

Delpech A., Girard F., Robine G., Roumi M., *Les Norias de l'Oronte,* Institut Français de Damas, distribué par Maisonneuve, 1997.

Dickinson, O., *The Aegean Bronze Age,* Cambridge University Press, 1994.

Dion R., *Histoire des levées de la Loire,* Paris, 1961.

Domergue C., « L'eau dans les mines d'or romaines du nord-ouest de l'Espagne », in *L'homme et l'eau en Méditerranée et au Proche Orient, III,* Maison de l'Orient Méditerranéen, Lyon, 1986.

Domergue C., Binet C., Bordes J.L., « La roue de Sao Domingos », *La revue du musée des arts et métiers*, n°27, pp. 49-59, 1999.

Driver G.R., Miles J.C., *The Babylonian Laws,* vol II, Oxford, 1955.

Ducelier A., Kaplan M., Martin B., *Le Moyen Age en Orient,* Hachette, 1990.

Dumont-Heusers M.F. « L'alimentation en eau de Vaison la romaine », in *Les aqueducs de la Gaule romaine,* p.179-192, PULIM, 1997.

Durand J.M., *Documents épistolaires du palais de Mari, tomes I et II,* Cerf, 1997 et 1998.

Eschebach H., « Die Innerstädtische gebrauchswasserversorgung dargestellt am beispiel Pompejis », *Journées d'étude sur les aqueducs romains, Mai 1977,* pp. 81-132, Les Belles Lettres, 1983.

El Faïz M. , *Les Maîtres de l'Eau- Histoire de l'Hydraulique Arabe,* Actes Sud, 2005.

G.Fabre G., Fiches J.L., Leveau P., Paillet J.L., *Le pont du Gard. L'eau dans la ville antique,* CNRS ed., 1993.

Fahlbusch H., « Remarks on Vitruv's colluviriae », *Proc. IAHR Congress*, Cagliari, Sept 1979.

Fahlbusch H., « Vitrivius and Frontinus - Hydraulics in the Roman period », in *Hydraulics and hydraulic research, a historical review,* Garbrecht ed. Balkema, 1987.

Fernandez Ordonez J.A., *Catalogo de noventa presas y azudes espanoles anteriores a 1900,* CEHOPU, Madrid, 1984.

Fiches J. L., « La stratégie du tracé », in *L'aqueduc de Nîmes et le Pont du Gard - Archéologie, Géosystème, Histoire*, edited by G. Fabre, J.L. Fiches, J.C. Paillet, Conseil Général du Gard, 1991.

Fichou J. C., Le Hénaff N., Mével X., *Phares - Histoire du balisage et de l'éclairage des côtes de France*, Le Chasse-Marée, 1999.

Finet A. *Le code de Hammurapi,* Cerf, 1996.

Fleury P., « Héron d'Alexandrie et Vitruve - à propos des techniques dites pneumatiques », in *Science et vie intellectuelle à Alexandrie,* G. Argoud ed., University of Saint Etienne, 1994.

Francfort H. P., « La civilisation de l'Indus aux rives de l'Oxus », *Archeologia,* 227, pp. 44-55, 1987.

Francfort H.P., *Fouilles de Shortughai - Recherches sur l'Asie Centrale protohistorique*, vol II, De Boccard, Paris, 1989.

Frontinus, *De Aquaeductu Urbis Romae - Les aqueducs de la ville de Rome,* original Latin text with a translation into French and comments by P. Grimal, Les Belles Lettres, 1961.

Gaibov V., Kochelenko G., « La Margiane », *Dossiers d'Archéologie,* n° 271, p 46-53, 2002.

Garbrecht G., « Die entwicklung der wasserwirtschaft Pergamons bis zur romischen kaiserzeit », *Journées d'études sur les aqueducs romains,* pp. 141-163, Les Belles Lettres, 1983.

Garbrecht G., « The water supply at Tushpa (Urartu) », *World Archeology,* vol II, n°3, pp. 306-312, 1980.

Garbrecht G., « Sadd-el-Kafara : the world's oldest large dam », *Water Power and Dam Construction,* july 1985, pp 71-76.

Garbrecht G., « Hydraulics and hydraulic concepts in antiquity » in *Hydraulics and hydraulic research, a historical review,* Garbrecht ed, Balkema, 1987.

Garbrecht G., « Water management for irrigation in Antiquity (Urartu, 850 to 600 BC) », *Irrigation and drainage systems,* 2, pp. 185-198, 1988.

Garbrecht D. « Talsperre und tunnel am hafen Seleukeia », *Historische Talsperren 2,* p 83-89, Wittner, 1991.

Garbrecht G., « Historical water storage for irrigation in the Fayum depression (Egypt) », *Irrigation and drainage systems,* vol 10, pp. 47-76, 1996.

Garbrecht G., Jaritz G., « Neue Ergebnisse zu altägyptischen Wasserbauten im Fayum », *Antike Welt*, 23, 4, pp. 238-254, 1992.

Gardin J. C., *Prospections archéologiques en Bactriane Orientale* , vol III , de Boccard, Paris, 1998.

Gashe H., Tannet M., Cole S., Verhoeven K., "Fleuves du temps et de la vie : permanence et instabilité du réseau fluviatile babylonien entre 2500 et 1500 avant notre ère », *Annales d'Histoire et Sciences Sociales,* 57-3, p 545-559, may-june 2002.

Gentelle P, « Les irrigations antiques à Shabwa », dans *Fouilles de Shabwa II, Rapports préliminaires,* ed. by J.F. Breton, *Syria*, vol LXVIII, Librairie Orientaliste Paul Geuthner, Paris, 1992.

Gentelle P., *Prospections archéologiques en Bactriane Orientale,* vol I, de Boccard, Paris, 1989.

Georgelin Y., Arzano S., « Pytheas le massaliote », *Neptunia,* n°209, pp. 4-14, 1998.

Gernet J., *Le monde chinois,* Armand Colin, 1990.

Geyer B., « Aménagements hydrauliques et terroir agricole dans la moyenne vallée de l'Euphrate » in *Techniques et pratiques hydro-agricoles traditionnelles en domaine irrigué* (Proc IFAPO Conference of Damas), vol I, pp. 63-84 , Geuthner, Paris,1990.

Geyer B., Besançon J., « Environnement et occupation du sol dans la vallée de l'Euphrate syrien durant le néolithique et le chalcolithique », *Paléorient,* vol 22/2, p. 5-15, 1997.

Ghirshman R., *Tchoga Zanbil (Dur Untash),* Proc Archeological Delegation in Iran, vol XL, Geuthner, 1968.

Gies F, Gies J. , *Cathedral, forge and waterwheels,* Harper Collins publisher, 1994

Gilgamesh, *The epic of Gilgamesh - an English version,* Penguin, Baltimore, c1960.

Goblot H., *Les qanats, une technique d'acquisition de l'eau.* Mouton ed - EPHESS, Paris, 1979.

Goyon J.C., « Ebauche d'un système étatique d'utilisation de l'eau : Egypte pharaonique de l'ancien au nouvel empire », in *L'homme et l'eau en Méditerranée et au Proche Orient - Aménagements hydrauliques, état et législation,* Maison de l'Orient, Lyon, 1982.

Goyon J.C., « Transports par voie d'eau et organisation étatique de la vallée du Nil à l'époque pharaonique », in *L'homme et l'eau en Méditerranée et au Proche Orient - l'eau dans les techniques,* Maison de l'Orient, Lyon, 1986.

Graham J.W., *The palaces of Crete,* Princeton University Press, 1987.

Granet M., *Danses et légendes de la Chine ancienne*, PUF, 1994 (firstly ed in 1926).

Granet M., *La civilisation chinoise,* Albin Michel, 1994 (firstly ed in 1929).

de Graeve M.C., *The ships of the ancient near east,* Departement Orientalistieck, Leuven, 1981.

Grenier A., *Manuel d'archéologie gallo-romaine,* tome IV, Picard, 1960.

Grimal N., *Histoire de l'Egypte ancienne*, Fayard, 1988.

Grimal, N., *A History of Ancient Egypt*, Translated by Ian Shaw, Blackwell Publishers, Oxford, 1992.

Guery R., Pirazzoli P., Trousset P., « Les variations du niveau de la mer depuis l'antiquité à Marseille », *Les dossiers de l'archéologie, n°50, pp. 8-27, 1981.*
Guillerme A., « La destruction des aqueducs dans le Nord de la France », *Les Dossiers d'Archéologie ,* 38, 1979, p 94-96.

Guillerme A., *Les temps de l'eau – la cité, l'eau et les techniques,* Champ Vallon, 1983.

Hadj Sadok M., *Al-Idrissi - le Maghreb au VIème siècle de l'Egire, (translation into French of the Arab origial text)* Publisud, 1983.

Haj Ibrahim, « L'irrigation dans les ghoutas de l'ouadi el-Majarr » in *Techniques et pratiques hydro-agricoles traditionnelles en domaine irrigué* (proc. IFAPO conf. of Damas), vol II, pp. 295-2312, Geuthner, 1990.

du Halde, *Description de l'empire de la Chine*, P.G. Lemercier imprimeur, Paris, 1735 (coll. ENPC).

Heers J., *La ville au Moyen Age en Occident,* Fayard, 1990.

Helms S., « Jawa, site chalcolithique », *les dossiers d'histoire et archéologie,* n°118, pp. 92-94, July 1987.

Helms S., « Paleo-Beduin and Transmigrant Urbanism », in *Studies in the Histoty and Archeolgy of Jordan, I,* pp. 97 - 113, Department of Antiquities, Amman, 1987.

Herodotus, *The History,* Translated by David Grene, University of Chicago Press, 1987.

Herodotus, *The History of Herodotus*, George Rawlinson Tr., D. Appleton and Company, New York, 1885.

Heurgon J., *La vie quotidienne des Etrusques*, Hachette, 1989

Hiebert F. T., « The oasis and the city of Merv (Turkmenistan) », *Archéologie Islamique*, 3, pp. 111-126, 1992.

Hill D. R., « Technologie », in *Histoire des sciences arabes*, Roshdi Rashed ed., vol III, Seuil, 1997.

Hills R. L., *Power from wind,* Cambridge University Press, 1994.

Hodge A. T., *Roman aqueducts and water supply,* Duckworth, 1995.

Homer, *Iliad,* XXI, 250, translated by Samuel Butler, http://darkwing.uoregon.edu/~joelja/iliad.html#b21; See also *The Iliad and Odyssey of Homer*, translated by Pope. Published London: printed for C. Bathurst, J. Buckland, W. Strahan, J. Rivington and Sons, T. Davies, 1783.

Hoffmann H., « L'hydrologie des sites bénédictins et cisterciens de l'Allemagne centrale », in *L'hydraulique monastique,* pp 99-109, Créaphis, 1996.

Huot J. L., *Les sumériens,* Errances, 1989.

Huot J. L., *Les premiers villageois de Mésopotamie. Du village à la ville.* Armand Colin, 1994.

Huot J. L., Thalmann J.P., Valbelle D., *Naissance des cités,* Nathan, 1990.

Al Idrissi, *La première géographie de l'Occident,* translation into french by Henri Bresc et Annliesse Nef, Flammarion, 1999.

Ibn Jubayr, « Relations de Voyage », translation into French of the Arab original text by Paule Charles-Dominique, in *Voyageurs arabes,* Gallimard, 1995.

Ibn Battûta, « Voyages et périples », translation into French of the Arab original text by Paule Charles-Dominique, in *Voyageurs arabes,* Gallimard, 1995. (see also Ibn Battûta, *Travels*, J. Murray, London (Johnson Reprint Company Ltd., New York), Translated by Samuel Lee, 1829).

Izarra (de) F., *Hommes et fleuves en Gaule romaine,* Errance, 1993.

Jaccottey L. « Le problème de l'approvisionnement de Besançon antique : l'aqueduc d'Arcier », in *Les aqueducs de la Gaule romaine,* p. 399-404, PULIM, 1997.

Jacobsen H., Lloyd S., *Sennacherib's aqueduct at Jerwan,* Oriental Institute Publications vol XXIV, The University of Chicago Press, 1935.

Janssen M., « Mohenjo-Daro : architecture et urbanisme », in *Les cités oubliées de l'Indus,* Musée National des Arts Asiatiques Guimet, 1988.

Jeancolas L., « Présentation des aqueducs antiques de Lyon », *Journées d'étude sur les aqueducs romains, Mai 1977,* pp. 179-205, Les Belles Lettres, 1983.

Joffe E. G. H., « Khattara and other forms of gravity-fed irrigation in Morocco », in *Qanat, Kariz and Khattara - Traditional Water Systems in the Middle East and North Africa,* pp. 195-209, P. Beaumont, M. Bonine, K. Mc Lachlan ed., Middle East & North African Studies Press, 1989.

Kamel M. W., « L'importance structurelle du qanaye et de la noria en Syrie » in *Techniques et pratiques hydro-agricoles traditionnelles en domaine irrigué* (proc. IFAPO conf of Damas), vol II, pp. 383-393, Geuthner, Paris, 1990.

Keller W., *The Etruscans,* Randon House, Inc., 1976.

Kerisel J., *Le Nil : l'espoir et la colère,* Presses des Ponts et Chaussées, Paris, 1999.

Kinder T. N., « Aménagement d'une vallée de larmes : les Cisterciens et l'eau à Pontigny », in *L'hydraulique monastique,* pp 383-399, Créaphis, 1996.

Knauss J. « Mykenische Talsperren in Arkadien und Beotien », *Historische Talsterren 2,* p 19-41, Wittner, 1991.

Kohl P. L., *Central Asia, palaeolithic beginnings to the iron age,* ed. Recherches sur les Civilisations, Paris,1984.

Kramer S. N., *History Begins at Sumer*, The University of Pennsylvania Press, Philadelphia, 1981.

Kramer, S. N., *Sumerian Mythology, A Study of Spiritual and Literary Achievement in the Third Millennium B.C.,* University of Pennsylvania Press, Philadelphia, 1944, revised 1961.

Lacordaire S., *Sources et fontaines de Paris,* Fayard, 1979.

Lafont B., « Nuit dramatique à Mari », *Florilegium Marianum, Mélanges M. Fleury*, p. 93-105, SEPOA, Paris, 1992.

Lagardère V., « Moulins d'Occident Musulman au Moyen Age (IX au XVème siècle) : Al-Andalus », *Al-Qantara, Revista de estudias arabes,* vol XII, facs. 1, pp. 59-117, 1991.

Lalouette C., *Textes sacrés et textes profanes de l'ancienne Egypte,* vol I & II, Gallimard, 1994.

Lambton A.K.S., « The origin, diffusion and functionning of the qanat », pp. 5-11, P. Beaumont, M. Bonine, K. Mc Lachlan ed., Middle East & North African Studies Press, 1989.

Landry P., « Eaux souterraines et qanâts d'après un livre arabe du XIème siècle » in *Techniques et pratiques hydro-agricoles traditionnelles en domaine irrigué* (proc. of IFAPO conf. of Damas), vol II, pp. 271-283, Geuthner, Paris, 1990.

Lane M. E., *Guide to the antiquities of the Fayyum,* The American University in Cairo Press, 1985.

Lasssalle V., « Le pont du Gard et l'aqueduc de Nîmes », *Dossiers de l'archéologie,* 38, pp.52-61, 1979.

Lassus J., « L'eau courante à Antioche », *Journées d'étude sur les aqueducs romains,*

may 1977, pp. 207-229, Les Belles Lettres, Paris, 1983.
Lecoq P., *Les inscriptions de la Perse achéménide,* Gallimard, 1997.

Lefebvre C., « L'aqueduc de Metz », in *Les aqueducs de la Gaule romaine,* p. 405-442, PULIM, 1997.
Lemaitre C., Allart E., « L'aqueduc de Lisieux », in *Les aqueducs de la Gaule romaine,* p. 467-518, PULIM, 1997.

Leveau Ph., « La construction des aqueducs », *Dossiers de l'archéologie,* 38, pp. 8-18, 1979.

Leveau Ph., Paillet J.H., « Alimentation en eau et développement urbain à Caesarea de Mauretanie », *Journées d'étude sur les aqueducs romains, may 1977,* pp. 231-234, Les Belles Lettres, Paris, 1983.

Li G., Du G., « An ancient hydraulic engineering in China : the Dujiangyan irrigation project », *proc. XXX IAHR Congress,* Aristotle University of Thessaloniki, theme E, p 331-335; aug. 2003.

Lloyd G. E. R., *Greek Science after Aristotle,* Chatto and Windus, London, 1973.

Lyonnet B., « Etablissements chalcolithiques dans le nord-est de l'Afghanistan : leurs rapports avec les civilisations de l'Indus », *Paléorient,* vol 7/2, p. 57-74, 1981.

Madani T., « Le réseau hydraulique de Fès », *Archéologie Islamique,* 8-9, p. 119-142, 1999.

Manley B., *Historical atlas of ancient Egypt,* Penguin books, 1996.

Margueron J.C., « Espace agricole et aménagement régional à Mari au début du IIIème millénaire », in *Irrigation and cultivation in Mesopotama, I, Bulletin on sumerian Agriculture*, p 49-60 , Cambridge, 1988.

Margueron J.C., *Les Mésopotamiens*, Armand Colin, 1991.

Margueron J.C., *Mari,* Picard, Paris, 2004.

Maspero H., *La Chine antique*, PUF, 1965

Masson V. N., « The bronze age in Khorassan and Transoxiana », in *History of cvilizations in Central Asia*, vol I, UNESCO publishing, 1992.

Mayerson P., « The port of Clysma (Suez) in transition from Roman to Arab rule », *Journal of Near Eastern studies,* vol. 55, n°2, p. 119-126, 1996.

Measson A., « Alexandrea ad Aegyptum », in *Science et vie intellectuelle à Alexandrie,*

G. Argoud ed., de University of Saint Etienne, 1994.
Messier R. A., « Sijilmassa. Five seasons of archeological inquiry. », *Archéologie Islamique,* 7, p. 61-92, 1997.
Micheau F., « Les Institutions Scientifiques dans le Proche Orient Médiéval », in *Histoire des Sciences arabes*, Roshdi Rashed ed., vol III, Seuil, Paris, 1997.

Monchabert .J.Y., « Réflexions à propos de la datation des canaux » in *Techniques et pratiques hydro-agricoles traditionnelles en domaine irrigué* (proc. IFAPO conf. of Damas), vol I, pp. 87-100 , Geuthner, Paris, 1990.

Moussly N., *Le problème de l'eau en Syrie,* Thèse, Lyon, 1951.

Motte T. W. *Imperial China,* Harvard University Press, 1999.

Mülle S., « Prospection archéologique de la plaine de Malia », *Bulletin de Correspondance Héllénique,* n°120 - 2, pp. 921-928, 1996.

Müller S., « L'organisation d'un territoire minoen », *Les dossiers d'archéologie,* n° 222, pp. 52-53, Avril 1997.

Needham, J., *The Shorter Science and Civilization in China,* Cambridge University Press, 1978.

Needham J., Ling W., *Science and Civilization in China,* vol IV, part II (Mechanical Engineering), Cambridge University Press, 1965.

Needham J., Ling W., Gwei-Djen L., *Science and civilization in China,* vol IV, part III (Civil Engineering and Nautics), Cambridge University Press, 1971.

Nordon M., *Histoire de l'hydraulique 1, l'eau conquise*, Masson, 1992.

Orrieux C., *Les papyrus de Zénon,* Macula, 1983.

Özis Ü, « Forchheimer and ancient hydraulic works in Turkey », *proc. Memorial Symposium*, XXVIII IAHR Congress, 1999.

Parisse M, ed. , *Atlas de la France de l'an mil - Etat actuel des connaissances*, Picard , 1994.

Pelletier A., « L'alimentation en eau à Vienne dans l'Antiquité » *Journées d'étude sur les aqueducs romains, may 1977,* pp. 293-307, Les Belles Lettres, Paris,1983

Pirenne J., « La route de l'encens, richesse de l'Arabie », *Les dossiers de l'archéologie,* n°33, p. 11-15, 1979.

Pirenne J., « Shabwa, capitale du royaume de l'encens », *Les dossiers de l'archéologie,* n°33, p. 74-79, 1979.

Pirenne J., « La juridiction de l'eau en Arabie du sud d'après les inscriptions » in *L'homme et l'eau en Méditerranée et au Proche Orient - aménagements hydrauliques, état et législation*, pp. 81-102, Maison de l'Orient Méditerranéen, Lyon, 1982.

Plato, *Critias,* Translated by Benjamin Jowett http://classics.mit.edu/Plato/critias.html

Platon N., *La civilisation Egéenne,* Albin Michel, 1981.

Pliny the Elder, *The Natural History,* with an English translation by H. Rackham, Harvard University Press, Cambridge, Massachusetts, 1969.

Poillon C., *Pompes et machines à élever les eaux*, Bernard et Cie, Paris, 1885 (coll. ENPC).

Provost A., Leprêtre B. « L'aqueduc gallo-romain de Carhaix (Finistère) », in *Les aqueducs de la Gaule romaine,* p. 525-548, PULIM, 1997.

Polo, M., *Le Devisement du Monde. Le livre des Merveilles* (text translated by A.C. Moule, P. Pelliot, L. Hambis), La Découverte, 1998. (see also Marco Polo, *The Travels of Marco Polo*, revised from Marsden's Translation, Liverright Publishing Corporation, New York, 1953).

Porter Y., « Adduction et stockage de l'eau à Delhi, aux XIIIème et XIVème siècles », *Archéologie Islamique,* 3, pp. 169-183, 1992.

Rachet G., *Civilisations et archéologie de la Grèce préhellénique,* Editions du Rocher, 1993.

Rakob F., « L'aqueduc de Carthage », *Dossiers de l'archéologie,* 38, pp. 34-42, 1979.

Reddé M., « Portus, le plus grand port du monde romain », *Les dossiers d'archéologie,* n°71, pp.54-61, 1983.

Redmount C.A., « The Wadi Tumilat and the canal of the pharaohs », *Journal of Near Eastern Studies*, vol. 54, n°2, p. 127-135, 1995.

Renger J., « Rivers, watercourses, and irrigation ditches and other matters concerning irrigation based on old Babylonian sources (2000-1600 BC) », *Irrigation and cultivation in Mesopotamia, Bulletin on Sumerian agriculture V,* p 31-46, Cambridge, 1990.

Renou A., *La civilisation de l'Inde ancienne,* Flammarion, 1950.

Ricci M., Trigault N., *Histoire de l'expédition chrétienne au royaume de la Chine (1582-1610).* Desclée De Brouwer, 1978.

von Richthofen F.F., *China*, 1877, edited by Akademische Druck u. Verlagsanstalt, Graz, 1971.

Riuju L., Zhaojin Z., Jialin H., « River training works throughout the history of China », in *Hydraulics and hydraulic research, a historical review,* Garbrecht ed., Balkema, 1987.

Rigal D. « L'aqueduc de Cahors », in *Les aqueducs de la Gaule romaine,* p. 311-332, PULIM, 1997.

Roaf M., *Cultural Atlas of Mesopotamia and the Ancient Near East*, Andromeda, Oxford, 1990.

Robert J.N., *De Rome à la Chine - sur la route de la soie au temps des Césars,* Les Belles Lettres, Paris, 1997.

Robin C., « Quelques épisodes marquants de l'histoire sudarabique », in *L'Arabie Antique, de Karib'il à Mahomet, (Revue du Monde Musulman et de la Méditerranée,* n°61), pp. 55-70, EDISUD, 1991.

Robin C., *Inventaire des inscriptions sudarabiques,* tome I, Académie des Inscriptions et Belles Lettres, De Boccard, Paris, 1992.

Roccati A., *La littérature historique sous l'ancien empire égyptien,* Cerf, Paris, 1982.

Rouillard J., « L'eau en Champagne du Sud et en Bourgogne du Nord : les abbayes cisterciennes de Pontigny et Vauluisant d'après les cartulaires », in *L'hydraulique monastique,* pp 363-381, Créaphis, 1996.

Roux, G., *Ancient Irak,* Allen and Unwin, London, 1964.

Rozhanskaya M., « La Statique », dans *Histoire des sciences arabes*, Roshdi Rashed ed., vol II, Seuil, 1997.

Ryckmans J., « Le barrage de Marib ou l'Arabie Heureuse », *Les dossiers de l'archéologie,* n°33, p.28-35, 1979.

Safadi C., « La foggara en Syrie », », in *Techniques et pratiques hydro-agricoles traditionnelles en domaine irrigué* (Proc. IFAPO Conf. of Damas), vol II, pp. 285-292 , Geuthner, Paris, 1990.

Saliby N., « Les installations hydrauliques à Qasr el Heir el Gharbi », in *Techniques et pratiques hydro-agricoles traditionnelles en domaine irrigué* (Proc. IFAPO Conf. of Damas), vol II, pp. 475-484 , Geuthner, Paris, 1990.

Sadler G., « Le terroir agricole de Diyateh, l'irrigation comme condition d'existence de ce terroir » in *Techniques et pratiques hydro-agricoles traditionnelles en domaine irrigué* (Proc. IFAPO Conf. of Damas), vol II, pp. 421-451, Geuthner, Paris, 1990.

Sanlaville P., « Changements climatiques dans la région levantine à la fin du Pléistocène supérieur et au début de l'Holocène. Leurs relations avec l'évolution des sociétés humaines », *Paléorient,* vol 22/1, p. 21-29, 1996.

Santa Maria Scrinari V., « Les nouvelles fouilles au port de Rome », *Les dossiers d'archéologie,* n°71, pp.47-53, 1983.

Sarianidi V., « Food-producing and other neolithic communities in Khorassan and Transoxiana », in *History of civilizations in Central Asia,* vol I, UNESCO publishing, 1992.

Sarrazin J.L., « Les Cisterciens et la genèse du marais poitevin », in *L'hydraulique monastique,* pp 111-119, Créaphis, 1996.

Sartre M., *L'Orient romain,* Seuil, 1991.

Sauvaget J., *Historiens Arabes,* Maisonneuve, 1988.

Scarre C., *The Penguin Historical Atlas of Ancient Rome*, Penguin Books, London, 1995.

Schnitter N.J., *A history of dams - the usefull pyramids,* Balkema, 1994.

Sellin R.J.H., « The roman water mill at Barbegal », *Proc. AIHR Congress, Cagliari, 1979.*

Shelmerdine C.W., « Review of aegean prehistory VI : the palatial bronze age of the southern and central greek mainland », *American Journal of archeology,* n°101, pp. 537-585, 1997.

Sima Qian , *Records of the Grand Historian of China,* translated from the *Shih chi* of Ssu-ma Ch'ien by Burton Watson, Columbia University Press, New York, 1961.

Smith N.A.F., *The heritage of Spanish dams*, pub by The Spanish National Committee on Large Dams, Colegio de Ingenieros de Caminas, Canales y Puertos, 1970.

Sonnet B., « Le canal de la Cent-Fonts », *Les Dossiers d'Archéologie,* n° 234, p. 38-39, 1998.

van Soldt W., « Irrigation in Kassite Babylonia », in *Irrigation and cultivation in Mesopotama, I, Bulletin on sumerian Agriculture,* p 104-120 , Cambridge, 1988.

Sollberger E., Kupper J.R., *Inscriptions royales sumériennes et akkadiennes,* Cerf, Paris, 1971.

Stordeur D., « El Kowm 2 Caracol et le PPNB », *Paléorient,* vol 15/1, p. 102-110, 1989.

Steens E., *La Chine Antique* , ed du Rocher, Paris, 1989.

Steinkeller P., « Notes on the irrigation system in third millenium south Babylonia », in *Irrigation and cultivation in Mesopotama, I, Bulletin on sumerian Agriculture,* p 73-92 , Cambridge, 1988.

Strabo, *The Geography of Strabo,* William Heinemann Ltd., London, 1932, translated by H. L. Jones.

Strabon, *Géographie,* (translation of F. Lasserre), Les Belles Lettres, 1981.

Strabon, *Le voyage en Egypte,* translated into French and commented by J. Yoyotte, P. Charvet, S. Gompertz, NIL ed, Paris, 1997.

Strassberg R.E., *Inscribed landscapes ; travel writtings from imperial China,* Univ. of California Press, 1994.

Taraqji A., Braemer F., Echallier J.C., « Khirbet el-Umbashi », *Catalog of the Syro-European Archeology Exhibition* , Nat. Museum of Damas, Institut Français d'Etudes Arabes de Damas, 1996.

Taylour W., *The Mycenaeans,* Thomas & Hudson, 1983.

Toueir K., « Le nahr el-Nil entre Raqqa et Heraqleh », in *Techniques et pratiques hydro-agricoles traditionnelles en domaine irrigué* (Proc. IFAPO Conf. of Damas), vol I, pp. 217-227, Geuthner, Paris, 1990.

Trenard L. ed. , *Histoire des Pays-Bas français* , Privat, 1972.

Vallet R., « Habuba Kebira ou la naissance de l'urbanisme », *Paléorient,* vol 22/2, p. 45-76, 1997.

Vercoutter J., « Journal de bord », in *Cités disparues,* Baud ed., p. 157-170, Autrement, 1991.

Verhulst A., *Précis d'histoire rurale de la Belgique*, University of Bruxelles, 1990.

Viollet P.L., Chabard J.P., Esposito P., Laurence D. *Mécanique des fluides appliquée,* Presses de l'Ecole Nationale des Ponts et Chaussées, Paris, 1998.

Viollet P.L. « The predecessors of European Hydraulic Engineers : Minoans of Crete and Mycenaeans of Greece (2100-1200 BC) », *proc. XXX IAHR Congress,* Aristotle University of Thessaloniki, theme E, Aug. 2003.

Viollet P. L. *Histoire de l'Energie Hydraulique,* Presses de l'Ecole Nationale des Ponts et Chaussées, Paris, 2005.

Viollet P.L. « Water management and Engineering in the Early Civilizations », *Proc. International Workshop "La Ingenieria y la Gestión del Agua a Través de los Tiempos", Alicante*, may 2006, University of Valencia.

Vitruvius, *De Architectura - De l'architecture,* books VIII & X, original Latin and translation into French with comments by L. Callebat, Les Belles Lettres, Paris, 1973-1986.

Vogel, A., "Die kupfersteinzeitlichen Dämme von Jawa in Jordanien", in *Historische Talsperren 2*, p 9-18, Garbrecht ed., Wittner, 1991

Wikander, O., « The water-mill », in *Handbook of Ancient Water Technology,* Wikander ed., pp. 371-400, Brill, 2000

Will P.E. , « Hangzhou », in *Dictionnaire de la civilisation chinoise (Encyclopaedia Universalis)*, p. 343-345, Albin Michel, 1998.

Wilson, A., « Machines, Power and the Ancient Economy », *Journal of Roman Studies,* vol XCII, pp 1-32, 2002.

Xenophon, *Xenophon in Seven Volumes III. Anabasis, Books I - VII,* trad. C. Brownson, Harvard University Press, 1958.

Zakri S., « Pratiques hydro-agricoles traditionnelles d'après les manuscrits arabes d'agronomie à l'époque médiévale » in *Techniques et pratiques hydro-agricoles traditionnelles en domaine irrigué* (Proc.IFAPO Conf. of Damas), vol I, pp. 35-49, Geuthner, Paris 1990.

Zangger E., « Landscape changes around Tiryns », *American Journal of Archeology,* n°98, p. 189-212, 1994.

Zettler A., « L'économie hydraulique dans les abbayes du lac de Constance et sur le plan de Saint-Gall », dans *L'hydraulique monastique,* pp 65-75, éditions Créaphis, 1996.

Zheng L., « Talsperren und where im alten China », *Historische Talsterren 2,* p 109-130, Wittner, 1991.

Zimansky P.E., « Ecology and empire : the structures of the Urartian state », *Studies in Ancient Oriental Civilization,* n°41, The Oriental Institute, Chicago, 1985.